Conspiracy Theories and the Failure of Intellectual Critique

CONSPIRACY THEORIES AND THE FAILURE OF INTELLECTUAL CRITIQUE

Kurtis Hagen

University of Michigan Press
Ann Arbor

Published in the United States of America by the University of Michigan Press
Printed and bound by CPI Group (UK) Ltd, Croydon, CR0 4YY

First published July 2022

A CIP catalog record for this book is available from the British Library.

Library of Congress Cataloging-in-Publication data has been applied for.

ISBN 978-0-472-13310-9 (hardcover : alk. paper)
ISBN 978-0-472-22034-2 (e-book)

Contents

Digital materials related to this title can be found on the Fulcrum platform via the following citable URL: https://doi.org/10.3998/mpub.12089461

Preface

The aim of this book is to make serious consideration of conspiracy theories respectable among the sophisticated. To some this aim may seem audacious, even ludicrous. But to most of the philosophers who have contributed substantially to the philosophical literature on conspiracy theories over the past two decades—who are at least arguably the most relevant authorities on the matter—it would be a welcome development, and long overdue. The philosophers I'm particularly thinking of include Charles Pigden, David Coady, M Dentith, Lee Basham, Juha Räikkä, and even Brian Keeley, whose seminal 1999 article, "Of Conspiracy Theories," is often misinterpreted as arguing that conspiracy theories are generally unwarranted.

To suggest that serious consideration of conspiracy theories ought to be respectable, of course, does not mean that *all* individual conspiracy theories ought, upon consideration, to be regarded as respectable. Rather, it is to say that conspiracy theories ought not be dismissed on account of their being conspiracy theories, and that each ought to be given a fair hearing, given the recognition that sometimes conspiracy theories have turned out to be true, and also that, of the many serious attempts to provide reasons to adopt a dismissive stance toward conspiracy theories generally (including Keeley's), none have survived scrutiny (as Keeley readily concedes of his own attempt).

Further, it turns out that much of the social science scholarship on the subject, especially the psychological study of conspiracy theorists, has turned out to be flawed and misleading. Those flaws are not *randomly* directed; rather, they consistently serve to disparage conspiracy theorists unfairly. This suggests that there may be a bias against conspiracy theorists in the academy, as, indeed, there seems to be among "sophisticated" people more generally. If sophisticated people become aware of such a bias, they should become cautious about their own attitude; otherwise they fail a test of sophistication. This then—apparently biased research on conspiracy theorists—is another reason to adopt a more respectful attitude toward conspiracy theories and the people who believe them, which, after all, is most of us in varying degrees.

Previously Published Material

Earlier versions of following chapters were first published as indicated below:

Chapter 4: "Is Infiltration of 'Extremist Groups' Justified?"
International Journal of Applied Philosophy 24.2 (Fall 2010): 153–168.

Chapter 5: "Conspiracy Theories and Stylized Facts."
Journal for Peace and Justice Studies 21.2 (Fall 2011): 3–22.

Chapter 7: "Conspiracy Theorists and Monological Belief Systems."
Argumenta 3.2 (issue 6, May 2018): 303–326.

Chapter 8: "Conspiracy Theories and the Paranoid Style: Do Conspiracy Theories Posit Implausibly Vast and Evil Conspiracies?"
Social Epistemology 32.1 (2018): 24–40. Available online at http://www.tandfonline.com/10.1080/02691728.2017.1352625

Chapter 10: "Conspiracy Theorists and Social Scientists."
In *Taking Conspiracy Theories Seriously*, edited by M R. X. Dentith (London: Rowman and Littlefield, 2018): 125–140.
The chapter published in *Taking Conspiracy Theories Seriously* is, in turn, a revised and substantially extended version of a short paper first published under the title "What Are They Really Up To? Activist Social Scientists Backpedal on Conspiracy Theory Agenda," in the *Social Epistemology Review and Reply Collective* 6.3 (2017): 89–95.

In addition, the booklet *Conspiracy Theory: A Philosophical Defense* (Burlington, VT: Fomite, 2018), contains some material from, or incorporated into, Chapters 8 and 9.

Acknowledgments and Backstories

Defending the view that conspiracy theories ought to be taken seriously, which flies in the face of the "common sense" of many academics, presented challenges. For example, I encountered barriers to publication (such as frequent desk rejections) unlike what I encounter when publishing on "harmless" topics in Chinese philosophy, even when I challenged dominant paradigms in that sphere. But the most significant barrier was always psychological. I seriously worried that I might be wasting my time on articles that would never be published, no matter how cogent. Giving up always seemed a rational option. In such a situation, the encouragement of respected friends and colleagues—so important in all cases—takes on a heightened significance. And so, I have some thanking to do, both for substantial help and for significant moral support.

In the early days, I received encouraging feedback from my colleagues at SUNY Plattsburgh, E. Thomas Moran and the late Martin Lubin in particular. I also received both encouragement and helpful proofreading and editing services from friend and retired philosophy professor Kent Simmons, who continues to provide aid and encouragement.

I sent a draft of my first paper, which addressed Cass Sunstein and Adrian Vermeule's proposal involving "cognitive infiltration," to the writer and political activist Marc Estrin. He was outraged at the proposed cognitive infiltration, of which he had not been aware. Apparently, few people were aware of it, though the offending article was published eight months earlier, in April 2009, and an early draft had been available online since January 2008. So the proposal had been publicly available for just about two years. (The situation reminds me a little of the Tuskegee Study in the sense that the results were being published in academic journals and yet "nobody" knew about it.) Estrin asked if I would mind if he wrote a blog about it, and I approved of the idea. He published a blog on his own site January 10, 2010, and on *The Rag Blog* the next day. Then suddenly, over the next several days, a number of sites reported on the proposal, including *Raw Story* (which was later credited in Project Censored's *Censored 2011*). On January 15, Glenn Greenwald got into the mix with an article on Salon.com called "Obama Confidant's Spine-Chilling Proposal" (2010). It seems that this little internet buzz alerted David Ray Griffin to the issue, and he quickly wrote a book on the subject, *Cognitive Infiltration: An Obama Appointee's Plan to Undermine the 9/11 Conspiracy Theory*, which I reviewed (Hagen 2011b). A casual glance at the respective copyright dates may *seem* to indicate that my first article (Hagen 2010) was published before Griffin's book (2011). But, for the record, I know that not to be true. Griffin's book was published first.

Incidentally, much later, upon publication of "Conspiracy Theories and the Paranoid Style," Estrin suggested that I rework that material for more of a lay audience and publish it as a booklet in a series that he publishes. That is the origin of *Conspiracy Theories: A Philosophical Defense*. Estrin's feedback during the process helped shape that work.

Back in early 2010, Estrin arranged for both of us to be interviewed about cognitive infiltration and our responses to it on Thomas Kiely's radio show, *INN World Report*. This put me in touch with Kiely, who provided a number of helpful content suggestions via email.

Kiely in turn put me in touch with Lance DeHaven-Smith and Matthew T. Witt, both of whom had written about "SCADs" (State Crimes Against Democracy), and Kiely also introduced me, via email, to the late Hank Albarelli, author of *A Terrible Mistake*, which is about the CIA's role in the dosing of Frank Olsen with LSD and his subsequent murder. My thinking on these and related matters was enriched by the ensuing series of emails among the five of us.

After the publication of my first article on this topic, Charles Pigden reached out to offer his congratulations. This was tremendously encouraging. Pigden, along with David Coady, had been leading the way with their defense of a position that later came to be called "particularism." They had established a proof of concept—that it was possible to publish papers offering a rational defense of conspiracy theories in academic journals (though I did do something even a bit more edgy by defending some controversial conspiracy theories, particularly 9/11 conspiracy theories).

Lee Basham, who was also an early contributor to the philosophy of conspiracy theories, reached out to offer his congratulations after my first article as well, and later contacted me again when he found out that I was pursuing a critique of the psychological literature on conspiracy theories, as he was himself. In fact, at that moment, we were both developing what was substantially the same critique of an article called "Dead and Alive" (Wood et al. 2012). We've remained in close contact ever since, and he has become the first person I turn to for feedback and idea-sharing on these matters.

More recently, Basham informed me of a then-upcoming special issue of the journal *Argumenta* that would focus on the ethics and epistemology of conspiracy theories. He encouraged me to contribute an article, which I did. It was included in the special issue, and Chapter 7 of this book is a lightly revised version of that article. The guest editor for the special issue was philosopher Juha Räikkä, who has also published on this topic. In addition to his comments on my submission, Räikkä has also reviewed some of the other chapters of this book and provided useful comments and encouragement.

M R. X. Dentith has—along with friendly encouragement—provided detailed commentaries and useful suggestions regarding several of the chapters of this book at various stages of their development. At one point, Dentith suggested that I may have misrepresented Brian Keeley's position. So, after taking another look and making some revisions, I contacted Keeley about this. After clarifying that issue, Keeley graciously offered to take a look at other parts of my manuscript, which was by then nearing completion, and he offered several constructive criticisms that have helped me sharpen several sections, especially in Chapter 2.

Brian Martin has provided helpful and wise feedback on most of the manuscript. And his own neutral way of addressing controversial issues has been an inspiration, though I have not always found myself able to follow his model.

Meredith Cargill, a rhetorician, had assisted me tremendously on another project, *Lead Them with Virtue*, and provided feedback again on a couple chapters of this book. I deeply appreciate all his efforts to help me improve both my arguments and their expression.

I add a word of appreciation for my wonderful wife, Loni, for making possible the conditions under which this project could be completed. I also thank my parents, George and Joann Hagen, for their support and feedback, as well as proofreading, along the way.

Finally, anonymous reviewers have also contributed significantly to improvements at various stages. Of course, any faults that remain are my own, and the above-mentioned people may not endorse all my conclusions.

Introduction

Conspiracy theories are often disparaged by academics. For example, in a recent essay, psychologist Nicholas DiFonzo characterizes conspiracy theories as "outlandish, unlikely, and paranoid tales" (2019, p. 257), "incredible and improbable tales" (p. 265), and "wildly false stories" (p. 257). And yet he recognizes that he must add, "and every once in a while a true one" (p. 257). There's the rub. On the one hand, he maintains, "To apply the label 'conspiracy theory' to a story is not just to say that it is in doubt, but that it is false, and wildly so" (p. 263). But on the other hand, he must immediately contradict himself by admitting that "some conspiracy theories turn out to be true" (p. 263). DiFonzo seems to think this contradiction is solved by simply declaring the true conspiracy theories to be "exceptions." But even if we accept that, how many exceptions are there? And which theories are the exceptions? It seems that once we have admitted, as even DiFonzo recognizes we must, that some conspiracy theories turn out to be true, we cannot simply dismiss them as "wildly false stories" just on the basis that they are conspiracy theories.

Philosophers have, over the last couple decades, come to appreciate the implication of this simple truth: serious people ought to take conspiracy theories seriously. In a recent anthology, *Taking Conspiracy Theories Seriously*, edited by M R. X. Dentith (2018), a number of philosophers press this case. Even *Conspiracy Theories and the People Who Believe Them*, edited by Joseph Uscinski (2019), which includes essays from some philosophers, but contains mostly essays by social scientists, and includes DiFonzo's essay, also contains a number of essays calling for more cautious and even-handed treatment of conspiracy theories.

In this book, I address both the philosophical literature and the social science literature. I support the emerging consensus among philosophers writing on this issue, and I criticize the work of social scientists, though some of them are now beginning to recognize that much of their work on conspiracy theorists is seriously flawed.

I.1. Overview of the Book

This book is divided into three main parts.

Part I addresses the state of the philosophical debate over the epistemological status of conspiracy theories. The debate is chiefly over whether or not it is reasonable to dismiss a theory *on the grounds that* it is a conspiracy theory. The first chapter addresses the meaning of the phrase "conspiracy theory." In the second chapter, "The Philosophers of Conspiracy Theory," I briefly review the various attempts to justify dismissiveness toward conspiracy theories. I frame this in terms of a debate between dismissive "generalists," on the one hand, and "particularists," who insist that each particular conspiracy theory ought to be judged on its own particular merits, on the other. In Chapter 3, I address the issue of "conspiracist ideation" and the role of conspiracy theorists in uncovering real conspiracies, and I suggest that some of the ostensible "epistemic authorities" writing about conspiracy theorists may not be as reliable as one may hope and expect.

Part II focuses on the many problems in an article by the acclaimed scholar Cass Sunstein and his Harvard law school colleague Adrian Vermeule. I give this article considerable attention partly because of Sunstein's prominence, and partly because of the outrageousness of their proposal. The first chapter in this section focuses on Sunstein and Vermeule's proposed "cognitive infiltration" and problems with their arguments relating to that. The next chapter focuses on the implausibility of their account of the "causes" of the spread of conspiracy theories, and the fact that their "cure" is based on misleading depictions of reality. The final chapter in Part II, "Were Sunstein and Vermeule Misunderstood?" addresses Sunstein's revised version of the article, in which he unconvincingly "clarifies" his intent.

Part III is composed of four chapters, three of which address flaws in social science research on conspiracy theories. Chapters 7 and 8, respectively, address the notion that conspiracy theorists operate within a "monological belief system" and exhibit the "paranoid style." Chapter 10 addresses a call by a number of social scientists for more funding for research to help "fight conspiracy theories effectively," and the response of these social scientists when challenged on the appropriateness of that idea. Chapter 9 is an interlude that explores examples related to a theme in Chapter 8, namely, that many significant conspiracy theories do not posit implausibly malevolent conspiracies and thus do not exhibit that aspect of the "paranoid style." Taken together, the arguments in Part III suggest that there is a significant bias operating in the social sciences that is distorting the research on conspiracy theorists. Note that at least some of the psychological study of conspiracy theorists is conducted under the rubric of "anomalous psychology." And yet, as Joseph Uscinski and Joseph Parent acknowledge, "[I]t is safe to say that almost everyone believes in at least one conspiracy theory and many of us believe more than one" (2014, p. 6).

At the end of the book I include an appendix, "9/11 and Epistemic Authorities," which addresses the reliability of the mainstream media regarding 9/11, as well as that of the National Institute of Standards and Technology, government officials, and the 9/11 Commission.

I.2. Argument to the Best Explanation

Here I address some preliminary issues that will help with the understanding of many of the arguments that appear throughout this book. Namely, I explain the key elements of the "argument to the best explanation," and the related issue of simplicity, as well as the limitations of Occam's razor as a heuristic for evaluating conspiracy theories.

The meaning of the phrase "conspiracy theory" will be discussed in detail in Chapter 1. For now, it is only important to notice that it is a particular type of *explanation*. This suggests that, when judging whether or not one should believe a particular conspiracy theory, one is essentially asking whether or not it is the *best explanation*. And so, it is important to understand how "the argument to the best explanation" works.[1]

An argument to the best explanation can be thought of as having two parts, explanatory power and prior probability. Explanatory power is a measure of success in rendering a particular phenomenon-to-be-explained relatively unsurprising. Consider the phenomenon: John has a stomachache. There may be competing hypotheses put forward to explain John's discomfort. Perhaps one attributes John's condition to the fish he had eaten earlier. This explanation—John is sick because the fish was bad—has some explanatory power: *if* the fish he ate was indeed bad, it would not be surprising that he feels sick. Whether this is the *best* explanation, however, will depend, in part, on the degree of explanatory power that alternative explanations can provide. I say "in part" because there is another consideration: prior probability. This is a measure of the likelihood of the hypothesis being true *prior to any consideration of the particular phenomenon in question*. So, in this case, how likely is it that the fish is bad, *independent of John's particular circumstances*? One important element of prior probability is simplicity. It is generally thought that all else being equal (that is, assuming the other elements of prior probability and explanatory power are a wash) the simpler hypotheses is more likely. (More will be said about simplicity in the section below.) When evaluating competing hypotheses, one needs to consider both prior probability and explanatory power.[2]

To understand the power of prior probability, it is useful to consider a far-fetched example. Consider the hypothesis that teeny aliens in a tiny ship flew into John's mouth on a perverse mission to cause havoc in his stomach. This is obviously too silly to be worth serious consideration, but it is important to understand *why* it is so obviously not the best explanation. It is *not* because it lacks explanatory power. If teeny aliens with adequate technology really were trying to make John sick, it would not be too surprising that he became sick. The problem is entirely one of prior probability. Independent of John's current stomach problem, our background knowledge suggests that such beings just do not exist.

Now, some people seem to think that conspiracy theories have such a low prior probability that they can be rejected without even taking explanatory power into consideration. However, unlike the case of teeny aliens, our background knowledge tells us that sometimes conspiracy theories turn out to be true. This consideration alone suggests that each conspiracy theory ought to be judged on its own combination of prior probability and explanatory power in comparison to known rivals, including official accounts. And the more real conspiracies one is aware of, the higher the prior probability one will (and should) assign to analogous theories. For example, as historian Kathryn Olmsted notes, after various anti-Castro schemes were revealed, "Many citizens wondered: if government officials could engage in these ludicrous plots, then what other crazy things had they done—and were they continuing to do?" (2019, p. 290). This is a perfectly reasonable question. The knowledge of "ludicrous plots" raises the prior probability one should assign to theories postulating "ludicrous plots." It doesn't mean they are true, of course, or even likely. It just means the assessment of their likelihood ought to be judged higher than it would have been if one knew of no real cases involving ludicrous plots.

I.3. Simplicity

Conspiracy theories are sometimes criticized for positing theories that are too simple. The idea is that by simply postulating intentional agents directing the course of events, conspiracy theorists ignore the real complexity of the world. This criticism has been succinctly captured in the phrase "Shit happens."[3] The idea is that, due to the real complexity of the world, no group of conspirators could have their way with it, let alone ensure that their manipulations remain secret.

However, conspiracy theorists are also criticized for positing theories that are too complex[4]—too *large* and complex. Critics may appeal to Occam's razor, which is a heuristic that says, "Do not multiply entities unnecessarily," or, as the journalist David Aaronovitch puts it in his critique of conspiracy theories: "Other things being equal, one hypothesis is more plausible than another if it involves fewer numbers of new assumptions" (2010, p. 6). Too often, when Occam's razor is invoked, the clause "other things being equal" (or the word "unnecessarily") is downplayed or forgotten. But that is absolutely critical. When we evaluate competing theories, *other things are not equal*, and sometimes it is *necessary* to "multiply entities" or in other ways make the theory more complex. In other words, when we take "other things"—the relevant particulars—into consideration, the relatively more complex theory may be the best one, all things considered.

As it applies to conspiracy theories, Occam's razor suggests that, among other things, we ought to prefer explanations of events that involve relatively fewer actors. So, for example, if "the lone gunman did it" adequately explains a political assassination, we ought to prefer that explanation over more complex explanations that involve the coordination of various individuals and organizations, such as the Mob and the CIA. (The key word in that sentence is "adequately.") At first it may seem that the presence of what seems to be excessive complexity is a reasonable basis on which to dismiss conspiracy theories: we should prefer official accounts because, compared to conspiracy theories, they invariably provide simpler explanations.

But Occam's razor, which is only a rule of thumb, can be misleading. Consider this exaggerated example (which I am appropriating and adapting from Lance deHaven-Smith 2013b, pp. 65–66): the simplest explanation of certain aspects of what I observe in my daily life is, or at least *seems to be*, that the world is roughly flat and still. Now, suppose someone proposed that the world is in fact spherical, traveling at great speed through space, and spinning. Far from being the simpler explanation, this seems to introduce entirely unnecessary complications that do nothing but present a host of new problems—not the least of which is explaining why we don't all fall off. Suppose the answer to that problem was "gravity"—an invisible force holding us to the great ball, spinning and whirling through space. Well, besides being just too crazy to be seriously considered, it is an ad hoc complication to resolve a problem with a theory that was already unnecessarily complicated. Or, at least, this is how it seems, from a particular perspective—a perspective prior to certain, rather significant, considerations.

So, what has gone wrong here? Einstein's famous comment gives us a hint. He reportedly said, "Things should be as simple as possible, but no simpler." In other words, sometimes an explanation *needs* to be relatively complex. After all, an explanation of something must actually *explain* it. Sometimes the simpler theory just doesn't do the job. If a proposed explanation, a hypothesis, can't account for some phenomenon—say, by being inconsistent with it—then that hypothesis lacks explanatory power with respect to that phenomenon. And that counts against the hypothesis. When comparing hypotheses, we need to weigh up all elements of explanatory power, as well as considerations of prior probability, the latter of which includes simplicity. In the case of the supposed flatness of the world, while that hypothesis does explain much of the *surface* phenomena we *casually* experience, it turns out not to explain lots of other phenomena, more esoteric phenomena—which are nonetheless just as real and in as much need of explanation as anything else. At the end of the day, the hypothesis that we are held down by gravity on a rotating ball flying through space turns out to do a much better job explaining all the relevant phenomena. And so it is rightly accepted as correct, even though it is neither the simplest nor the most obvious position.

Similarly, the hypothesis that there was some insider complicity in the events of September 11 may well (for all we know before investigating) offer the best explanation of the available phenomena. This is true even if, considering its relative complexity, it is *not* the best explanation of the *surface* phenomena, that is, the phenomena that are repeatedly emphasized in mainstream media reports. So, to be perfectly clear, I am saying that flippant appeals to Occam's razor in cases like this often reflect a lack of sophistication. One must become aware of the phenomena that are put forward as requiring an explanation before one can decide what is the simplest *adequate* explanation. Occam's razor may still be relevant in the final analysis, but it can only do its job properly *after one has investigated the particulars*; it cannot be legitimately appealed to as a way of cutting off the need for such an investigation.

Consider political assassinations again. It is always simpler, in a shallow sense, to blame an assassination on a lone gunman than to posit a conspiracy. But when there is evidence of more bullets being fired than the purported lone gunman could have fired (as in the case of the Robert Kennedy assassination), a priori preference for simpler theories does not make that evidence go away. There must be a tug of war between the *degree of complexity* entailed by an alternative hypothesis and the *strength* of the evidence that there were too many shots fired, as well as any other indications of conspiracy (such as an autopsy report that is inconsistent with the fatal shot being fired from the witnessed position of the accused shooter). There is no formulaic solution other than this: one must consider competing arguments.

PART I

The Philosophy of Conspiracy Theories

This part focuses on the philosophical literature on conspiracy theories. Chapter 1 evaluates proposed definitions of the phrase "conspiracy theory." In Chapter 2, I discuss the issue of whether or not conspiracy theories, by their nature *as* conspiracy theories, are epistemically suspect. I address various attempts to answer affirmatively, citing and amplifying criticisms of those attempts. Chapter 3 focuses more narrowly on the problems with a single article, but for the purpose of making a number of more general points, involving (1) the concept of "conspiracist ideation," (2) the potential role of conspiracy theorists in aiding in the revelation of real conspiracies, and (3) the sometimes problematic and prejudicial treatment of conspiracy theorists by academics. This analysis shows that ostensible epistemic authorities sometimes fail to live up to expectations as epistemic authorities, which is a theme that reoccurs throughout the book.

ONE

What Is a Conspiracy Theory?

Let us never tolerate outrageous conspiracy theories concerning the attacks of September the eleventh.

—President George W. Bush (November 10, 2001, address to the United Nations)

But honestly, I mean, conspiracy theories . . .

—Tony Blair, in reference to a leaked memo that suggested that President Bush had discussed the bombing of an Al Jazeera TV station[1]

Theories suggesting that, for example, elements within the US government were behind the assassination of President Kennedy or in some way facilitated the attacks of September 11 are paradigmatic conspiracy theories, which generally conflict with official accounts. Regarding the attacks of September 11, for example, the official account is that Osama bin Laden, nineteen hijackers, and at least some other members of the al-Qaeda network with which they were affiliated, concocted and carried out a surprise attack. Although this theory clearly involves a conspiracy, it is not regarded as a "conspiracy theory." Conspiracy theories regarding the attacks, in one way or another, generally suggest some level of complicity on the part of US government officials and/or intelligence operatives. In some circles, such theories are dismissed *because they are conspiracy theories*. Should they be?

This book is mostly about what our attitude toward conspiracy theories ought to be. But we first need to understand more clearly what we are talking about when we are talking about conspiracy theories. This chapter seeks primarily to clarify that, but secondarily to begin the argument that a conspiracy theory ought not be dismissed simply because it is a conspiracy theory. I begin with a rough taxonomy of various types of conspiracy theories and examples. I then consider several proposed definitions and ultimately provide a working definition that captures most clear cases, without too many false positives, and has other advantages as well. Chiefly, it helps us understand why government officials, mainstream media personalities, and even academics would be inclined to express dismissiveness toward conspiracy theories. At the same time, it enables an analysis—which unfolds more fully in the succeeding chapters—that explains why this dismissiveness is actually inappropriate.

1.1. Preliminaries

1.1.1. A Rough Taxonomy of Conspiracy Theories

As suggested above, I take theories that imply insider complicity in the assassination of President Kennedy and in the events of September 11, 2001, to be particularly paradigmatic cases of the category "conspiracy theory." Cases that are very similar or highly analogous to either of those would usually count as "conspiracy theories," unless there was a particularly salient difference. There are a variety of other relatively clear cases that are significantly different. We need not insist on perfect agreement regarding what cases should be on a (necessarily incomplete) list of conspiracy theories for such a list to be useful. In that spirit, I offer the following rough taxonomy of cases and categories of theories that are often discussed as examples of conspiracy theories.[2] I've included separately cases that are similar but either acknowledged or at least relatively well established. Some "established cases" may not strictly count as "conspiracy theories." But their similarity to genuine conspiracy theories is nonetheless noteworthy.

Political Assassinations

These include allegations against members of US law enforcement or intelligence agencies in the assassination of John F. Kennedy, Bobby Kennedy, or Martin Luther King Jr. Similar theories about other political assassinations (or attempted assassinations) would also count, even if they are less often discussed. Examples include the attempted assassinations of Ronald Reagan and Gerald Ford. Certain theories involving plane crashes may be placed in this category, such as those involving the deaths of Senator Paul Wellstone and JFK Jr.

Established cases: Of course, the CIA *has* engaged in assassination. For example, the Phoenix Program, during the Vietnam War, involved CIA-led assassinations (see section 8.7, below). And the CIA attempted to assassinate Fidel Castro and played a role in the abduction and execution of Patrice Lumumba of the Democratic Republic of the Congo as well. But there has never been an admission that the CIA or any other US government agency has assassinated anyone domestically. Nevertheless, the murder of the Black Panthers leader Fred Hampton appears to be a pretty clear case in which the FBI, along with police, played a role in what was essentially a domestic quasi-political assassination.

False Flag Terrorism

These are allegations that a terrorist event was not caused by the prima facie culprits, who have been framed by the real culprits. For example, certain interpretations of September 11, 2001, the London bombings of July 7, 2005, and the Moscow apartment bombings of 1999 hold that these acts were perpetrated by the governments of the people who were attacked, in order to blame the attacks on someone else as a purported enemy. The Syrian Ghouta chemical attack of 2013 and the Khan Shaykhun chemical attack of 2017 have also been alleged to be false flag attacks.

Generally acknowledged cases: the Mukden Incident and Operation Himmler, which began the Pacific and European sides of World War II, respectively. Also included are terrorist attacks associated with Operation Gladio.

Stand-Downs, Provocations, and Exaggerations or Mischaracterizations Used Propagandistically to Promote War

Allegations that an attack either was allowed to happen, encouraged to happen, or did not really happen as depicted by official Western authorities. Examples include foreknowledge of, or provocations leading to, the Japanese attack on Pearl Harbor, and the reported cause of the sinking of the *Maine*.

Established cases: the Gulf of Tonkin and the coordinated propagandistic attempt to use the September 11 attacks to justify war with Iraq.

Alternative Accounts of Mass-Casualty Events

Allegations that mass-casualty events are not what they seem, either because there are additional unpursued suspects or because the event itself was exaggerated or faked. In the United States, the alleged political motive is often gun control. Examples include the Oklahoma City bombing, the Boston Marathon bombing, and the Le Penca bombing of 1984, as well as the Columbine and the Sandy Hook school shootings.

Large-Scale Election Fraud

Allegations of political insiders rigging elections, especially in the 2000 and 2004 US presidential elections. (The deadly plane crash of Republican IT specialist Michael Connell is often alleged to be connected.) More recently, Democrats (mostly) have alleged that Republicans colluded with Russians to help secure Trump's 2016 election, and Republicans have alleged that Democrats stole the 2020 election for Biden.

Elite Pedophilia

Allegations of elite pedophilia or a cover-up on behalf of prominent serial child abusers. Elite pedophilia conspiracy theories often involve allegations of political blackmail targeting sexually compromised politicians. Examples include Pizza-gate, the Finders, and the Franklin scandal. (The latter often includes the small plane crash of investigator Gary Caradori.) Suspicions have also been raised regarding the billionaire and convicted sex offender Jeffrey Epstein,[3] and Bill Clinton, who has been flown around on Epstein's jet and visited Epstein's private island. Epstein's friendliness with Donald Trump and Britain's Prince Andrew has also raised some eyebrows.[4]

Established cases: "serial child molester" Dennis Hastert, former Speaker of the House (see section 1.4.2, below) and British media personality Jimmy Savile. The investigation into allegations of child abuse against Savile was called "Operation Yewtree." Dentith and Keeley explain: "As the recent Operation Yewtree investigation in the UK has shown, prominent Britons (and at least one Australian) in the seventies and eighties not only successfully sexually predated on young men and women, but the attempts to expose this scandal were at best ignored, and at worse, covered up by influential members of British society" (Dentith and Keeley 2019, p. 292).

Other cases that are generally recognized as (or as *having been*) "conspiracy theories" include Watergate and the Iran-Contra affair, both of which have now-official versions as well as alternative accounts. The PRISM program, and various aspects of COINTELPRO count as established conspiracy theories involving illegal government surveillance of US citizens, by the National Security Agency and the FBI, respectively.

Except for those listed as "established cases," this list is not intended to suggest that these cases are warranted. They are listed and grouped to provide a sense of the dimensions and the contours of the category, which I ultimately attempt to capture in my definition. But first I will explore other definitions, following a brief discussion of the relevance of the definition of conspiracy theories to their epistemic status.

1.1.2. A Challenge to Conventional Wisdom

Charles Pigden, a philosopher who has contributed much to the understanding of the epistemic issues surrounding conspiracy theories,[5] describes the "conventional wisdom" regarding them as follows: "Conspiracy theories should be neither believed nor investigated" (2007, p. 219). Rejecting this, Pigden has thrown down a challenge:

> If you wish to vindicate the conventional wisdom, you must do two things. First you must give an interpretation of the term "conspiracy theory" with roughly the right extension. (Most of the theories castigated as "conspiracy theories" must qualify as such, and most of the conspiracy-postulating theories that conspiracy skeptics believe in must not.) You must then show that on this interpretation, the strategy of neither investigating nor believing in conspiracies makes epistemic sense. Until this is done, the idea that conspiracy theories as such are intellectually suspect is a superstition that can be safely dismissed. (2007, p. 230)

I do not attempt to answer Pigden's challenge, for I agree with what it implies. I don't think a definition or interpretation of the term that fits the bill is possible. I discuss several attempts at defining the phrase, and ultimately provide my own suggestion for a definition (which is strongly influenced by Pigden and by David Coady). This is, at least roughly, the sense of "conspiracy theory" that I have in mind when, throughout this book, I argue, in essence, that Pigden is right to reject the "conventional wisdom" regarding conspiracy theories.

It turns out that defining "conspiracy theory," even if not trying to answer Pigden's challenge, is not easy. I start with a critique of two proposed definitions: a simple and highly inclusive "constituent parts" definition preferred by a number of philosophers who (like me) are inclined to defend conspiracy theories as potentially true and important, and also a question-begging definition preferred by some dismissive critics. Then I explore difficulties with some seemingly innocuous descriptive definitions. In the end, I endorse the view that, as the phrase is typically used, conspiracy theories contrast with official accounts, and typically imply scandalous conduct on the part of Western governments or elites (perhaps extending to democratic allies such as Japan and South Korea).

I am not endeavoring to provide a perfect set of necessary and sufficient conditions, as I don't imagine that "conspiracy theory" corresponds to a Platonic Form. Nevertheless, while a fully adequate definition may prove elusive, exploring the difficulties with various proposals can help us sharpen our understanding. In any case, it may not be necessary to have a perfect definition, so long as we keep in mind paradigmatic cases whose family resemblances allow us to discuss relevant matters intelligibly.

1.2. Attempts at Definition

1.2.1. The "Constituent Parts" Definitions

To conspire is to "whisper together," that is, to secretly scheme. And a "conspiracy theory," it may be thought, is just a theory that posits a conspiracy. M Dentith, in particular, has championed this idea, which I will call the "constituent parts" definition. Dentith writes:

> [A] conspiracy theory—if we break down the term into its constituent parts—is just a theory about a conspiracy; to wit, a theory about two or more people working together in secret towards some end. This general, non-pejorative definition has been defended by philosophers like Brian L. Keeley (Keeley 1999), Charles Pigden (Pigden in press),[6] David Coady (Coady 2012), Lee Basham (Basham [2018a]) and myself (Dentith 2014). (2019b, p. 2244; cf. Dentith 2019a, p. 102)

However, most of these philosophers acknowledge that this definition does not capture what people generally mean *when they speak dismissively* of "conspiracy theories." And when the phrase comes up in the mainstream media, or in the academy, or issues forth from the mouth of a politician, it generally is used dismissively. Coady and Pigden both offer suggestions for better describing what is often meant by the term. These suggestions will be discussed below, each in their own section.

It may be that ordinary people (as opposed to mainstream media personalities, politicians, and academics) use the phrase nonpejoratively in roughly the way the constituent parts definition suggests. Basham, for example, takes this position, though he prefers causation to be included in the formulation: "[A] conspiracy theory is any explanation of events that includes a conspiracy as a salient cause" (Basham 2016, p. 6). I remain unconvinced that the constituent parts definition represents ordinary usage. Here I agree with philosopher Juha Räikkä, who writes, "Obviously, in ordinary language 'conspiracy theory' does not refer to *all* explanations that include [the] term 'conspiracy'" (2018, p. 208). However, there is evidence that ordinary people are not *influenced* to respond negatively to an allegation on account of its association with the term "conspiracy theory" (Wood 2016). Citing Wood, Basham remarks, "'[C]onspiracy theory' possesses no negative connotation except as residue among certain academic, media, and political elites" (2018b, p. 40). While significant, Wood's finding does not fully answer the question of the meaning in popular usage. And, in any case, mainstream media personalities, politicians, and academics are influential, and *their* usage is important. Further, I am addressing the arguments of those who believe that conspiracy theories can be dismissed. To challenge them directly and avoid charges of equivocation, I'm seeking a working definition that approximates *their* usage, recognizing that there will be gray areas and that usage and understandings may vary.

Let's consider a critique of the "constituent parts" definition. Philosopher Patrick Stokes describes a questionable defense of conspiracy theories that relies on this definition.

> [Equating "conspiracy theories" with conspiratorial explanations] is the first step in what has become a standard move in the epistemological literature on conspiracy theory: define "conspiracy" in a very formal and minimal way, and then show how there is nothing intrinsically irrational, or even unreasonable, about explanations of that form. Conspiratorial activity is at least sometimes, perhaps even often, the best available explanation to infer to.[7] (2016, p. 35)

And yet, as most parties seem to accept, "[T]his definition clashes with the ways we *generally* talk about conspiracy theory" (Stokes 2016, p. 35). And so, showing that a "conspiracy theory" in this broad sense should be treated as a potentially good explanation does not show that *what we generally mean by "conspiracy theory"* should be treated likewise. A group of social scientists succinctly put the criticism this way: "[A]sserting that a conspiracy theory is any kind of thinking or explanation that involves a conspiracy . . . seems like a premature attempt to settle the issue" (Dieguez et al. 2016, p. 22).[8] The issue is whether or not theories that are generally thought to count as "conspiracy theories" may be warranted, or whether we are entitled to a dismissive attitude about *those* theories. While this book is largely a defense of conspiracy theories as potentially warranted, I agree with the critics of the "constituent parts" definition on this: appeal to this definition is not the way to do it.

Those who argue that we can dismiss conspiracy theories based on generalities about them have been labeled "generalists." Following philosophers Joel Buenting and Jason Taylor (2010, pp. 568–569), Dentith and Keeley state: "According to the *Generalist*, the rationality of conspiracy theories can be assessed without considering particular conspiracy theories. On this view, conspiratorial thinking *qua* conspiracy thinking is itself irrational" (2019, p. 286). In Chapter 2, I criticize generalist perspectives at length. But refuting them convincingly will be made more difficult and confusing if we accept the "constituent parts" definition of conspiracy theories, because generalists who reflect on the obvious fact that people are convicted of "conspiracy" all the time—with "beyond a reasonable doubt" as the standard—will be quick to deny that they mean to include such things as petty conspiracies in their understanding of "conspiracy theories." In so doing, they will be denying that they believe that "conspiratorial thinking *qua* conspiracy thinking is itself irrational." While some generalists (such as David Aaronovitch, discussed in section 1.2.2, below) do at least seem to suggest that *conspiratorial thinking* is irrational or unwarranted, most do not mean to conflate "conspiracy theories" with "conspiratorial thinking," as the constituent parts definition encourages. If one wants to show that the more reflective generalists are wrong (and they *are* wrong), one has to show that they are wrong about the cases they actually intend to refer to. When people say, "Conspiracy theories are unwarranted," they do not mean to assert anything about indictments against miscreants charged with petty crimes, or about the official account of 9/11, or about surprise parties. The fact that such things also involve conspiratorial activities does not show they have all the qualities necessary to count, according to common usage (or at least dismissive usage), as "conspiracy theories."

In other words, the conspiratorial activity alleged in what counts as a "conspiracy theory" is a small fraction of conspiratorial activity more generally. Although it may not be possible to achieve a perfect definition of "conspiracy theory," the "constituent parts" definition surely captures far too many cases, and in so doing invites confusion.

Supporters of the "constituent parts" definition seem intent to highlight just how common conspiring is. Indeed, the ubiquity of conspiratorial activity is important to notice. Otherwise, we may mistakenly regard conspiracy theories as less plausible, prima facie, than they really are, *simply because we have underestimated the ubiquity of conspiratorial activity*. But it is not necessary to *define* "conspiracy theory" to include this ordinary conspiratorial activity in order to make the point that such activity is normal and pervasive. We just need to mention, as Pigden does, that "Conspiracy . . . is endemic in political life, a fact that should be fairly obvious to anyone who bothers to read the newspapers or watch the nightly news" (2006, p. 162).

If we accept an overbroad definition, then, to borrow the words of Pigden once again, "[E]very politically and historically literate person is a big-time conspiracy theorist" (2007, p. 222). However, this defense of conspiracy theories is likely to be unpersuasive to skeptics, who will respond, "It is only a *particular kind* of theory about conspiracies that I mean by 'conspiracy theory.'" In order to mount a valid defense of the kind of theories treated dismissively under the rubric "conspiracy theories"—to argue that they ought to be investigated and treated fairly rather than summarily dismissed—a more targeted definition is needed, one that makes some sense of, even if it doesn't truly justify, the dismissive attitude of intellectuals and elites. Below, I propose just such a working definition, based on suggestions by Coady and Pigden. I argue that, even with such a definition, a strong defense of conspiracy theories as potentially warranted can still be mounted. For even under more narrowly circumscribed definitions, politically and historically literate persons will at least be "small-time" conspiracy theorists, admitting that *at least some* conspiracy theories, on their own understandings of the term, have turned out to be true.

1.2.2. Question-Begging Definitions

Another problematic way of defining "conspiracy theories," this one favored by those who are dismissive of conspiracy theories, is the question-begging strategy of defining conspiracy theories as false (unreal/imaginary), unwarranted, or implausible. As Alfred Moore puts it: "Conspiracy theories are often defined as beliefs that are (among other things) untrue or unwarranted or unfalsifiable" (2019, p. 111).

Daniel Pipes, for example, makes a distinction between "conspiracies, which are real, and conspiracy theories, which exist only in the imagination" (1997, p. 20). He does so under the heading "Terms and Concepts," and immediately after stressing the importance of "exact words and precise concepts," so we can be sure he means exactly what he says here. In another formulation, he writes, "A *conspiracy theory* is the fear of a nonexistent conspiracy" (p. 21), and then he associates conspiracy theories with "unwarranted belief" (p. 21).[9] But how do we know, given any particular theory, whether it describes a real conspiracy or merely an imaginary "conspiracy theory"? It seems we would have to do some serious investigation into the relevant evidence before we could even know how to refer to the theory. Creating such an obstacle to speech is not particularly helpful. Further, since it is obvious that people use the term "conspiracy theory" prior to any serious consideration of the particular merits of the theory in question, defining "conspiracy theory" in this way can only be misleading and prejudicial in practice.

The journalist David Aaronovitch has a similar strategy. He defines conspiracy theory as "'the attribution of deliberate agency to something that is more likely to be accidental or unintended.' . . . [O]ne might add 'the attribution of secret action to one party that might far more reasonably be explained as the less covert and less complicated action of another'" (2010, p. 6). "[A] conspiracy theory is the unnecessary assumption of conspiracy where other explanations are more probable" (Aaronovitch 2010, p. 6). Focusing on Aaronovitch's contrast between conspiratorial explanations and "other" (nonconspiratorial) "accidental or unintended" explanations, Charles Pigden points out that many of the conspiracy theories that Aaronovitch chooses to focus on are unwarranted not because there is a *nonconspiratorial explanation* that is more plausible in each case, but because there is a different and (at least seemingly) better *conspiratorial explanation*—such as the official account of 9/11.[10] In making this point, Pigden is emphasizing that Aaronovitch seems, quite confusedly and mistakenly, to think there is something wrong with conspiratorial explanations as such.[11]

I want to focus on a different problem with Aaronovitch's definition, a problem that Aaronovitch recognizes in Pipes's definition. For Aaronovitch rejects Pipes's definition, asking, "How . . . can Pipes prove categorically that a conspiracy is 'nonexistent'?" (2010, p. 5) Indeed, he can't. He can only define it to be so, which, as I've suggested, simply shifts the problem to knowing when something really is a "conspiracy theory." However, by the same token, how can Aaronovitch prove, *prior to* any investigation, that a theory is less likely, all things considered, than an alternative (conspiratorial or not)? Relative likelihood of an explanation can only be established on a case-by-case basis, considering all the relevant particulars. One doesn't know *in advance* that any theory generally thought to be a conspiracy theory (such as those listed in section 1.1.1, above) is "unlikely" or unwarranted. By *defining* conspiracy theories as "unlikely," Aaronovitch faces the same problem as Pipes: especially in controversial cases, one never really knows whether one is talking about a genuine "conspiracy theory" or not. Uncertainty in life is unavoidable. But it is one thing to be unsure whether something is true, warranted, or likely; we are used to that. It is worse to not even know what term to use to refer to what one wants to talk about. The term ought to have a descriptive content that helps us know what kind of explanation we are referring to, and then we can ask whether *that* is true, warranted, or likely.

In an article titled "Towards a Definition of 'Conspiracy Theory,'" psychologist Robert Brotherton adds some descriptive content to his definition, but also includes the idea that conspiracy theories are, as part of their definition, not very plausible. He writes, "[A] conspiracy theory is an unverified and sensationalistic[12] claim of conspiracy which contradicts a more plausible account" (Brotherton 2013, p. 12).[13] Spelling out more detail, Brotherton writes:

> I define conspiracy theory as an unverified claim of conspiracy which is not the most plausible account of an event or situation, and with sensationalistic subject matter or implications. In addition, the claim will typically postulate unusually sinister and competent conspirators. Finally, the claim is based on weak kinds of evidence, and is epistemically self-insulating against disconfirmation. (Brotherton 2013, p. 9)

There is a lot wrong with this. For it is not clear that even the most paradigmatic of conspiracy theories meet these criteria. Are JFK conspiracy theories based on "weak kinds of evidence"? It is my experience that members of the JFK research community appeal to evidence of all sorts. And regarding 9/11, conspiracy theorists are constantly appealing to physical laws—hardly the weakest form of evidence. Also, do JFK conspiracy theories generally "postulate unusually sinister and competent conspirators"? The answer is no, as I explain in detail in Chapters 7 and 8. In any case, here I am most concerned with the phrase "not the most plausible account," if we take it to mean the account that comes out as the most credible after thorough investigation, not just the most plausible hypothesis prior to any investigation. The theory of a lone gunman is certainly simpler than an elaborate conspiracy, which may make it seem more of a plausible explanation at first. But is it a fact that JFK assassination conspiracy theories have turned out to be less well supported by the evidence than the Warren Commission account? That seems pretty hotly contested and has been for half a century. Lots of people who are extremely knowledgeable about the relevant particulars take the side of conspiracy, and I wouldn't want to have to debate them.[14]

And yet those who define conspiracy theories as unlikely or unwarranted, or the like, do not hesitate to name particular theories as "conspiracy theories." Brotherton et al. (2013), for example, mention a number of examples, starting with 9/11, 7/7 (the London bombings), and JFK conspiracy theories. Then they remark, "Such beliefs are usually unsubstantiated and implausible" (2013, p. 279). But if conspiracy theories are *defined* as relatively implausible, why are they only "usually" implausible?

Most philosophers seem to agree with me that the definition of "conspiracy theory" should be nonpejorative. This seems necessary if we ever wish to make the legitimacy of any such theory a topic of investigation rather than a foregone conclusion. The definition should *describe* some category without prejudging the warrant or truth-value of individual members. Once it is clear *what* we are talking about, we may proceed to evaluate it. If we simply *define* conspiracy theories to be unwarranted, then we would have to argue over the details of each theory, to adjudicate its degree of warrant, *before we could even know whether or not it is a "conspiracy theory"*! In other words, if we define conspiracy theories as unwarranted, those who support counternarratives and those supporting official accounts will not be able to agree upon terms. The trick with the pejorative definition, of course, is that people *assume* that certain theories count as conspiracy theories, and then try to write them off as unwarranted on that basis, without ever actually carefully examining their individual merits.

1.2.3. Some Relatively Neutral Definitions

Various other definitions have been offered which go a bit beyond the simple "constituent parts" definition, and yet attempt to be neutrally descriptive (as they should). Here are three such definitions, provided by law professors, a philosopher, and social scientists, respectively:

> [A] conspiracy theory can generally be counted as such if it is *an effort to explain some event or practice by reference to the machinations of*

> *powerful people, who attempt to conceal their role (at least until their aims are accomplished).* (Sunstein and Vermeule 2009, p. 205)

> A conspiracy theory is a proposed explanation of some historical event (or events) in terms of the significant causal agency of a relatively small group of persons—the conspirators—acting in secret. (Keeley 1999, p. 116)

> For *conspiracy theory*, we use a standard definition: an explanation of historical, ongoing, or future events that cites as a main causal factor a small group of powerful persons, the conspirators, acting in secret for their own benefit against the common good. (Uscinski and Parent 2014, p. 32; cf. Uscinski 2019, p. 48)

At first glance these seem relatively straightforward, with seemingly minor differences from each other and from the "constituent parts" definition. Two of them mention that the conspiracy is small, two mention the conspirators are powerful, one mentions the event at issue is historical, and so on. None of them, however, gets any closer than the constituent parts definition does to the meaning intended by those who dismiss all conspiracy theories in general, or to identifying the kinds of cases of most concern to them. Their dismissive attitude does not hinge on whether the alleged conspirators are powerful, whether the group is small, or whether the events are historical.

Also, various aspects of these definitions are questionable or unclear. (1) First, there is no "standard definition," as the third definition claims to be; the definition is very much contested, as should be becoming clear. (2) The second and third definitions both refer to *small* groups. This is odd, since many critiques of conspiracy theories object to their purported *vastness* (see Chapter 8). And, as Coady notes, "[I]t is natural to think that a conspiracy theory according to which there are many conspirators is *ipso facto* more of a conspiracy theory than one which postulates fewer conspirators. Someone who believed that absolutely everyone else was plotting against her would, I suggest, be a conspiracy theorist *par excellence*" (Coady 2003, p. 200). (3) The first and third definitions make reference to *powerful* people, but it is not entirely clear who is included and who is excluded by this qualification. And, (4) the phrase "acting in secret" in the second and third definition could be more nuanced. The first definition handles this aspect better. Alternatively, using the phrase "by partly secret means" (as does Pigden 2007, p. 222) allows for the realistic possibility that some parts of a plot may be carried out in the open, or at least be not entirely hidden.

Further, the third definition, which posits that conspiracy theories are "against the common good," seems to assume that "the common good" is uncontroversial. Problems with this assumption are discussed at length in Chapters 8 and 9. It might be better to follow Pigden once again by replacing "against the common good" with a phrase like "morally suspect" (see Pigden 2006, p. 157).[15] Or, if "morally suspect" seems too weak, I would suggest "appalling to a significant segment of the most relevant population." After all, many significant conspiracies, though appalling to many, may plausibly be justified, or rationalized, by the (alleged) conspirators. These nuances are important because they significantly impact considerations of prior probability. It is more likely, prima facie, that some group of people conspired to do what they can at least rationalize as "good" than that they are self-consciously bent on wrongdoing.

1.3. Approaching a Better Definition

1.3.1. Contrast with Official Accounts

Philosopher David Coady has suggested that conspiracy theories, as that phrase is typically used, contrast with "official stories." This is clear from the last sentence of this description:

> A conspiracy theory is a proposed explanation of an historical event, in which conspiracy (*i.e.*, agents acting secretly in concert) has a significant causal role. Furthermore, the conspiracy postulated by the proposed explanation must be a conspiracy to bring about the historical event which it purports to explain.[16] Finally, the proposed explanation must conflict with an "official" explanation of the same historical event. (Coady 2003, p. 201)

Philosophers Joel Buenting and Jason Taylor accept Coady's definition and emphasize the final part. They write, "[A] conspiracy theory is an explanation of an event that is contrary or opposed to the received, official explanation . . . the *official story*" (Buenting and Taylor 2010, p. 569). Philosopher Juha Räikkä agrees: "Official explanations can be theories and they can refer to conspiracies, but they cannot be conspiracy theories" (2009, p. 187), at least not according to the "ordinary meaning" of the phrase (2009, p. 188). Räikkä calls this the "conflict criterion" when he reaffirms this position in

2018. There he adds a nuance, writing, "Usually an explanation is called a conspiracy theory only if it conflicts *or has conflicted* with a received explanation of the same political event" (2018, p. 210, emphasis added; citing Coady 2003, p. 199). Likewise, philosopher Keith Harris understands conspiracy theories as positing explanations that are "alternative to the official account of the event(s)" (2019, p. 236), and philosopher Susan Feldman writes, "[C]alling an account a 'conspiracy theory' usually signals that it is a counternarrative, running contrary to the received account" (Feldman 2011, p. 15). Similarly, political scientist Joseph Uscinski suggests that "conspiracy theories conflict with establishment accounts" (2019, p. 20), explaining: "Conspiracy theories are competitors in the marketplace of ideas. Their prime competition stems from the establishment. . . . For example, Kennedy assassination conspiracy theories conflict with the Warren Report" (Uscinski 2019, p. 109; cf. p. 20).[17]

Not everyone agrees with this,[18] and Coady himself would prefer that we stop using the phrase "conspiracy theory" altogether, rather than try to rehabilitate it. Nevertheless, on this issue, I side with Buenting, Taylor, Räikkä, Harris, Feldman, and others in regarding Coady's suggestion as important. Indeed, this nuance is particularly relevant when discussing arguments suggesting that conspiracy theories are implausible or unwarranted because (ostensible) "epistemic authorities" (supposedly) invariably disagree with them. I address the arguments of philosopher Neil Levy in this regard in Chapter 2.

For now, consider a feature of Brotherton's above-cited definitions: that conspiracy theories are "unverified." By "unverified," or alternatively, "unsubstantiated" (2013, p. 5), Brotherton means "not regarded as verified by legitimate epistemic authorities . . . [and] invariably at odds with the mainstream consensus among scientists, historians, or other legitimate judges of the claim's veracity" (Brotherton 2013, p. 12). However, there are issues that need to be addressed regarding who counts as a genuine or legitimate epistemic authority, and about situations in which epistemic authorities disagree. Should we simply declare a minority position among experts as "implausible" simply because it does not conform to the current consensus? On the one hand, that is not how scientific or other academic debates are generally resolved. And, on the other hand, the progress of science, in the real world, is less than ideal; reigning paradigms tend to continue *not* until the accumulating evidence sufficiently favors an alternative, but until a sufficient number of prominent supporters of the status quo have died.[19] Both of these considerations suggest that simple appeal to consensus among epistemic authorities may not be appropriate. This would be particularly true in cases where there may be strong biases and perverse incentives.

Further, the "JFK research community" and the "9/11 truth movement" both include various types of experts. Are these not "legitimate judges" of veracity? Like them or not, they cannot be easily dismissed. But how exactly should their views be weighted? How do we compare the hundreds of architects and engineers who favor a conspiracy theory about 9/11, having presumably spent some time critically considering the relevant physical evidence, with the (presumably) vastly greater number of architects and engineers on the other side of the issue who have presumably directed less effort to considering the issue? (How do we validate the numerous presumptions here, and measure the extent of the relevant discrepancies, as would be needed before we could even begin such a comparison?) How should we weigh the relatively small number of experts at the National Institute for Standards and Technology (NIST), who have had access to the most information, compared to the (perhaps) larger number of critics who are denied access to some evidence, and have to do the best they can with publicly available information? And to what degree do we discount the opinions of experts who find themselves embedded in a biased institutional structure?[20] (For NIST employees, pressing the case that a particular piece of evidence suggests controlled demolition could be career limiting.)[21] These questions do not have easy answers.

The larger question is this: Does the fact that a theory is "at odds with the mainstream consensus among scientists, historians, or other [official or quasi-official] judges of the claim's veracity" really mean that it is at odds with *the most competent* judges of its veracity, that is, those *most deserving to be* epistemic authorities? Brotherton seems to think so. This book, in contrast, is filled with reasons to be skeptical—not so much by extolling the merits of conspiracy theorists, but by exploring the inadequacies of those who *should be* more reliable, but too often are not.

As far as definitions are concerned, I am willing to stipulate (for the sake of argument) that for an explanation to count as a "conspiracy theory," it must contrast with an official story, or at some point in time have done so. But whether or not that implies such a theory is unwarranted or implausible is something that will have to be argued (see the section on Neil Levy in the next chapter). It is not something that follows straightforwardly from the definition. All that follows from this criterion in the definition is that anyone defending a conspiracy theory faces the added burden of proof to overcome the presumption in favor of the already dominant official account. It does not imply that this burden can never be met.

Of course, if we define conspiracy theories as opposed to the view of "properly constituted," unbiased, critical, and informed epistemic authorities—that is, *genuine* epistemic authorities—then we simply have a question-begging definition in disguise. To say that a conspiracy theory conflicts with the view of *genuine* epistemic authorities is essentially the same as declaring it "unwarranted." But to say simply that it is in opposition to an official account leaves the question of warrant open. For official accounts are not always warranted.

Further, if we were to accept the view that conspiracy theories are defined as being in conflict with *genuine* epistemic authorities, then in order to decide whether or not a conspiratorial account of the assassination of President Kennedy, for example, even counts as a "conspiracy theory," we would have to figure out *who the better epistemic authorities really are*. One way to do this would be to look at the relevant evidence and see which side deals with it most honestly and reliably. It seems a bit of a roundabout approach. Nevertheless, I argue in the appendix that, regarding 9/11, many of the sources we ought to be able to trust as part of a properly constituted set of epistemic authorities are in fact less than fully reliable. Indeed, a theme that runs through Parts II and III is also that scholars are unreliable in a clearly lopsided way in their treatment of conspiracy theories and conspiracy theorists, which strongly suggests that a pervasive bias is limiting the efficacy of the checks and balances that should have prevented clear errors and fallacies from becoming widely accepted in certain circles in the academy.

1.3.2. Appalling Conduct by Western Governments

Although Pigden has professed a personal preference for the "constituent parts" definition of *conspiracy theory*, he nevertheless recognizes that the term is not most typically used in this sense. His observations are revealing:

> [T]he concept of a conspiracy theory as it is commonly employed is a chauvinist construct. It is not to be understood in terms of governments generally, but in terms of *Western* governments, and recent Western governments at that. When people say or imply that conspiracy theories ought not to be believed, what they actually mean (in so far as they have a coherent idea) is that we should not believe theories that postulate evil[22] schemes on the part of recent or contemporary Western governments (or government agencies) and that run counter to the current orthodoxy in the relevant Western countries. (Pigden 2007, p. 229)

Pigden does not seem to approve of this usage, labeling it a "chauvinist construct."

Philosopher Patrick Stokes, who remains "wary of conspiracy theorizing as a practice" (2016, p. 37), nevertheless acknowledges the point. He writes:

> Philosophers acknowledge [that the constituent parts] definition clashes with the ways we *generally* talk about conspiracy theory. They like to remind us that, according to this definition, the "official" explanation for the 9/11 attacks is itself a conspiracy theory; that is, it explains the attacks as the outcome of a conspiracy on the part of al-Qaeda. Yet when we think of "conspiracy theories" we don't generally think of such "accepted" explanations as falling under that heading. We don't typically group officially sanctioned beliefs about al-Qaeda flying planes into buildings or Russian FSB agents murdering Kremlin opponents with polonium-laced tea with beliefs about the New World Order or the "Clinton Body Count." Yet there's nothing *structural* that differentiates the first set of beliefs from the second. If there is a formal difference between "Putin murdered Alexander Litvinenko" and "Bill Clinton murdered Vince Foster" it is hard to see what it might be. Appeals to the official status of one story but not the other don't work, because an officially sanctioned story in one society might be considered a conspiracy theory in another. If we attempt to force that sort of solution we end up, as Pigden points out, with a blatantly gerrymandered and chauvinistic definition according to which a conspiracy theory is "a theory which posits a secret and morally suspect plan *on the part of Western governments or government agencies* to influence events by partly covert means" [Pigden 2006, p. 164]. We have obvious reasons to look askance at any definition of conspiracy theory that entails that conspiracies are something only *other* societies do. (Stokes 2016, p. 35)

This is all very illuminating, but there is something just a little off about the last sentence. There is nothing necessarily wrong with a *definition* of "conspiracy theories" that is exceptionalist. If that is how the term is used, then that, in a significant sense, is *what it means*. And (putting subtle philosophical arguments aside) it is not inappropriate for a definition to try to capture what people mean by a term. The problem is not with the *definition*, but with the *term* being defined. It is not Pigden's *definition* that is chauvinistic, as Stokes says; it is rather *the term* that is chauvinistic, as

Pigden says. And it is only a real problem for people who want to use the term dismissively. The problem is that such "a blatantly gerrymandered and chauvinistic" concept is unlikely to separate legitimate accusations of conspiracy from unfounded ones.

I would like to add one caveat to the idea that the phrase "conspiracy theories" applies to theories that accuse Western governments. First, nobody, to my knowledge, explicitly claims that that is what he or she means when using the term—unless "Western" is understood not geographically but rather as shorthand for democratic countries, or "open societies." For it is not uncommon for those critical of conspiracy theories to make an explicit distinction between open societies and closed societies, suggesting that conspiracies that conflict with official stories are less likely to be true in open societies (see Chapter 4 for a discussion of this idea). And so perhaps it would be better to conceive "conspiracy theories" as involving sufficiently disturbing allegations against governments of countries that are generally regarded as "open," namely, democracies with a "free press." This would include countries such as Japan and South Korea, who are, in addition, allies of the West.

The inclusion of South Korean conspiracy theories is potentially significant given the scandals associated with the downfall of President Park Geun-hye. Interestingly, a nontraditional journalist and "noted liberal pundit," Kim Eo-jun, whose perceived credibility has increased in the wake of the various revelations, has been accepted into the mainstream Korean media, even though he is still referred to as a "conspiracy theorist" (*eummo ilonga* 음모 이론가), though seemingly without much stigma—or so my trusted Korean sources inform me. Unfortunately, the Korean scandals are not covered well in the English-language press, so I leave it as a promising area for future research for someone fluent in Korean.

1.3.3. Putting It Together

We are now ready for a working definition. When the term *conspiracy theory* is used in a dismissive way, in its descriptive aspect it seems to refer to a theory that expresses a contrarian attitude toward officials, governments, and other authorities in Western or open societies by searching for and claiming to have found evidence of appalling, secret actions by elites in those societies who steer both events and the official accounts of those events. I do not claim that this definition will cover every use of the term in public, political, legal, and philosophical discourse. Nor do I claim that every future use of this term ought to conform to this definition. I claim only that it includes most of the cases intended by those who are dismissive of conspiracy theories in general, without too many false positives. This book supports the position that such conspiracy theories cannot reasonably be assumed to be unwarranted. They cannot be presupposed to be unwarranted before any investigation, and they cannot be presumed to be unwarranted simply because they fit the category.

Note that the condition "appalling conduct" has a subjective side to it, for to be appalled is to *regard* something as horrible. However, "appalling" is not *wholly* subjective. The idea is that it would be reasonable to expect that many if not most people in the most relevant community would be appalled. For example, if Americans found out that Vice President Cheney, along with a cabal of other insiders, orchestrated or enabled the 9/11 attacks to serve as a pretext for war, they can reasonably be expected to be appalled, even if the conspirators thought they were somehow justified.

However, sometimes critics magnify the degree to which conspiracy theories allege something appalling in order to make those theories seem unrealistic. For example, Brotherton seems to imply that conspiracy theories are relatively implausible because "they assume everything is intended, with malignity" (Brotherton 2013, p. 5). In other words, "almost nothing happens by accident, only by agency" (p. 11)—by "evil" agency (p. 11). This, as we will see in detail in Chapters 8 and 9, is a misleading exaggeration. Conspiracy theories often do not posit a malevolent intent, and they do not posit that everything that occurs was intended by shadowy but powerful conspirators, a view that implies "preternatural effectiveness," which is an aspect of "the paranoid style." This also conforms to "the conspiracy theory of society," a theory in which almost no one believes.[23] (The "conspiracy theory of society" is discussed in section 2.3, and "the paranoid style" is the focus of Chapter 8.)

A definition satisfactory to everyone will probably never be found, nor should any one definition be expected to serve all purposes, but the above characterization serves well enough to indicate the reasons that many people feel uncomfortable with conspiracy theories. Conspiracy theories suggest that (1) ostensible epistemic authorities are not truly reliable, and that (2) *our* leaders, or people in positions of authority in social or political structures that we associate with, would engage in appalling conduct. These are the aspects of conspiracy theories that underlie skepticism toward them, more than the mere suggestion of conspiratorial activity. Thus, the two features that constitute the only criteria in the constituent parts definition turn out to be not even the most important criteria. And so, if one wants conspiracy theories to be taken more seriously, as I do, it is these other considerations that must be addressed.

1.4. Further Considerations

1.4.1. Political Conspiracy Theories

There is one sense in which the concept of "conspiracy theory" that I am concerned with in this book differs from some usages of the term. I am not trying to provide a comprehensive descriptive definition covering all the ways the term actually gets used in public. I'm focusing on what could be called "*political* conspiracy theories," a subset of theories commonly dismissed under the rubric of "conspiracy theories." While what counts as "political" is arguable, there are certain theories that are often characterized as "conspiracy theories" that are so different from the more paradigmatic (political) conspiracy theories that their political aspect, if one exists at all, is less salient than other features. I'm thinking particularly of theories involving such entities as Bigfoot and UFOs, which are often lumped together with conspiracy theories involving only human activities (often, it would seem, to suggest guilt by association). It is as if "pointing to Enron's conspiracy to manipulate the supply of electricity to inflate rates in California is akin to asserting the existence of little green men," as sociologists Marin Orr and Ginna Husting put it (2019, p. 82). I also consider skepticism of mainstream scientific orthodoxies as a distinct, albeit related, phenomenon.

Although these other topics may have political and conspiratorial components, and some more than others, they are more fundamentally about other matters. Theories involving aliens from other worlds, for example, must posit the visitation of Earth by, well, *aliens from other worlds*! While such visitation, for all I know, may indeed be occurring, it is such a different issue that it seems unfair to group it with less fantastic, merely terrestrial theories. As for skepticism regarding scientific orthodoxy, many of these skeptics do not posit, or at least need not posit, a significant level of conspiratorial activity. At the very least, conspiratorial activity is not usually central—even if exaggerated critiques of these theories often suggest that it is. Further, I would be inclined to argue that the better versions of skeptical arguments regarding scientific orthodoxies do not typically involve appalling conduct among Western elites. While I find all of these issues interesting, especially UFOs and skepticism regarding scientific orthodoxies, and while they share some similarities with, and sometimes overlap with, political conspiracy theories, they are different enough to deserve separate treatment, and will not be discussed in these pages (except where unavoidable), nor included in what I mean by "conspiracy theories."

In addition, relatively innocuous counternarratives, such as the theory that Elvis faked his own death, are at best marginal cases.[24] And there may be good reasons to exclude them in our reasoning about the plausibility of conspiracy theories, or at least be aware of relevant considerations. Partly to capture the idea that "conspiracy theories" that are generally of interest are ones that have important implications in the world, but also because of certain epistemic implications, Lee Basham has coined the phrase "ambitious conspiracy theories" (2018a, p. 274). Alternatively, Basham also uses the adjective "momentous," which I think excludes too much. For example, the case study below, regarding a rape accusation against Bill Clinton, involves a conspiracy theory that I would regard as "ambitious" because it levels a very serious allegation against a prominent and powerfully connected person, but I would not classify it as "momentous." It should be noted, nevertheless, that Basham's *point* regarding momentous conspiracy theories is well taken. His point is this: the more "momentous" a conspiracy, the more its exposure will involve revealing what Basham calls "toxic truths." These are truths that, in a certain sense, nobody has an interest in revealing because they are too destabilizing. This leads to what is essentially a wide-scale cover-up, and denouncement of any "conspiracy theory" that gets too close, though no explicit *conspiracy to cover up* is required.[25] As Basham elsewhere explains, "The more damaging a conspiracy theory to the established order, the greater the reason every element of the hierarchy has to crush it, regardless of whether or not they are directly involved in the conspiracy itself" (2011, p. 69).[26]

It might be objected that by excluding these subcategories I have merely defined what I myself happen to be interested in. There is some truth in that, though I don't think that is the whole truth. In any case, it is not a wholly unreasonable thing to do, to stipulate how one intends to use one's key terms. I am trying to defend certain things and not others. Some issues—such as UFOs and scientific controversies—just require a different defense and are beyond the scope of what I can provide here. (And I'm not sure of the degree to which I want to provide one at all.) Other issues I have no interest in defending, either because I don't think they deserve to be defended or because I don't think they are important enough to matter.

1.4.2. Case Study: The "Bill Clinton Is a Rapist" Conspiracy Theory

Juanita Broaddrick has accused Bill Clinton of having violently raped her in 1978, when Clinton was the attorney general of Arkansas and running for governor. Broaddrick, at the time, was a nursing home administrator and Clinton supporter. Does the idea that Clinton raped Broaddrick count as a conspiracy theory? If so, why? There is no conspiracy implied in the theory of the *event itself*. But let's suppose that the theory also involved the idea that Hillary Clinton knew something of the event (whether or not she believed it to be consensual) and helped, in small ways, to cover it up. (This is not an arbitrary supposition, for it seems to be, at least roughly, what Broaddrick believes. See Broaddrick 2017, p. 45.) Many people would regard this as a conspiracy theory, not just in the "constituent parts" sense—a theory involving a conspiracy—but in the sense that people mean when they speak dismissively about conspiracy theories—a theory implicating elites in a Western society engaged in appalling conduct denied in official accounts.

What is it about this case that makes it seem worthy of dismissiveness? Surely it is not that it alleges rape. Nor is it the allegation that there was at least some minimal collusion in the covering up of the rape. What makes this a "conspiracy theory" is that the allegation is against *Bill Clinton*. And yet that is kind of strange. Why would such an accusation against Bill Clinton in particular be regarded as any less credible on its face (that is, before considering the evidence) than other accusations of rape, which are not generally dismissed? After all, unlike most people, Bill Clinton has a history that justifies heightened suspicion that allegations of sexual misconduct may be true. His sordid encounter with Monica Lewinsky is well known. And though the truth about the Paula Jones case is not clear, the fact that he paid $850,000 to settle the case does not support the notion that he deserves *more benefit of the doubt* than others who are accused of sexual misconduct.

Surely the real reason Broaddrick's allegations are treated dismissively is because of Clinton's *status*. At the time of the alleged incident Clinton was attorney general of Arkansas and running for governor, and, obviously, he went on to become president of the United States. Isn't *that* really the reason that the accusation is called a conspiracy theory?

But, again, that is not a good reason to be dismissive about such an accusation. We know that elites sometimes behave badly—we even know that Clinton himself is not above engaging in sexual misconduct and lying about it. And other examples abound: President Kennedy was hardly a saint in this regard (though I don't know of allegations of rape). And then there is, at the congressional level, the outrageousness of Anthony Weiner. Let's not forget how Senator John Edwards disgraced himself during the 2008 presidential primary race (which was particularly disappointing to me). To be fair to both sides of the aisle, I should here repeat that Dennis Hastert, former Speaker of the House, was "a serial child molester," according to District Judge Thomas M. Durkin, who sentenced Hastert to fifteen months in prison for a related financial crime.[27] And recently, Supreme Court justice Brett Kavanaugh has been credibly accused of sexual misconduct in his youth. The accusations of Christine Blasey Ford, who was only one of three prominent accusers, were so credible that even Republican senators who went on to vote in favor of Kavanaugh's appointment to the Supreme Court acknowledged her credibility and shied away from any criticism beyond suggesting that it may have been a case of mistaken identity. (I'm sorry if I missed your favorite example; this list is not meant to be exhaustive.)

It may be that people who have significant power and prominence (or are related to powerful people) are *more* likely to engage in sexual misconduct, including rape. It is hard to say. On the one hand, they have more to lose, so they *should* be especially careful. On the other hand, they may have influence over levers of power that emboldens them. Such people may also have oversized egos, and some may have a heightened propensity to take risks. In any case, there is little reason to think that they are morally better than the rest of us. And even if they were, on average, better than us, we are dealing with individuals, not averages.

So, it seems that part of what makes something count as a "conspiracy theory" is that it *involves an accusation against an elite*. That is the main point here regarding what "conspiracy theory" means. But let's pursue this example a bit further to clarify what is wrong with that.

While this is not the place to adjudicate the *truth* of Broaddrick's accusation, we do want to gain some purchase on whether a person is *justified in dismissing* such an accusation with little or no knowledge of the pertinent details. It seems that if one has *no* knowledge of any details, one should not regard this accusation much differently than any other rape accusation. But in cases like this, especially when the accused is a prominent person, it is not uncommon for those who are dismissive to have *some* information. For example, in this case, one may be aware that (1) Broaddrick did not tell her then-husband after the alleged incident, that (2) she later attended another Clinton event (or at least was present at the location shortly before the event), and that (3) when called to give a deposition in the Paula Jones case, Broaddrick denied that the rape took place.

Are such considerations sufficient to warrant dismissiveness regarding such a charge? No. There may well be reasonable explanations for these facts (and, for all one with limited information really knows, those details might not be entirely correct). In this case, according to Broaddrick: (1) Her marriage was already in trouble, and she neither spoke to her husband much nor believed he would be sympathetic (2017, p. 42). (2) Prior to the alleged rape, she was involved in setting up the later event. After the incident, she made an excuse for not being able to attend the event, but instead swung by before the event to drop off checks (2017, p. 44). In other words, she was fulfilling obligations stemming from before the purported rape, while trying (unsuccessfully) to avoid further involvement with the Clinton campaign. (3) She was drawn unwillingly into the Paula Jones case and did what she needed to do (lied under oath) to prevent herself from being drawn in further; it was an act of self-protection (2017, pp. 63–64).

I am not arguing that these are *adequate* explanations. Rather, they are meant to provide concrete examples of the *kinds* of explanations that *may* exist, and which, for all we know before investigating, *may* be adequate. In addition, there may be positive evidence that bolsters the allegations. In this case, though Broaddrick did not tell her husband, she did tell credible others. One of these, Norma Rogers, was with Broaddrick shortly after the alleged incident and could testify to her condition at that time (see Broaddrick 2017, pp. 39–43). And when Broaddrick was again dragged unwillingly into legal proceedings, this time by Kenneth Starr, she reported that the rape *did* take place, but that Clinton never asked her to lie about it. (It was obstruction of justice that Starr was pursuing so he did not pursue her case further.)[28]

It would take much more evidence than the brief sketch above to substantiate Broaddrick's claims—that is not the objective here. But it would also take much more evidence to be justified in dismissing them—*that* is more to the point. If there are good reasons to reject Broaddrick's claims, then *that specific set of reasons* is the basis for rational rejection, not the notion that the claim fits into some *type* of belief. And yet to use the "conspiracy theory" label as a term of abuse discourages the exploration of this evidence, and thus does a disservice to both truth and justice.

1.4.3. What to Do about the Label "Conspiracy Theory"?

As is clear from the preceding discussion, using the phrase "conspiracy theory" as a term of abuse is inappropriate; that all-too-common practice should stop. (And people who take themselves to be sophisticated, in particular, should be more careful.) But there are two general lines of thought about *how* it should stop. (1) We could just stop using the phrase altogether (as David Coady recommends).[29] Perhaps we could substitute a nonpejorative alternative, such as "State Crimes Against Democracy (SCADs)," as Lance deHaven-Smith, professor emeritus of public administration and policy at Florida State University, has recommended. Or, (2) the phrase could be rehabilitated—we could keep using it but stop using it *as a pejorative*.

I tend to favor rehabilitation. And so I will continue to use the phrase, though in a nonpejorative sense (except in reference to other people's usage). I do, however, also think that the introduction of the term SCAD can be useful. Further, decreased general use of the term, rehabilitated use, and the use of alternatives are all compatible. We can refrain from using it when it is easily avoidable, use it in a nonpejorative sense when it is not, and use alternatives when suitable.

There will be occasions when one would like to single out conspiracy theories that are unwarranted. In such cases, one should be explicit, calling them "*unwarranted* conspiracy theories." This change has practical significance because it makes it harder for anyone to pretend that all she has to do to show that an idea is unwarranted is to show that it counts as a conspiracy theory. It will be transparently clear that she will have to give reasons to think that it is, in addition, an *unwarranted* one.

TWO

The Philosophers of Conspiracy Theory

Are Conspiracy Theories Epistemically Suspect?

[W]e are entitled to an attitude of *prima facie* skepticism to the theories propounded by conspiracy theorists.

—Steve Clarke (2006, p. 129)

[A] "conspiracy theory" is an explanation that conflicts with the account advanced by the relevant *epistemic authorities*. . . . [I]t is almost never rational to accept such a conspiracy theory . . . [A]ccepting the official story is almost always rational.

—Neil Levy (2007, pp. 181, 190)

Sometimes, conspiracy theories turn out to be right—Watergate really was a conspiracy—but mostly they are bunkum.

—Quassim Cassam (2015)

[T]here is indeed something deeply epistemically wrong with conspiracy theorizing.

—Pete Mandik (2007, p. 205)

[T]he idea that there is something suspect about conspiracy theories is one of the most dangerous and idiotic superstitions to disgrace our political culture.

—Charles Pigden (2006, p. 139)

The errors of excessive willingness and excessive unwillingness to believe conspiracy theories correspond to two epistemic vices . . . "paranoia" and "naivety."

—David Coady (2006, p. 10)

> The real problem . . . is determining what a relatively conspiracy-free society would look like, that we would be *well justified in believing is relatively conspiracy-free*. One thing is clear: It would not look like ours.
>
> —Lee Basham (2003, pp. 100–101)

> [I]f we accept that conspiracies are not unlikely . . . surely they can—in a range of cases—feature in our best explanations.
>
> —M Dentith (2016, p. 582)

In this chapter, I address the arguments that philosophers (and would-be philosophers)[1] have offered in various attempts to justify a dismissive attitude toward conspiracy theories. I also cite responses that other philosophers have given to these arguments, often amplifying those responses with my own. In the end, I hope to persuade rigorous thinkers that a dismissive attitude toward conspiracy theories is not intellectually defensible. At least, despite considerable effort, nobody has succeeded in defending it well. So the burden of proof rests squarely with those who wish to adopt such an attitude.

2.1. Generalism and Particularism

Since the 1990s, primarily, philosophers have been arguing over the epistemic status of conspiracy theories. As the debate developed, two distinct camps became distinguishable. Some of these scholars have sought to justify the rather dim view of conspiracy theories that is generally assumed in the academy. Others have argued that the bad reputation that conspiracy theories have in such circles is undeserved. (This is not to say, of course, that the bad reputation of *any particular* conspiracy theory is necessarily undeserved.) These two camps have come to be referred to as "generalists" and "particularists"—at least by the particularists. I should acknowledge here that I am in the particularist camp, and that one reason I highlight the distinction between particularists and generalists is that it clearly exposes the implausibility of the generalist project. The chief problem for generalists is the fact, which they usually admit, that at least some conspiracy theories are both true and warrant belief. Generalists, understandably, will thus be motivated to criticize the distinction, or claim that they are not actually generalists, or both (for example, see section 10.2, below). Nevertheless, the continuing prevalence of generalism in the academy is suggested by the stated purpose of a European Union research network, COMPACT, which seeks "better understanding of conspiracy theories in order to develop an effective response to them" (according to the organization's website).[2] Citing membership in that organization, M Dentith writes, "I am quite aware that generalism still is the dominant force in the academic sphere, with only faint lip service paid to particularism when people say 'Well, of course *that* conspiracy theory turned out to be true, but . . .'" (2018d, p. 65).

The labels "generalist" and "particularist" stem from an article by Joel Buenting and Jason Taylor.[3] They explain, "According to the *generalist view*, the rationality of conspiracy theories can be assessed without considering particular conspiracy theories" (Buenting and Taylor 2010, p. 568). Conspiracy theories, according to this view, may safely be regarded as unwarranted based on some general considerations about the nature of conspiracy theories, or about the conspiracy theorists who support them. Particularists deny this. They argue that each conspiracy theory ought to be judged on its own merits, and cannot be summarily dismissed as unwarranted or implausible simply because it is a conspiracy theory. As Buenting and Taylor put it, "[A]ssessing the rationality of a conspiracy theory should be done on a case-by-case basis" (2010, p. 570).

When particularists say that conspiracy theories have to be judged on a case-by-case basis, we do not mean that no generalities can be taken into consideration in evaluating cases, as one group of social scientists (discussed in Chapter 10) has implied (Dieguez et al. 2016, p. 23 n. 4). Certainly, considerations of prior probability,[4] which are general in nature, are relevant. What particularists deny is that there is any *decisive* general consideration about the class of ideas referred to by the term "conspiracy theories" that obviates the need for consideration of particulars when evaluating such theories.

In addition to Buenting and Taylor, the list of philosophers who have argued in favor of a particularist view regarding conspiracy theories is growing (though not all have *explicitly* embraced the term "particularist"). Lee Basham and M Dentith sum it up this way:

> In the last two decades philosophers like Charles Pigden, Brian L. Keeley, David Coady and, yes, ourselves, have taken a close look at conspiracy theories, and the news is in: belief in conspiracy theories is not irrational and the conspiracy theorist, despite the opprobrium expressed towards her, has emerged as good a thinker as you or us.

> Their theories are intriguing, and often constructed with a careful eye to the standards of both logic and evidence that we all share. "They" are just like us. In fact, "they" are us. (2015)

I think it is fair to add Juha Räikkä,[5] as well as myself, to this list. In large measure, our work has been a response to generalist arguments. Together, this work shows that serious academic attempts to justify dismissiveness toward conspiracy theories have all failed.

The category of "generalist" is not without its problems. Nobody is a generalist in the strict sense of ruling out all conspiracy theories on the basis of some generalization—because virtually everyone participating in the debate admits that some conspiracy theories are true. Still, there are a number of scholars who, in one way or another, can be viewed as justifying a dismissive attitude toward conspiracy theories on general grounds (as the epigraphs by Clarke, Levy, Cassam, and Mandik at the top of this chapter suggest). Though imperfect, the distinction between particularists and generalists provides a convenient way of referring to two distinct approaches to the philosophical study of conspiracy theories. Below, I briefly describe the arguments and positions of philosophers (or would-be philosophers) most of whom may be usefully characterized as generalist, at least in spirit. But first I want to mention two prominent philosophers who were also conspiracy theorists.

2.2. Two Twentieth-Century Conspiracy Theorists: Bertrand Russell and John Dewey

As I've indicated, conspiracy theorists have a bad reputation among intellectuals. Conspiracy theorists, many academics seem to think, just aren't as sophisticated as they themselves are. And yet, one of the most renowned British philosophers of the twentieth century, Bertrand Russell, was clearly unconvinced by the official account of the assassination of President Kennedy, as evidenced by his 1964 essay, "16 Questions on the Assassination," which was highly critical of the Warren Commission. His questions, and the reasoning behind them, had all the hallmarks of conspiracy theory—not that there is anything wrong with that. He asked, for example:

- Why were all the members of the Warren Commission closely connected with the US government?

- Why was the president's route changed at the last minute to take him past Oswald's place of work?
- How did Oswald manage to shoot the president in the front from behind?[6]

Let's consider one more of Russell's questions, this time including his reasoning:

> The Dallas police took a paraffin test on Oswald's face and hands to try to establish that he had fired a weapon on November 22. The Chief of the Dallas Police, Jesse Curry, announced on November 23 that the result of the test "proves Oswald is the assassin." The Director of the F.B.I. in the Dallas–Fort Worth area in charge of the investigation stated: "I have seen the paraffin test. The paraffin test proves that Oswald had nitrates and gunpowder on his hands and face. It proves he fired a rifle on November 22." Not only does this unreliable test not prove any such thing, it was later discovered that the test on Oswald's face was in fact negative, suggesting that it was unlikely he fired a rifle that day. Why was the result of the paraffin test altered before being announced by the authorities?

Over fifty years later, as I write this, Russell's point is still a good one. The paraffin test on Oswald's cheek was negative, which does tend toward exoneration (though not conclusively), and the police and the FBI, who clearly should have known better, dramatically mischaracterized this evidence. It is worth remembering that when people dismiss conspiracy theories as a whole, they are, among other things, dismissing *this*: the blatant mischaracterization of significant evidence by ostensible epistemic authorities regarding matters of great importance. Is that really wise? Russell *may* have been wrong about the strength of the case against Oswald, when taken as a whole—I'm not judging that here. The point is that the questions Russell raises are, and should be, concerning. There really should be no controversy about *that*. Yet it all gets swept under the rug when suggestive questioning and skepticism regarding official accounts gets dismissed as mere "conspiracy theory."

Another acclaimed philosopher of the same period, John Dewey, headed up what came to be known as the "Dewey Commission," which investigated what are now known as the Moscow Show Trials. In contrast to official opinions of the times, Dewey concluded that the trials were, well, *show trials*. Dentith provides a summary:

> The result of the commission was a report that included the claim that the verdicts in the Moscow Trials were a sham—the result of a conspiracy by senior members of the Communist Party. Stalin and his cronies countered that the Commission's report was a campaign of disinformation. . . . However, after Stalin's death Nikita Khrushchev, the new leader of the USSR, admitted that the Dewey Commission was largely right: Stalin *had* wanted the alleged Trotsky sympathisers to be found guilty to prove the existence of a (non-existent) conspiracy by Trotsky to return to Russia. Stalin and his cronies were the crazy conspiracy theorists, while the conspiracy theory put forward by the members of [the] Dewey Commission turned out to be warranted. . . . At the time, the Dewey Commission's claims were not just pooh-poohed by the USSR but also by the governments of the USA and the UK. Britain and America trusted Stalin's assurances that the trials were free and fair. (Dentith 2015, pp. 37–38)

One would think that a case like this should be a cause of some circumspection regarding how one should regard theories that conflict with official accounts. But the fact that Dewey had been both a conspiracy theorist and right to be one appears to have had little effect.

2.3. Karl Popper versus Charles Pigden

Instead, it was another prominent twentieth-century philosopher who exercised significant influence, at least in academic circles, regarding the status of conspiracy theories: Karl Popper. Popper's critique of what he called the "conspiracy theory of society" made dismissal of conspiracy theories seem to be the most intellectually respectable posture. So it is perhaps fitting that the recent debate that has occurred in the philosophical literature begins with a refutation of Popper—or, at least, a refutation of what Popper is commonly *thought* to have shown (Pigden 1995).

Popper's notion of the "conspiracy theory of society" involves "various powerful men and groups—sinister pressure groups, who are blamed for having planned the great depression and all the evils from which we suffer" (Popper 1972; reprinted in 2006, p. 13). According to Popper, "The conspiracy theory of society is very widespread, and has very little truth in it" (p. 13). In Popper's view, the "conspiracy theory of society" is the secular equivalent of believing that the schemes of Greek gods explain the events of history. Powerful individuals are thought to have engineered the major events of history: the great wars, the Depression, and so on. However, the conspiracy theory of society cannot be right because, as Popper points out, "*nothing ever comes off exactly as intended*" (p. 13). There are always unintended and unwanted consequences that cannot be explained by conspiracy. Although Popper does admit exceptions, naming "Hitler's war" (p. 15), he is clearly a generalist about "*the conspiracy theory of society*"—theories *of this kind* can be safely dismissed on the general grounds just stated, without consideration of any particulars.

Pigden (1995) points out that the central problem with Popper's argument is that it isn't an argument against *conspiracy theories*. It is an argument about something different, the "conspiracy theory of society," in which almost no one believes. Popper's assertion that this view is "very widespread" notwithstanding, most conspiracy theorists do *not* assume that everything worked exactly as planned. And the observation that "nothing ever comes off exactly as intended" has no force against any theory that does not assume this sort of perfection to begin with. (See Chapter 8 for examples.) That leaves just about all interesting conspiracy theories, certainly the much-discussed ones, unscathed by Popper's arguments. As Pigden puts it:

> [S]mug denials of this or that conspiracy on the grounds that the conspiracy theory of history [or society] is false are based on a confusion. The falsehood of the conspiracy theory (as understood by Popper) does not entail that *individual* conspiracy theories are false. It does not even create a presumption to that effect. (1995, p. 8)

Of course, that doesn't mean that all those theories are good ones. Whether they are or not has to be judged case by case—or so the particularists insist. My arguments in Chapter 8 can be seen as mirroring those of Pigden, especially those applying to "preternaturally effective" conspirators. For "the conspiracy theory of society" is a version of the paranoid style in which the notion of "preternaturally effective" conspirators is taken to its logical extreme. Essentially nobody asserts the extreme version. As for less extreme versions (which are thus somewhat less implausible), most conspiracy theorists do not assert or imply them either. Rather, most conspiracy theorists seem to hold a fairly commonsensical view of the imperfect ability of powerful people, some heading hierarchical organizations, to attain the cooperation of others in order to influence events, and to do so with some measure of success, at least some of the time. Some of this cooperation is thought to be conspiratorial, some merely bureaucratic, and some stemming from a confluence of interests, or from perceived incentives and disincentives. In other words, life is complicated, and collusion is *part* of it. People act intentionally, sometimes cooperatively, within a complex system of moving parts. Sometimes people work together secretly to achieve specific objectives, even if the process produces numerous uncontrollable loose ends.

The main point here is simply that Popper's critique of "the conspiracy theory of society" does not give us a reason to dismiss any specific conspiracy theory unless that theory assumes godlike powers of control. Few conspiracy theories do, and almost no one believes in them.

2.4. Jeffrey Bale: Distinguishing Bogus Conspiracy Theories from Plausible Accounts

Here I interject a leading expert on terrorism, Jeffrey Bale, whose (attempted) contribution to this issue is philosophical—his strategy is to introduce an incisive distinction. Bale's intention is, from my perspective, admirable. He seeks to remedy a bias in the academy against all things conspiratorial. He notes that "the word 'conspiracy' seems to set off an internal alarm bell that causes scholars to close their minds" (Bale 2007, p. 47). He elaborates:

> [Even when] belief in the historical importance or efficacy of political conspiracies . . . is expressed in a very cautious manner, limited to specific and restricted contexts, supported by reliable evidence and hedged about with all sorts of qualifications, apparently it still manages to transcend the boundaries of acceptable discourse and to violate unspoken academic taboos. (p. 47)

Bale usefully highlights a number of genuine and significant political conspiracies, such as the Loggia Massonica Propaganda Due (aka "P2")[7] (which is related to Operation Gladio),[8] and the Afrikaner Broederbond.[9] His explicit point here is that even hard-to-believe accusations of conspiracy are sometimes true.

However, Bale thinks that there is a class of conspiracy theories that are not plausible. His purpose is to establish "a clear analytical distinction between actual conspiratorial politics and 'conspiracy theories' in the pejorative sense of that term" (2007, p. 45). In other words, while Bale admits that conspiratorial activities do occur, real conspiracies, he believes, are distinguishable in a generalizable way from "'conspiracy theories' in the pejorative sense." The latter, as Bale describes them, have the characteristics of the paranoid style, which I discuss at length in Chapter 8. For now, let's just consider Bale's characterization:

> In the first place, conspiracy theorists consider the alleged conspirators to be Evil Incarnate. . . . Second, conspiracy theorists perceive the conspiratorial group as both monolithic and unerring in the pursuit of its goals. . . . Third, conspiracy theorists believe that the conspiratorial group is omnipresent, at least within its own sphere of operations. . . . Fourth, the conspiratorial group is viewed by conspiracy theorists as virtually omnipotent. . . . Finally, for conspiracy theorists, conspiracies are not simply a regular feature of politics whose importance varies in different historical contexts, but rather the motive force of all historical change and development. . . . Everything is the result of secret plotting in accordance with some sinister design. (Bale 2007, pp. 51–53)

Bale suggests that making this distinction will enable academics to address plausible claims of conspiracy without such projects suffering under the stigmatization properly belonging to "conspiracy theories" in the pejorative sense—those having the above-enumerated features.

Notice that these features include the characteristics of "the conspiracy theory of society," which, as already mentioned, almost no one believes in. Bale adds other elements of "the paranoid style" as well which, contrary to the assertion of many academics, most interesting conspiracy theories do not posit—a point elaborated at length in Chapters 8 and 9. What is noteworthy here is that Bale, like so many others who appeal to similar distinctions, fails to apply his "clear analytical distinction" strictly. If he did, only theories that actually had the objectionable qualities would be considered "conspiracy theories" (in the pejorative sense in which he uses the phrase). Yet he seems to regard the suggestion that anyone other than al-Qaeda was behind the 9/11 attacks as "nonsense" (Bale 2007, p. 46 n. 1).[10] Bale does temper his remarks, writing, "This is not meant to suggest, of course, that serious researchers should refrain from independently investigating and, if necessary, challenging problematic aspects of the 'official' version, either of the 9/11 operation or other recent terrorist attacks" (p. 46 n. 1). Nevertheless, he seems to assume that "serious research" must remain comfortably within the "al-Qaeda did it" paradigm, without seriously considering that radically alternative accounts might not actually be on the wrong side of his analytical distinction. Similarly, when it comes to JFK conspiracy theories, Bale is quick to dismiss elements or versions of such theories without rigorously questioning whether or not they truly fit his criteria. For example, Bale scoffs at the notion that the man standing with an open umbrella in front of JFK's car as shots rang out could possibly be serving as a signalman in a conspiracy to kill Kennedy. Bale remarks, "It is precisely this totalistic, all-encompassing quality that distinguishes 'conspiracy theories' from the secret but often mundane political planning that is carried out on a daily basis by all sorts of groups, both within and outside of government" (p. 53). In a footnote, Bale does suggest that he is open to some level of critique of the Warren Commission's "comforting official reconstruction." He writes:

> It should, however, be pointed out that even if the "umbrella man" was wholly innocent of any involvement in a plot, as he almost certainly was, this does not necessarily mean that the Warren Commission's comforting official reconstruction of the Kennedy assassination is accurate. As always, the fact that certain parties continue to promote unfounded and absurd conspiratorial scenarios should not lead serious researchers to neglect problematic issues that have been ignored or glossed over in official accounts. (p. 53 n. 5)

Here I mostly agree with Bale (though I doubt I would share his assessment of what is "unfounded" and his intuition regarding what is "absurd"). However, *if* umbrella man is "almost certainly" innocent, it is because the totality of the relevant evidence tends toward exoneration (assuming that is the case, which I don't pretend to know);[11] it is *not* because his involvement as a signalman (or as a distraction) would require a "totalistic, all-encompassing" conspiracy, *as it most certainly would not.*

Bale has, in effect, suggested that conspiracy theories (in the pejorative sense) be understood as those theories that exhibit the paranoid style. But then he employs a looser and more conventional understanding of what counts as a "conspiracy theory," presumably according to his own prejudices, unfairly dismissing theories that do not, or at least *may* not, actually meet his criteria. He further suggests, without evidence, that "the overwhelming majority" of those that present themselves as conspiracy theory authorities deserve the pejorative implications of that term:

> [T]he overwhelming majority of the self-appointed "experts" who concern themselves with alleged conspiracies are in fact "conspiracy theorists" in the negative sense outlined above. They seriously and passionately believe in the existence of vast, preternaturally effective conspiracies that successfully manipulate and control historical events behind the scenes. (pp. 55–56)

The analysis given in Chapters 8 and 9 provides reasons to suspect that this is, at best, an overestimation. More importantly, I argue that conspiracy theories in the neutral sense (such as nonconventional theories regarding political assassinations or purported false flag events) that many academics dismiss as conspiracy theories in the pejorative sense, have versions that cannot reasonably be dismissed in this way because they do not actually have the qualities that are flippantly attributed to them.

2.5. Brian Keeley and "Unwarranted Conspiracy Theories"

After Popper's critique of "the conspiracy theory of society" and Pigden's response to it, the next significant philosophical treatment of conspiracy theories was a 1999 essay published in the *Journal of Philosophy* by Brian Keeley called "Of Conspiracy Theories."[12]

Buenting and Taylor count Keeley among the generalists. However, Keeley does not actually *endorse* generalism. Indeed, in the end, he rejects it. However, he did *explore the viability* of generalism. His "initial motivation" in writing this article was to "present an analysis of conspiracy theories in the spirit of Hume's analysis of miracles" (Keeley 1999, p. 126). This is clearly a generalist project. As Keeley notes, "For Hume, miracles are *by definition* explanations that we are never warranted in believing" (p. 126). However, Keeley ultimately finds this kind of analysis of conspiracy theories not to be viable. In a 2003 article, Keeley puts it this way: "After exploring a number of possibilities, I conclude that, unlike Hume's analysis of miracles, no such *a priori* analysis is available in the case of conspiracy theories" (Keeley 2003, p. 105).

Unfortunately, however, Keeley's original 1999 article is often cited as though it shows that conspiracy theories generally have features that render them unwarranted.[13] Perhaps the reason Keeley's article is so misunderstood is his emphasis on the analysis of what he calls "unwarranted conspiracy theories [UCTs]," which is a *subset* of conspiracy theories, and should not be conflated with the larger category of conspiracy theories.

According to Keeley, UCTs have the following characteristics:

(1) A UCT is an explanation that runs counter to some received, official, or "obvious" account.
(2) The true intentions behind the conspiracy are invariably nefarious.
(3) UCTs typically seek to tie together seemingly unrelated events.
(4) [T]he truths behind events explained by [unwarranted] conspiracy theories are typically well-guarded secrets, even if the [purported] ultimate perpetrators are sometimes well-known public figures.
(5) The chief tool of the conspiracy theorist is . . . *errant data* [i.e., data that either conflict with these received account, or are unaccounted for by it]. (Keeley 1999, pp. 116–117)

And yet, Keeley admits, "These criteria . . . do not distinguish UCTs from all conspiracies we are warranted in believing. Both Watergate and the Iran-Contra Affair meet all of these criteria, yet the belief in these conspiracies seem prima facie warranted" (Keeley 1999, p. 118). In other words, even the subset of conspiracy theories that Keeley labels "unwarranted conspiracy theories" *are not necessarily unwarranted.*

To understand why that is so, let's briefly consider each of these five criteria in turn. First, since not all official accounts turn out to be true, the mere fact that a particular account conflicts with the official account of some event does not mean it is unwarranted. Second, although many, if not most, conspiracy theories do *not* posit nefarious motivations (see Chapters 8 and 9), nefarious motivations do sometimes lurk behind real conspiracies. The Nazi "Final Solution" is a striking example.[14] Admittedly, conspiracies *this nefarious* are rare. But the term "nefarious" is imprecise, and Keeley himself seems to think that the motivations behind Watergate and the Iran-Contra affair would count. In any case, the more extreme the interpretation of "nefarious" one requires, the more this criterion would weigh against a theory, but the fewer the conspiracy theories to which it would apply. Third, ordinarily, being able to tie together seemingly unrelated facts is an epistemic virtue. This is, after all, a hallmark of a good scientific theory, as well as detective work. So it would be very odd for this criterion to count *against* conspiracy theories. The real question is: *How successful* is the theory in its attempt to tie things together? To answer that we have to look at the particulars, not just notice, vaguely, that the theory seeks to tie together seemingly unrelated events. Fourth, we know that people do sometimes conspire. And we know of some secrets that were maintained for a long time. And we *can't* know of any secrets that were so well kept that they never got out at all, obviously. So this criterion cannot be used to judge a theory to be unwarranted.

Finally, regarding criterion 5, appealing to errant data is a *perfectly legitimate* epistemic move. Recently Keeley himself has stated explicitly that, regarding the appeal to anomalies or errant data, "this is far from a feature that distinguishes conspiracy explanations from scientific and religious ones. . . . [O]n this point, conspiracy theory is on par with other forms of explanation" (Keeley 2019, pp. 425, 426). After all, if there really are data that conflict with, or are unaccounted for by, a received account, so much the worse *for the received account*, not for its rival. Whether or not this is "the chief tool" makes no difference to the question of warrant. Further, Coady has argued that Keeley (and Steve Clarke) are just wrong to suggest that the use of errant data by conspiracy theorists distinguishes them from supporters of official accounts (Coady 2003, pp. 202–204). Coady cites a book by Gerald Posner, *Case Closed* (1993), which attempts to refute conspiracy theories regarding the JFK assassination by appeal to data that is "errant" in relation to those theories—which is a perfectly legitimate move on Posner's part. The relevant question is not: What *kind* of evidence was used? It is rather: Who has made the most convincing case, all things considered?

Keeley himself recognizes his criteria do not succeed in differentiating warranted conspiracy theories from unwarranted ones. Indeed, he doesn't think any set of criteria will. He writes, "There is no criterion or set of criteria that provide a priori grounds for distinguishing warranted conspiracies from UCTs" (1999, p. 118).[15] So Keeley is not, in the end, a generalist in the sense at issue here. Indeed, he recently stated explicitly, "In general, I will take a particularist approach to conspiracy theories. As with scientific theories they vary in their plausibility" (Keeley 2019, p. 424). However, his emphasis on (purported) problems with what he calls "unwarranted conspiracy theories" reflects an interest in determining whether generalism can work for some *subsets* of conspiracy theories.

There is one other aspect of Keeley's article that deserves mention (and will come up again in the next section). Räikkä regards it as "Keeley's main argument." It is, in Räikkä's words, "an argument that stresses the importance of *trust* in belief-formation" (2009a, p. 189). This is the "public trust approach," according to which, writes Keeley, "we ought to recognize such theories [i.e., conspiracy theories] as embodying an almost nihilistic degree of skepticism about the behavior and motivations of other people and the social institutions they constitute" (1999, p. 126). Räikkä notes that this idea has been much criticized—by Pigden (2006), Clarke (2002), Basham (2001, 2003), and Coady (2003). While Räikkä doesn't find these criticisms fully convincing, he offers his own straightforward one: if one actually looks at collections of conspiracy theories, one finds that this is not, in fact, typical. Räikkä writes:

> [The view] that *most* conspiracy theorists end up making claims of larger and larger conspiracies . . . is empirically false. Typical conspiracy theories do not begin with skeptical assumptions, nor do they end up with them. Keeley seems to recognize this, as he writes that "*some* mature conspiracy theories" entail "pervasive skepticism of people and public institutions" (my [i.e., Räikkä's] emphasis). Anyone who thinks that the most favored "explanatory stratagem" of conspiracy theorists is to expand and expand their theories should read at least one book—there are many of them—that describes various conspiracy theories. Expanding one's theory has been only one strategy among many, and it has not been very popular. The relatively common view that a runaway expansion "is characteristic to many conspiracy theories" is simply false. (2009a, p. 195)

Finally, something should be said about another of Keeley's contributions. Eight years after the initial publication of his "Of Conspiracy Theories," Keeley published an article titled "God as the Ultimate Conspiracy Theory" (2007). Keeley there accepts the reasonableness of agnosticism toward conspiracy theories as a premise in an argument for the reasonableness of agnosticism about God. Except for narrowly circumscribed subsets of conspiracy theories, Keeley seems to be as much opposed to scholarly attempts to discourage the study of evidence related to particular conspiracy theories by appealing to generalizations about them as other self-professed particularists. And he succinctly points out the problem with generalism:

> The chief problem is that there is a class of quite warranted conspiracy theories about such events as Watergate, the Iran-Contra Affair, etc., and that there is no principled way of distinguishing, *a priori*, the two classes from one another. . . . [P]rior to any investigation, we ought to adopt an agnostic attitude with respect to conspiratorial claims. (Keeley 2007, p. 137)

2.6. Pete Mandik's Absurdism

Keeley's work seems to have influenced Pete Mandik, who, in an article entitled "Shit Happens," "defend[s] the view that there is indeed something deeply epistemically wrong with conspiracy theorizing" (Mandik 2007, p. 205). Keeley had quoted the slogan "Shit happens" as an epigraph at the beginning of his article, contrasted with Hegel's idea that "Reason rules the world" (1999, p. 109). This (false) dilemma is the framework in which Mandik sets up his argument, "embrac[ing] . . . the absurdist horn of the dilemma" (2007, p. 205).

Mandik defines conspiracy theories in terms of five elements:

> Conspiracy theories postulate 1. explanations of 2. historical events in terms of 3. intentional states of multiple agents (the conspirators) who, among other things, 4. intended the historical events in question to occur and 5. keep their intentions and actions secret. (2007, p. 206)

He maintains that "If something has all five elements of my working definition [of a conspiracy theory], then it is unwarranted, or at least no more warranted than a declaration of 'shit happens'" (Mandik 2007, p. 213).

Regarding theories that involve the first four elements but not the fifth, Mandik writes, "Examples would include just about any historical explanation involving multiple people" (2007, p. 216). This seems both correct and unproblematic. As for the first three, surely we must accept the existence of multiple agents with intentional states, and that explanations of historical events must somehow relate to some of them. And assuming we don't have an overly strong interpretation of the fourth element of Mandik's definition[16] (which Mandik does not suggest), that also seems both common and unproblematic. At least some historical events occur because, in part, certain individuals intended something approximating them. Man did not walk on the moon because "shit happened."

Mandik, however, does not seem to agree. He writes, "My own view of the matter, and the case I have tried to make in the current paper, is that the elements prior to the fifth one create most of the trouble and the prospects for resisting absurdism are quite slim" (2007, p. 217). So, according to Mandik, it is the first four elements, the ones that are common among both conspiracy theories and other historical explanations, that suggest that conspiracy theories are unwarranted. So too then would be nearly all historical explanations.

However, it is perhaps best to ignore Mandik's "own view of the matter" and focus on the supposed problem with the fifth element of his definition: secrecy—since this is the one element of his definition that sets conspiracy theories apart from other historical explanations. The kind of secrecy Mandik has in mind is the *keep your mouth shut* variety, not the *don't get caught* variety.[17] Mandik points out that, since conspiracy theories assume that the alleged conspirators are liars, their testimony cannot be relied on. This certainly reduces the epistemic resources that we have at our disposal in determining the truth. But it doesn't reduce it to zero. Notice that the problem that Mandik raises applies also to petty conspiracies, which are successfully prosecuted all the time based on all sorts of other kinds of evidence. It seems the problem would apply also to nonconspiratorial crimes in which the suspect, who can't be trusted, denies involvement—as Oswald did, for example. Should those who believe that Oswald was the lone gunman who intended to kill JFK change their view to "shit happens"?

Mandik's arguments are not likely to convince anyone with common sense. The fact that "shit happens" does not conflict with the equally clear fact that intentional agents, sometimes in collaboration with others, do influence events. Sometimes this influence is strong and direct enough to reasonably call it the cause of the event, even though a variety of unintended consequences complicate the situation. Most conspiracy theorists understand and acknowledge both realities: Not everything that happens was intended (by anybody, let alone by a particular set of conspirators), and yet *some* things that happen *were* intended, sometimes by conspirators. And sufficient evidence sometimes exists to warrant belief about what happened in such cases, even without, or before, admission by the guilty parties.

Mandik does, moreover, make one observation that is particularly relevant to "conspiracy theories and epistemic authorities." He writes, "[T]he more we lend credence to conspiracy theories . . . the more we are pushed to a kind of skepticism about any of our institutions" (2007, p. 205). Unfortunately, I have to admit that Mandik is correct about this. It is "unfortunate" because being able to rely on our institutions is desirable, and the more the better. But the fact that it would be advantageous to have reliable institutions does not show that our institutions are, in fact, reliable (which is, of course, a matter of degree). And even if we find our institutions to be less reliable than we would like, or if we now regard them as less reliable than we previously did, that need not lead to Keeley's above-mentioned "almost nihilistic degree of skepticism," to which Mandik alludes.

2.7. Cass Sunstein and Adrian Vermeule's Implicit Generalism

Cass Sunstein and Adrian Vermeule, who are the focus of Part II of this book, attempt to circumvent the problem of true conspiracy theories by claiming that their focus is only on conspiracy theories that are "demonstrably false" (2009, p. 206). However, they do not indicate how such falsity is to be demonstrated. As a result, whether or not Sunstein and Vermeule have actually given a generalist argument against conspiracy theories is somewhat ambiguous. On the one hand, their criterion for rejection of conspiracy theories, demonstrable falsity, suggests that they are at heart particularists. For it is presumably only through the analysis of such particulars that such falsity is demonstrated. On the other hand, they make almost no effort to deal with particulars or cite others who have dealt with particulars, and instead seem to assume that official accounts *must* have been based on expert assessment of all the relevant particulars and are therefore reliable. But this seems to be a kind of generalism, at least in spirit, insofar as it implies that conspiracy theories can be safely dismissed *simply because* they conflict with official stories, which are assumed to be backed by epistemic authorities. And thus, common folk need not trouble their small minds with particulars. The reason this may be thought of as a form of generalism is that it suggests that conspiracy theories may be safely dismissed based on a simple generality: (conventionalist) experts are usually right. Mainstream (conventionalist) experts are assumed to be unbiased, and dissenting experts are essentially ignored. This book challenges the notion that experts are unbiased and reliable enough to accept Sunstein and Vermeule's inference. In fact, as I will show in Chapters 2 through 4, Sunstein and Vermeule are themselves unreliable.

2.8. Neil Levy and Epistemic Authorities

While the generalist argument I have attributed to Sunstein and Vermeule is largely implicit, philosopher Neil Levy has made this argument explicitly. In an article entitled "Radically Socialized Knowledge and Conspiracy Theories," Levy argues that conspiracy theories typically conflict with the views of epistemic authorities and that such theories should be regarded as highly unlikely on that account. In his words, "[A]ccepting the official story is almost always rational" (Levy 2007, p. 190). As in the case of Sunstein and Vermeule, while this is perhaps not *strictly* generalist—because it assumes that the particulars have been addressed adequately by the "authorities"—

Levy's position is generalist in spirit. In his view, even though particulars (the facts) ultimately determine what is warranted (of course), we should trust expert judgment regarding them. And, it is assumed, this is what the official account is based on. So *we* need not worry over the particulars. For even though experts are *sometimes* wrong, they are usually right. Thus, Levy concludes, we should "cultivat[e] only our own garden: that relatively small sphere in which we can claim some expertise" (2007, p. 191). This is generalist in spirit because it suggests that individuals ought not consult specifics of any particular conspiracy theory (except if they happen to have relevant expertise, but even then their exploration of the particulars should be narrow). Rather, ordinary people, and even "the intellectual," should rely on the judgment of their epistemic betters. Levy writes:

> The typical explanation of an event or process which attracts the label "conspiracy theory" is an explanation that conflicts with the account advanced by the relevant *epistemic authorities*. I argue that both for the layperson and for the intellectual, it is almost never rational to accept such a conspiracy theory. (2007, p. 181)

The key point is that "epistemic authorities, when properly constituted, are far better positioned to explain events than are isolated agents" (Levy 2007, p. 182).

Coady has responded to Levy in an article that immediately follows Levy's in a special issue of the journal *Episteme*. I'll just reiterate the bare bones of Coady's counterargument. First, Coady acknowledges that there is a sense in which Levy's position is true. However, in *that* sense, it "is little more than a tautology" (Coady 2007, p. 198). The problem is that Levy conflates "official stories" with what would be supported by a network of properly constituted epistemic authorities. But this can hardly be taken for granted. By "properly constituted," Levy means they have the "right kind of structure." He explains:

> The right kind of structure is that exemplified by science: knowledge claims are the product of a socially distributed network of enquirers, methods and results are publicly available (especially, but not only to other members of the network), inquirers are trained in assessing knowledge claims according to standards relevant to the discipline, and rewards are distributed according to success at validating new knowledge *and* at criticizing the claims of other members of the network. (Levy 2007, p. 188)

Unfortunately, as Coady points out, the various elements constituting the "right kind of structure" are "nothing more than ideals" (2007, p. 199). Coady characterizes "official stories" as accounts that are "propagated by an institution which has power to influence what is widely believed at a particular time and place" (2007, p. 200). This is not equivalent to an ideal network of genuine epistemic authorities.

Levy also suggests that conspiracy theorists risk a kind of epistemic suicide by doubting (supposedly) authoritative sources of knowledge—disconnecting themselves, as it were, from the network of knowledge acquisition and validation. But this is at best an exaggeration. As Coady remarks, "Conspiracy theorists, understood as people predisposed to be sceptical of official stories, are not (or need not be) any more sceptical about epistemic authority or expertise *as such* than other people. Rather, they merely have a particular view about who the epistemic authorities or experts are" (2007, p. 201).

In addition, Levy fails to appreciate how significant the qualification "prima facie" is. One formulation of Levy's position is this: "A conspiracy theory that conflicts with the official story, where the official story is the explanation offered by the (relevant) *epistemic authorities*, is *prima facie* unwarranted" (Levy 2007, p. 182). To say that a theory is "*prima facie* unwarranted" is to make a claim about its *prior probability*. But generally theories are not, and *should* not be, judged simply on their prior probability. One *also* ought to consider the relevant evidence for or against the theory. In other words, to say something is prima facie unwarranted is just to say that one should not believe it is true, *prior to consideration of the evidence*. But conspiracy theorists do not ask for people to believe their theory prior to considering the evidence. So even if it is true that counternarratives should be considered prima facie unwarranted, that tells us little about how they should be regarded *after* considering relevant evidence.

When we consider another of Levy's formulations, the same types of objections are applicable. Levy writes, "Though it is not epistemically irrational to reject official stories *per se*, there is a class of official stories that, other things being equal, we ought to accept. Responsible believers ought to accept explanations offered by *properly constituted epistemic authorities*" (2007, p. 187). As mentioned above, "properly constituted epistemic authorities" exist only very imperfectly; quite often who counts as genuine epistemic authorities is contested, as is the neutrality of "epistemic authorities" favoring official stories. Further, other things are *not* generally equal—there may be, and often is, relevant evidence that conflicts with official accounts. This is particularly clear when one considers the details of real examples, such as critiques of official inquiries like the Warren Commission or the 9/11 Commission.

Basham's concept of "toxic truths" is particularly relevant here. "Toxic truth" is a convenient label for an idea that has been vaguely discerned by conspiracy theorists for some time: some truths would be so disruptive to acknowledge that epistemic authorities would be inclined to ignore them or even cover them up—even without being told to. As Basham puts it, "[S]ome conspiratorial scenarios, even if well evidenced, are too 'toxic' for our usual institutions of public information to disseminate to the public, or even investigate. Cover-up by intentional neglect, not descending control, is the easily predictable consequence" (2018a, p. 271).[18] At the very least, as political scientist Joseph Uscinski points out, "[I]nvestigating alleged conspiracies vigorously could cost journalists access or respect" (Uscinski 2018, p. 238).

We can conclude, then, that Levy has provided inadequate justification for judging conspiratorial counternarratives to be unwarranted, *all things considered*. To say that they are nevertheless "*prima facie* unwarranted" has little if any practical significance, even if that were true.

But my critique of Levy is not yet over. As the title of his article suggests, Levy's argument rests on the rather controversial premise that knowledge is "radically socialized." Levy remarks, "[K]knowledge production is deeply social, *in a way that few epistemologists have recognized*" (2007, p. 183, emphasis added). That's right: Levy's claim that counternarratives can be safely dismissed simply because they conflict with the position of supposed epistemic authorities rests on a theory that, Levy acknowledges, enjoys little support from the relevant epistemic authorities.

But surely Levy considers himself the right kind of epistemic authority to challenge the greater number of epistemic authorities aligned against him. Fair enough. If that is not allowed, after all, how could science progress? But Levy fails to see that conspiracy theorists are doing something similar. He wants to disqualify them as nonexperts, but it's not quite that simple. They have their own "expert witnesses," so to speak. It is only reasonable and fair to hear what the dissenting experts, of all kinds, have to say. But according to Levy, in a case like 9/11, there are no such experts. He writes:

> It is not because the government tells us that the attacks of 9/11 were carried out by Al Qaeda operatives that we dismiss rival explanations, according to which the government itself was behind the attacks, as just a conspiracy theory. It is because the relevant epis-

temic authorities—the distributed network of knowledge claim gatherers and testers that includes engineers and politics professors, security experts and journalists—have no doubts over the validity of the explanation that we accept it. (Levy 2007, p. 187)

But it is not true that no such epistemic authorities have doubts about the official explanation of what happened on 9/11. More than three thousand architects and engineers have signed a petition indicating doubts regarding the cause of the destruction of the Twin Towers and Building 7.[19] And perusal of the website patriotsquestion911.com reveals that hundreds of various kinds of epistemic authorities have doubts about some aspect of the official story (though the degree to which these people have relevant expertise varies considerably).[20]

But we haven't gotten to the radical part yet. In order to reach the conclusion that conspiracy theories are bunk, Levy argues that our ideas aren't in our own heads. That's at least partly what makes "radically socialized knowledge" so radical. If this was supposed to help people come to the understanding that conspiracy theories can be safely dismissed, it doesn't seem like much of a shortcut. I'm not sure which would be more difficult, sorting out all the minutiae involved in 9/11 conspiracy theories, or trying to figure out if the theory of radically socialized knowledge, which "few epistemologists have recognized," is right. And even if it were found to be right, there would be questions about its applicability, such as those mentioned above. I tend to think it makes more sense to just look at the relevant evidence for conspiracy theories on a case-by-case basis.

In sum, the idea that we should just accept the conclusions of ostensible epistemic authorities ignores a host of complicating factors. And, as if that were not enough, there are also practical problems. As Coady notes, "Not only is the claim that we should almost always believe official stories false, it is dangerously self-defeating, since, to the extent that it gains widespread acceptance, official stories will be less subject to scrutiny and hence less likely to be true" (2007, p. 200).

2.9. Quassim Cassam on Bad Thinkers

Instead of focusing on the problem with conspiracy theories, some philosophers (as well as social scientists) focus on the people who espouse them. The philosopher Quassim Cassam is a good example. In a 2015 online magazine article called "Bad Thinkers," Cassam imagines a character he calls "Oliver" who believes that 9/11 was an inside job. Cassam then psychoanalyzes Oliver:

> Oliver believes what he does because that is the kind of *thinker* he is or, to put it more bluntly, *because there is something wrong with how he thinks*. . . . [H]is beliefs about 9/11 are the result of the peculiarities of his intellectual constitution—in a word, of his intellectual character. . . . [H]e thinks that 9/11 was an inside job because he is gullible in a certain way. He has what social psychologists call a "conspiracy mentality." (Cassam 2015)

As Basham and Dentith point out in response, "The problem for Cassam's argument is that it relies on characterising belief in conspiracy theories solely along the lines of theories many of us *already think are unwarranted*" (Basham and Dentith 2015).[21] In other words, Cassam has done nothing to *show* that Oliver's beliefs are unwarranted; he just asserts that they are, and assumes many of his readers will agree, and so panders to them. Although Cassam admits that sometimes conspiracy theories are true, he declares Oliver, and the conspiracy theorists he is supposed to represent, to have a defective intellectual character, an epistemic vice. But it is not clear that this is correct.

Oliver may well have been epistemically *more responsible* than most of his conventionalist counterparts, *even if* he is wrong. As Basham and Dentith point out, people who accept the official story of 9/11 may well do so for poor reasons, not because they have informed themselves of the facts, done any clear thinking, or in any way exhibited exemplary epistemic character. For example, they may have simply fallen for "the oft-repeated, common sense wisdom that conspiracy theories are bunk" (Basham and Dentith 2015). At least Oliver *made an attempt* to inform himself and to think critically about the issue. And further, it is at least possible that, even if there are good answers to the questions that trouble Oliver, both official reports and mainstream media coverage failed to provide those answers; instead they address those questions in evasive and inadequate ways, if at all (see appendix). Oliver's position could therefore have been justified given the information available to him.[22] And further, Oliver may have acquired *more and better* information than most of his conventionalist counterparts. Given the assumptions here, Oliver could be faulted, we must admit, for failing to ferret out the information that adequately addressed his concerns (assuming such information exists). Confirmation bias may well have played a critical role in that failure. However, confirmation bias is not an epistemic fault that distinguishes Oliver from his conventionalist counterparts.

In 2016, Cassam gives the example of Oliver again in an academic paper published in *The Monist*. This time he cites the *9/11 Commission Report* and also the NIST report. Cassam declares that it is okay to be dismissive regarding Oliver's sources, but it is not rational to be dismissive about, for example, the *9/11 Commission Report* (2016, p. 165). Although Cassam doesn't say why he thinks this, presumably what he regards as "reputable studies" are assumed to be the product of genuine epistemic authorities. On the basis of appeal to these official documents, Cassam declares, "[Oliver's view] is (I take it) not just false but demonstrably false" (2016, p. 162). (The words "I take it" seem to be doing much of the work here.) Cassam further declares, "In fact, there were no explosive residues in the debris of the WTC towers and reputable studies have shown that aircraft impacts could have caused the towers to collapse" (2016, p. 162). Regarding the former claim, Cassam cites nothing, and he does not address the charge that NIST admitted that it didn't look for evidence of explosives, or to reports that independent scientists have found what they take to be explosive residues. Regarding the latter claim—that aircraft impacts could have caused the towers to collapse—he cites Lew, Bukowski, and Carino 2005, which is part of NIST's report on the collapse of the Twin Towers. Notice that even if the latter claim is true, it doesn't refute the controlled demolition hypothesis. Indeed, the idea that planes could not cause the buildings to collapse is not even one of the most cited reasons to favor the demolition hypothesis, which are generally more nuanced than that. Nevertheless, Cassam concludes that Oliver's views are "conclusively refuted" (p. 162) and "baseless and false" (p. 162). Cassam may well be right regarding the falsity of Oliver's beliefs. But it is not clear that they are *demonstrably* false, baseless, or refuted. To show that would require more analysis and less question-begging.

Further, and perhaps most importantly, even if there were an utterly conclusive demonstration of the falsity of Oliver's beliefs, it doesn't follow that Oliver was more epistemically irresponsible than his conventionalist counterparts, that he "should know better" (p. 162), and that he ignores critical evidence, relies on untrustworthy sources, and jumps to conclusions, as Cassam suggests (p. 164). Not every responsible detective will follow the same trail of evidence and focus on the same suspects. Oliver may have, despite reasonable efforts to consider counterarguments, simply failed to find the keys to unlock the case, partly for reasons that are random or circumstantial. Until one proves otherwise, it is wise to assume the best about Oliver, and engage him with reasons—answer his questions thoughtfully and explain to him this "demonstration" that so clearly shows him to be wrong. After all, it is a complex issue, and, contrary to Cassam's assertions, Oliver's views are not baseless—at least not any more so than some of Cassam's claims.

Some of Cassam's descriptions of Oliver appear to be little more than stereotypes. For example, Cassam says that "[Oliver] relies on dodgy websites, paranoid talk radio stations, and a narrow circle of eccentric, conspiracist friends and acquaintances" (p. 164). That may be true of some 9/11 conspiracy theorists, but I know of no reason to think it is true of David Ray Griffin or Graeme MacQueen. Cassam does allow that Oliver "tries, in his own way, to find things out and to extend his knowledge by carrying out investigations directed at answering certain questions." However, Cassam declares, "[H]is investigations are blighted by his intellectual vices. Because he is gullible, dogmatic, closed-minded, cynical, prejudiced, and so on, he ignores important evidence which bears on his questions, relies on unreliable sources, jumps to conclusions and generally can't see the wood for the trees" (p. 164).

Regarding at least some of this, Cassam thinks he can respond to the potential objection that Oliver is merely a contrived "fictional case" (Cassam 2016, p. 170)—an objection that was in fact made by Basham and Dentith 2015. He does so by appealing to the "fairly extensive psychological literature on this topic" (p. 171). Indeed, a number of psychologists have made concerted efforts to suggest that there is something significantly peculiar, anomalistic, and pathological about the psychology of conspiracy theorists. Part III of this book is largely a critique of this literature. Suffice it to make a couple points here. First, Cassam repeatedly suggests that conspiracy theorists are dogmatic and closed-minded. Regarding this claim, not only is Cassam potentially vulnerable to the charge of hypocrisy, but the psychological studies are mixed and unclear regarding the closed-mindedness of conspiracy theorists (Wood and Douglas 2019, p. 247), and dogmatism seems either a redundant category—a synonym for closed-minded—or is something that has not been, to my knowledge, studied in relation to conspiracy theories.

Second, Cassam uncritically accepts the finding of a seriously flawed study. Cassam writes elsewhere, "Psychologists talk about a 'conspiracy mentality,' a personality trait that disposes those who have it to believe conspiracy theories, even obviously contradictory conspiracy theories" (2018). Unfortunately for Cassam, the finding that conspiracy theorists are disposed to believe "obviously contradictory conspiracy theories" is a result of faulty thinking *on the part of the psychologists who did the study*—this is explained in detail in Chapter 7. Why did Cassam fail to recognize the rather obvious flaw in the study to which he alludes? Would it be unfair to explain it by saying he himself suffers from the epistemic vice of gullibility? I think so. His oversight, in my view, is more likely attributable to bias—but then that's just a guess. I can't prove a psychological reason for his mistake, just as he hasn't proven one about the people Oliver is supposed to represent. But at least I *have* painstakingly shown (in Chapter 7) that Cassam's error *is* an error. Cassam, in contrast, hasn't given any significant reason, other than the most sweeping question-begging generalities, to think that Oliver is wrong. And so, though I think impugning other people's intellectual character is not generally a good idea, there is perhaps a stronger case against Cassam than against Oliver.

2.10. Steve Clarke, Part I: The "Fundamental Attribution Error" Error

Like Cassam, Steve Clarke has suggested that conspiracy theorists have a psychological problem, and that for this reason we are entitled to "an attitude of prima facie skepticism towards the theories propounded by conspiracy theorists" (2002, p. 133). According to Clarke's 2002 article, the problem with conspiracy theorists is that their reasoning about conspiracy theories involves the "Fundamental Attribution Error" (FAE). That is, they overemphasize "dispositional" explanations and underemphasize "situational" ones. In other words, conspiracy theorists allow themselves to be unduly influenced by the *character* that they attribute to the purported conspirators, when they should be looking more at the *situation* in which actions are taken or decisions are made.

The first clue that there is something wrong with Clarke's analysis is this: Clarke suggests that there is something problematic about explaining things in terms of dispositions, and yet he explains the persistence of conspiracy theories by saying that conspiracy theorists have a disposition to explain things in terms of dispositions.[23] In any case, Clarke suggests that this disposition causes conspiracy theorists to cling to theories even when they are losing epistemic support. In the philosophy of science, research programs that are failing in this way are referred to as "degenerating research programs."[24] (More on that in section 2.11 below.)

Clarke discusses an experiment in which seminary students, en route to give presentations, pass by a person (an actor) who appears to be in need of assistance. It turns out that, when the seminarians were told that they were late, they were much less likely to stop to help the person (the percentage that stopped to help fell from 63 percent to 10 percent). This result differed greatly from the expectations of people who were asked to predict the results of the experiment. This experiment purportedly shows that the situation, not the character of the individual, best explains what people do, and that people systematically get this backward.

Coady points out, however, that the situation can only influence people in so far as they have certain dispositions, such as a "disposition to be punctual" (Coady 2003, p. 207). So the study could be read as showing that people underestimated the relative strength of the disposition to avoid being late (especially when being late would be embarrassing). In addition, Pigden points out that many of the dispositions that conspiracy theories posit are simply *human* dispositions, dispositions that we all *have* (desire, greed, etc.),[25] and further, that it is not the case that we all have exactly the same dispositions. The latter is presupposed by our judicial system: if we all had the same dispositions, there would be no need, at least no *practical* need, for "correctional facilities," since no one would be more likely to commit a crime than anyone else except insofar as they may be so prompted by circumstance.

And, still further, even if "situation" is *more* important than "disposition," assuming that they can even be cleanly separated, it is still the case that disposition is a factor. The experiment seems to confirm this. In addition, the experiment leaves open the possibility—indeed it is common sense—that *sometimes* disposition is *the decisive factor*. Even some official stories rely on this. For example, James Earl Ray is said to have assassinated Martin Luther King out of racial animus. So it is not a fallacy to include disposition in one's theory, so long as it is reasonable to suppose that the individuals in question are likely to have that disposition, either simply because it is common or on the basis of past behavior. In addition to conspiracy theorists, police detectives and prosecutors are likely to find dispositional considerations significant. Dispositions are real and they matter.

But are there good reasons to think that conspiracy theorists consistently draw wrong conclusions because they overestimate dispositions? Just providing a couple examples, as Clarke does, does not establish that conspiracy theories always, or even generally, involve this error to a greater degree than official accounts. (If only *some* conspiracy theories are grounded on the error, we cannot dismiss the entire class, but must at least first distinguish those that commit it from those that do not.) We should also bear in mind that sometimes official accounts rely heavily on appeals to disposition: think of "the evil empire," the "axis of evil," Saddam the *brutal* dictator, as well as the supposed motive of "lone gunmen" and the "evildoers" purportedly behind the 9/11 attacks, in addition to James

Earl Ray's supposed racial animus. Further, for the FAE to weigh strongly against conspiracy theorists, the theorists would not only have to commit this error, but have to *rest their argument on it.*

Do "conspiracy theories" that suggest high-level government complicity in the attacks of 9/11 really rely on attribution of some peculiar disposition, rather than on "situational" factors? If one looks at their arguments, they seem mostly about facts and situations, even though they do address the question of motive—just as any prosecutor would. Do they stress motive or disposition more so than defenders of the official account? It is not clear that they do. The most absurd dispositional account, it seems to me, is that al-Qaeda flew planes into buildings because they "hate our freedoms" (Bush 2001).

Facing various criticisms, Clarke has (sort of) retracted his argument in a short essay entitled "Appealing to the Fundamental Attribution Error: Was It All a Big Mistake?" (2006).[26] In that essay, Clarke tries to avoid answering the question posed in the subtitle with a clear yes, even though that answer seems inescapable. Let's take a closer look.

Clarke here admits, contrary to his original position, "I now hold that . . . we are not entitled to conclude that the FAE pervades lay social cognition" (2006, p. 131). And he acknowledges that this is a problem because he hasn't shown that *conspiracy theorists* are particularly prone to this error. He writes:

> However, if the FAE is less widespread than I had earlier thought then it seems less plausible, given available evidence, to think that conspiracy theorists are more in its grip than most people, because the postulated gap between the ordinary people and conspiracy theorists is now widened and the evidence we currently have about the presence of the FAE is derived from evidence about the behaviour of ordinary people (and seminarians), not from evidence about the behaviour of conspiracy theorists. (Clarke 2006, p. 131)

Nevertheless, he maintains:

> All that I require for the purposes of Clarke (2002), however, is the conclusion that conspiracy theorists, who are a distinct minority of the population (at the moment) are in the grip of the FAE to a great extent and this *possibility* is compatible with the FAE being less present in the general population than I had earlier suggested. (p. 131, emphasis added)

So, in the end, the best that Clarke could do is to appeal to the "bare possibility of [his] opinion," to borrow a phrase from David Hume. Clarke concludes:

> It *may be* that there is some factor or combination of factors, other than the FAE, that underpins the psychology of conspiracy theorists and which could serve to explain why they are motivated to hold on to conspiracy theories at times when other people have abandoned such theories. It would, of course, strengthen my argument if these factors were articulated, but as long as it is *credible* to think that there are such psychological factors, the overall argument remains defensible. (p. 132, emphasis added)

So was appealing to the FAE a big mistake? It seems that Clarke's answer amounts to this: "Yes, mistakes were made. But it is still *possible* that there are psychological factors that may justify dismissiveness." The question now becomes: Are there compelling reasons to think that there *are* psychological factors that explain belief in conspiracy theories, and which thereby justify a dismissive attitude toward conspiracy theories? This question can be broken into two parts.

One: *If* there are psychological factors that explain belief in conspiracy theories, does that mean that a dismissive attitude toward conspiracy theories would be warranted? The answer is no, not unless (implausibly) it could *completely* explain *all* conspiracy theories. For even if many conspiracy theorists are plagued by irrational biases, it would still not be safe to assume that all official accounts are true, and that conspiratorial counternarratives must be false.

Two: Are there *in fact* psychological factors that "explain why [conspiracy theorists] are motivated to hold on to conspiracy theories"? There has been a fair amount of work in the social sciences directed toward showing that there are such psychological factors. (Much of this work cites Clarke 2002 as if Clarke had proven something.) Some of the social scientists engaged in this work speak of a "conspiracy mentality," as mentioned in the previous section, which causes some (apparently defective) people to believe in conspiracy theories. In Chapters 7, 8, and 10, I will argue that much of this work is not sound and that we do not, in fact, have strong reasons to conclude that the thinking processes of conspiracy theorists, in general, are any less reliable than those of conventionalists.

When thinking about this issue, we should always remember the following. First, the "psychological factors" under consideration are matters of degree and are not uniformly distributed. Even where there is a statistically significant difference between conspiracy theorist and conventionalist, it may be marginal in magnitude and may apply to only a subset of conspiracy theorists. Second, psychological factors often have both positive and negative sides to them, such as being "disagreeable/critical." Third, there may be other psychological factors that push people toward conventionalism, or even toward neutral positions. Psychologists who are partial to conventionalist interpretations of events may be inclined to study, and in so doing pathologize, the psychological factors associated with conspiracy theorizing, while ignoring those factors that might be applicable to conventionalists. Indeed, this seems to be what is happening.

As long as these psychological factors are marginal, or have epistemic merits as well as demerits, or are counterbalanced by psychological factors that influence others in opposite ways, then even if they are real, they can do little to help us evaluate the warrant of any specific conspiracy theory. And, as I have already suggested, even if they turn out to be one-sided and large (which does not seem to be the case), this will not tell us which conspiracy theories were true or warranted. And so, the psychological study of conspiracy theories, even if it were to reveal pathological thinking, would not thereby undermine particularism. It would not show that any particular conspiracy theory could be safely dismissed as unwarranted prior to investigating the particulars. To evaluate the warrant of a conspiracy theory, the actual evidence for it must be considered, along with all relevant prior probability considerations.

2.11. Steve Clarke, Part II: Conspiracy Theories Aren't Specific Enough to Be Warranted

Clarke has also published an article, "Conspiracy Theories and the Internet: Controlled Demolition and Arrested Development," which argues that "the hyper-critical atmosphere of the internet has slowed down the development of conspiracy theories, discouraging conspiracy theorists from articulating explicit versions of their favoured theories, which could form the hard core of Lakatosian research programmes," that is, research programs that can be evaluated as either progressive or degenerative (Clarke 2007, p. 167). There are essentially two claims here. First, in the internet age, conspiracy theories have become significantly less specific regarding their positive claims, their "hard core" commitments. Second, this lack of specificity renders these newer conspiracy theories mere "proto-theories"

(p. 175) that are not proper Lakatosian research programs, that is, not amenable to being judged according to whether they are progressive or degenerating. Clarke seems to think these considerations imply that we should not accept postinternet conspiracy theories such as those suggesting that the WTC towers were brought down by controlled demolition. The first claim, regarding decreasing specificity of conspiracy theories, is interesting and plausible. However, I will argue, it is not well established. I will further argue that, even if both claims are true, there are no particularly interesting implications regarding the epistemic status of contemporary conspiracy theories.

Let's start with the claim that postinternet conspiracy theories are importantly less specific than preinternet conspiracy theories. As an example, Clarke contrasts the relatively nebulous phrase "domestic terrorists," used on at least one occasion by a 9/11 conspiracy theorist, with the precision of Elvis Presley's death being faked by Elvis himself—as if he wouldn't almost *have to* be in on it—and an accomplice. On its face, the Elvis example seems like a poor choice for comparison. JFK assassination theories, however, seem much more suitable for this purpose. Regarding these, Clarke writes:

> [The assassination of JFK] has been alleged to have been orchestrated by the CIA, the FBI, the KGB, the Mafia, Lyndon B. Johnson, Richard Nixon, Fidel Castro, Cuban exile groups, the US military and some combinations of the above. All of these are specific groups or individuals, with particular motives, unlike the nebulous "domestic terrorists" that Griffin (2005) blames for the World Trade Center collapse. (Clarke 2007, p. 176)

The fact that Griffin at some point used a nebulous phrase is not particularly significant, unless he never went beyond that. The problem is that he *did* go further, rather conspicuously. It is hard to see how Clarke could have missed this. Clarke's bibliography includes Griffin's 2006 book, *Christian Faith and the Truth behind 9/11*, which contains a chapter entitled "Bush Administration Responsibility for 9/11: From a *Prima Facie* to a Conclusive Case," and also a chapter entitled "Imperial Motives for a 'New Pearl Harbor.'" Further, that book is divided into two main parts, "Part 1: Evidence That 9/11 was a False-Flag Operation" and "Part 2: A Christian Critique of 9/11 and American Imperialism." The "Bush Administration" is named as being responsible.[27] And the administration's motive is clearly implied. Griffin is conveying, with the very chapter titles and part titles in this book, that the events of 9/11 constituted a false flag operation, conducted by the Bush administration in support of an American imperialist agenda, more specifically the agenda described in the Project for the New American Century document *Rebuilding America's Defenses*, to which the phrase "A New Pearl Harbor" (which is also the title of his first book on the subject) unambiguously refers.[28] As for the specificity of "who done it," it is not clear that the "Bush Administration" is substantially less specific than the earlier accusations against "the CIA" or "the Mob"—usually meaning unspecified elements within those organizations.

And if we look a little closer we find even more specificity. For example, also in *Christian Faith and the Truth behind 9/11*, Griffin writes:

> The neocons had during the 1990s provided an agenda for a more militarist, unipolarist, imperialist foreign policy and had helped shape a cultural-political context in which the implementation of this agenda would be easier than it otherwise would have been. The attacks of 9/11 allowed this agenda to be implemented. From these facts, however, we cannot legitimately infer that the various neocons whose writings are discussed here were in on the plans for 9/11. This inference can be made only about those who had become part of the new administration in 2001, and probably only some of them, *especially Cheney and Rumsfeld*. (2006, p. 90, emphasis added)[29]

Further, Kevin Ryan has published a book called *Another 19: Investigating Legitimate 9/11 Suspects* (2013), which names quite a few specific suspects. Though it came out well after Clarke's article, it nevertheless stands as evidence against the idea that conspiracy theorists nowadays tend not to be very specific about perpetrators. (In any case, since the claim was one of degree, citing a few examples cannot be conclusive either way.) And further, conspiracy theorists will be damned if they do and damned if they don't. If they name specific suspects without overwhelming evidence, even if circumstantial evidence is suggestive, they will surely be charged with "accusation without warrant" (Stokes 2016, pp. 36–37).[30] (See section 2.12 on Stokes below.)

All this notwithstanding, it may be true that conspiracy theorists nowadays tend to be more cautious than they used to be about putting forward specific positive theories, preferring to stay *mostly* on the safer ground of criticism of official stories. I have no firm opinion on this, as it is somewhat plausible and may be true in some degree, though it has not been convincingly demonstrated. In any case, it is not clear this would be a bad thing, or that it should reflect negatively on today's conspiracy theorists. It may be a move in a sensible direction. After all, conspiracy theorists don't have all the resources available to the backers of official stories. It is more reasonable, for example, to expect agencies like NIST to be able to explain, in specific detail, the collapse of the Twin Towers and Building 7 than it is to expect conspiracy theorists to do so—they just don't have the same kind of access. (Further, explaining the collapse of the World Trade Center towers was quite literally NIST's job.) Conspiracy theorists may, nevertheless, have access to *sufficient* relevant information to find serious enough inconsistencies and other problems with the official account to significantly undermine it. If that is the best that they can do given the epistemic circumstances, why should that be held against them? After all, many conspiracy theorists, including most 9/11 conspiracy theorists, explicitly argue that the evidence they provide suggests that *a more robust investigation* is warranted, one that they are not able to conduct by themselves. It is *that investigation*, they argue, that would likely provide a more detailed account.

But Clarke seems to think that the supposed lack of specificity in contemporary conspiracy theories makes a significant difference—so much so that many conspiracy theories, due to the lack of the right kind of specificity alone, cannot warrant belief. In service to his argument, Clarke asserts:

> A particular conspiracy theory needs at least to involve the identification of a specific conspiratorial group and to involve the specification [of] at least one motive to explain that group's conspiratorial activities before the research programme formed around that theory can be assessed as progressive or degenerative. (2007, p. 170)

As my comments above suggest, it is not clear that the 9/11 conspiracy theorists that Clarke has chiefly in mind fail to meet this standard. But is this even an appropriate standard? These particular requirements seem arbitrary, and the latter one, especially, seems unnecessarily onerous.[31] Suppose there are two competing theories, one attributing a death to suicide, the other to murder (which may involve a conspiracy). It is just not the case that one always needs to identify specific murder suspects before one can evaluate the relative merits of these vague but nevertheless meaningfully distinct competing explanations, especially if there are multiple plausible murder suspects, and the "suicide" explanation seems particularly plagued by difficulties. This suggests that Clarke was wrong to assert a requirement for a specific suspect for a Lakatosian analysis of conspiracy theories—if evidence of controlled demolition is sufficiently strong, that may be compelling evidence of conspiracy even if we don't know "whodunit." Or it may be that Lakatosian analysis is just not well suited for judging conspiracy theories, at least not by itself.

Further, Clarke seems too optimistic about the difference that naming perpetrators makes. Clarke supposes that if controlled demolition theorists had named specific perpetrators, then testable predictions could have been made. But this is largely implausible. Many conspiracy theorists have considered Dick Cheney to be a suspect (one is quoted above). What testable predictions follow from that? Could we really "predict when and where [he] can be expected to conduct follow up operations" (Clarke 2007, pp. 174)? It is just not clear that much predictive value is to be gained from naming individuals—*unless they are then faced with an invasive investigation.*

My own view is that Lakatosian analysis may sometimes be useful (and that this may not require naming specific suspects). It provides *one* consideration that particularists should be interested in as they investigate the particulars. But this type of evaluation must be done with sensitivity to the asymmetries that may exist between conspiracy theorists and defenders of official narratives, and to the particular challenges that arise when suspects have special forms of influence (see Keeley 2007 for more on the latter issue).[32] Theories involving human agents cannot be treated as if they were the same as theories about the invariable laws of nature. In any case, an adequate Lakatosian comparison between the official account and controlled demolition theories would be a significant undertaking and is beyond our scope here. But I suspect that people like David Ray Griffin and Steven Jones, two of Clarke's chief targets, would have substantive answers if challenged to defend their theories in these terms. Griffin has, in fact, argued that the official account of 9/11 is itself, in several respects, a degenerating research program (2011, pp. 72–86).

Now, let's suppose we accept Clarke's suggestion that, due to its vagueness in certain respects, the case that the WTC towers came down as a result of controlled demolition is just a "proto-theory." Clarke suggests this entails that it is not amenable to Lakatosian evaluation. Clarke explains, "Once we have [a sufficiently specific] theory we will be in a position to decide whether it is progressive or degenerative" (2007, p. 175)—and, it would seem, not before. But that does not mean there is *no way* to assess the "proto-theory" at all. And it does not mean that it can be safely dismissed either. So, it would seem, Clarke's claim that accusations involving controlled demolition are mere "proto-theories" tells us little if anything about how we ought to regard those accusations. And yet there is a strong rhetorical suggestion that we should reject such "proto-theories" and stick with official stories, at least until such time that a Lakatosian analysis can be applied. (It is as though Clarke has but one tool in his epistemic toolbox.)

Clarke seems, though not quite explicitly, to suggest that lack of specificity in an explanation is sufficient to warrant dismissal of that explanation. This will tend to render unlikely the kind of inquiry that may provide justification for details that Clarke insists must be presented up front. (Just imagine if law enforcement officers limited themselves in this way when investigating crimes.) In any case, here is Clarke's implicit argument, in a nutshell:

1. Conspiracy theories (nowadays) do not offer sufficiently specific positive theories, but rather concentrate on criticizing official accounts.
2. Criticisms of this sort "cannot be sufficient to overthrow a received view in favour of a [alternative] theory," unless that alternative theory provides "specific explanations of the events in question" (Clarke 2007, p. 177).[33]
3. The implied conclusion is that unless and until such specific explanations are produced, conspiracy theories may be safely dismissed as unwarranted.

We have already shown that the first premise is questionable, but let's grant it for the sake of argument. The implied conclusion doesn't really follow. The failure of Theory A to "overthrow" Theory B, at any particular point in time, does not imply that Theory A may be safely dismissed. It may be nearly on par with Theory B, and we ought to be open to the possibility of changes in epistemic status.

Further, premise 2 is false. One can reasonably favor a vague theory (or "proto-theory") or a family of theories (hedging our bets regarding which will pan out) over a more specific alternative. In judging competing theories, specificity is relevant, but we can't assume that other considerations will not outweigh whatever advantage greater specificity offers. For it is easy enough to imagine an elaborately detailed and plainly false theory. It wouldn't be rational to cling to it and reject its chief rival just because that rival was underdeveloped, especially if we assume there are at least *some* reasons to think the rival was, more or less, on the right track.

Clarke's own example, theories positing the controlled demolition of the WTC towers, can serve to illustrate: imagine Julia encounters the fact that the NIST report on Building 7 concludes that it fell at free-fall acceleration for over one hundred feet (NIST 2008a, pp. 45, 48), "primarily due to fires" (p. xxxv). Julia is incredulous; she recognizes that that would mean that the building offered *no* resistance during that portion of the collapse. After considering some additional evidence, Julia reaches the conclusion that Building 7, and the Twin Towers as well, were brought down by controlled demolition, somehow involving some type of thermite. Julia cites the orange flow of what looks like molten iron flowing out of the South Tower, the iron-rich microspheres found in abundance in the WTC dust, multiple reports of molten steel (which Julia now thinks may have been iron—a by-product of a thermite reaction), and red/gray chips found in the dust, which a group of scientists have judged to be unreacted nanothermite.[34] Based on this, Julia hypothesizes that thermite was somehow involved in the demolition of all three buildings. She adds, for other reasons, that the planes that hit the buildings must have been flown by computer or by remote control, at least at the end, not piloted by amateurs.[35] So she does have *some* relatively specific claims. But there are lots of details that Julia has no clear idea about. She doesn't know "whodunit." She doesn't know *exactly* what caused the orange flow—though she does suspect it was the product of a thermite reaction. And she doesn't know exactly *how* the red/gray chips figure in. She just thinks that *somehow* they ought to be accounted for. She emphasizes that those chips *should not be there* if the official story is right (if her understanding of the properties of the chips is correct—see section A.2.3 for reasons to think it is not). In other words, as Clarke has suggested, she emphasizes what appear to her to be *problems with the official story*, and she hopes that experts will look carefully and objectively at the issues that she has raised. For if these issues are honestly and rigorously attended to by those with the proper expertise and access, Julia suspects, more specific and definitive answers can thereby be found.

When asked *who* exactly she thinks was behind it, and what their motives were, Julia suggests that, for various reasons, the vice president ought to be considered a suspect, and that other high-level, like-minded neoconservatives were probably in on it with him. As for motive, she points to the PNAC document *Rebuilding America's Defenses*. Julia doesn't want to get sued, so she couches her accusations in somewhat tentative language, and isn't real specific about all the people she suspects (knowing also that she cannot yet prove her case), but it is nevertheless pretty clear, at least roughly, whom she has in mind. As for those who performed the hands-on work, Julia admits she has no idea. And as for *their* motive, she thinks money can go a long way. Besides, the low-level accomplices probably only had a very limited understanding of what was to transpire. And since they were complete unknowns, they may well have been quietly silenced after the event or, depending on their role, even before the event. But she doesn't have any specific ideas about this.

Let's add, just to complete the picture, that Julia has written to NIST about her concerns, and asked that organization how it explains the one-hundred-foot free fall of Building 7 and other technical questions. But NIST has largely dodged her questions. She has tried to get her academic friends to investigate the issues for themselves. But, for the most part, they can't be bothered—it would take some effort, after all. And besides, its just a crazy "conspiracy theory." Poor Julia, what a waste of energy on her part! After all, according to Clarke, she should just stick with the received view *simply on the grounds that her alternative lacks sufficient specificity*.

But this can't be right. Julia's relatively vague theory *may* be wrong, but Clarke's heuristic does not show that it is irrational, unwarranted, or false. *If* Julia's conclusion is actually unwarranted, it is because she is mistaken regarding the substance of her arguments, not because such considerations, even if valid, could be safely ignored on account of the insufficient specificity of her theory. The question is not whether or not Julia's thesis is elaborate enough. It is whether or not her analysis holds up. But in order to judge that, one has to look quite closely at the particulars. Bogus heuristics like that suggested by Clarke aren't helpful; they are misleading. They provide illegitimate excuses for not considering the merits of critiques of official stories.

There is one other aspect of Clarke's argument that deserves mention. Clarke suggests that *if* conspiracy theorists were to venture more specific positive accounts (which would have to be based only on the information they had at the time), those accounts would be found to be "degenerative," that is, evidence would tend to undermine them over time, making attempts to resuscitate them increasingly desperate. This is largely a speculative claim, but likely to be true. After all, if one demands more specificity than the available evidence is able to support, then the specifics guessed at are not likely to turn out to be right. The more specific a theory of any kind, the more likely it is to be wrong, at least in some aspect. However, that doesn't mean that the more general theory was on the wrong track.

Finally, there is something not entirely fair about the way Clarke motivates the idea that any elaboration will be degenerating. Clarke writes, for example, it would be "hard to believe that [a theory positing controlled demolition] will not be degenerative in at least one way" (2007, p. 175), namely, that it will be difficult to explain why nobody reported the "strange events" that Clarke assumes would have had to have taken place to prepare the building for demolition. But Clarke has simply ignored the ubiquitous efforts of conspiracy theorists to address this concern. For example, they often point to actual elevator repair work that could have served as cover, providing access to the core columns, without arousing suspicion. I am not arguing that these efforts to explain how the buildings could have been discretely prepared for demolition are adequate. I merely want to stress that *determining their adequacy* would require a treatment of the evidence that goes far beyond Clarke's sketchy and selective description.

2.12. Patrick Stokes's Reluctant Particularism

> [P]articularism may involve us in ethically problematic practices. The very act of entertaining the conspiracy theory as a *worthwhile hypothesis for investigation* may come at a serious moral cost, both in licensing socially harmful practices and in violating the attitude of trust that is, I argue, a precondition of ethical life. The position I advocate then is somewhere in the messy middle: a "defeasible generalism" or "reluctant particularism."
>
> —Patrick Stokes (2018a, p. 26)

Philosopher Patrick Stokes, who was "Once a conspiracy theory skeptic of a fairly conventional kind," according to Pigden (2018, p. 203), has now come to see that what he calls "blithe generalism" is not tenable. Stokes writes:

> Quite astonishingly, something like a broad consensus has emerged: regarded simply as explanations, conspiracy theories are not intrinsically irrational, and believing in conspiracy explanations is not necessarily unwarranted. (2018a, p. 25)

The fact that this is regarded as "astonishing" to a philosopher and public intellectual like Stokes might itself be a more legitimate cause for astonishment. For it should have been relatively obvious, as Stokes himself makes clear as he continues:

> After all, there are conspiracies we all agree happened—Watergate, Iran-Contra, etc.—and the search for a (non-question begging) definition of "conspiracy theory" that would exclude belief about these accepted conspiracies, but include those wacky theories we want to dismiss, appears doomed to failure. (Stokes 2018a, p. 25)

And yet, Stokes can't quite adopt a straightforward particularism, as he explains in two related chapters of a recent anthology.[36] He writes:

> [M]y position is not generalist—that is, it does not *dismiss* any given conspiracy theory out-of-hand *just because* it is a conspiracy theory. But it does give us a strong reason to be deeply suspicious of and reluctant to entertain explanations of the conspiracy style, lest we find ourselves obliged to level accusations out of loyalty to the theory rather than fidelity to the evidence. (Stokes 2018b, p. 200)

The difference between being dismissive and "being deeply suspicious and reluctant to entertain" is hard to discern. As Basham points out, "If 'reluctant' means we will not immediately embrace a theory, but seek significant evidence for or against, then this is simply the particularist position" (2018b, p. 43). However, if "reluctant" means something stronger—if it means that we refrain from conspiracy theorizing, then, as Dentith points out, "it is hard to work out how we can accept the former claim (the particularist position) if we also accept the latter (the position of reluctance)" (2018d, p. 61). Further, it is not clear why Stokes assumes there is a slippery slope between entertaining conspiracy theories and not being faithful to evidence. It would seem, on the face of it, that being reluctant to even entertain a certain explanation is close to refusing to consider evidence. So it is Stokes's own position that risks what he seeks to avoid, namely, failing to be faithful to the evidence.

Stokes's critique of conspiracy theorizing involves framing it as a tradition and practice. This practice is part of a "tradition of explanation" and makes use of "pre-existing templates" (Stokes 2018a, p. 28). Citing Richard Hofstadter,[37] Stokes suggests that conspiracy theorizing has a "recognizable 'style'" (p. 28), that it involves "historically conditioned recurring tropes and causal narratives" (p. 28) and "reflexive explanations" (p. 29) which are "specific explanatory tropes deployed across different events" (p. 29).

You may guess where this is going—connecting conspiracy theorizing with anti-Semitism (p. 29). And Stokes does have a point. Some conspiracy theorizing is anti-Semitic, or is in other ways prejudicial, which is obviously not good. Even some conspiracy theorizing that is not blatantly prejudicial may nevertheless be faulted, for all sorts of reasons, some of which Stokes describes (although some of his particular claims are contested in Basham 2018b). These considerations form the basis for Stokes's criticism of particularism. Stokes writes:

> The particularist, by insisting on reviewing each conspiracy theory solely on its own merits, requires a certain indifference to the contingent cultural, historical, and rhetorical explanatory repertoires from which conspiracy theorists typically draw their hypotheses. (Stokes 2018a, p. 29)

But that's not true at all. As Dentith points out, "[P]articularism is not indifferent to the social context of conspiracy theorizing" (2018d, p. 63). In other words, the considerations Stokes is worried about are included in the particularities that particularists consider when evaluating a particular theory. If the theory smacks of anti-Semitism, the particularist takes that into consideration. By itself that doesn't prove the theory false, but it is a bright red flag. However, the existence of an anti-Semitic argument should not be used to undercut the consideration of a non-anti-Semitic argument to a similar conclusion. As a close analogy, consider that critics of early JFK conspiracy theorists, in the years following the release of the Warren Report, pointed out that "rightists" were making some of the same claims.[38] At the same time, an infamous CIA memo, which provides advice for "countering and discrediting the claims of the conspiracy theorists," suggests that its media assets "Point out also that parts of the conspiracy talk appear to be deliberately generated by Communist propagandists."[39] In this light, it seems that Stokes's suggestion that we should avoid some theories because of their purported relation to other theories and traditions is not a new one; it is itself an anti-conspiracy theory trope, with its own suspect history.

Consider another way Stokes expresses the issue. He writes:

> There may be no a priori epistemic reason not to believe in a given conspiracy theory simply *because* it is a conspiracy theory. But nonetheless we can ask whether we have ethical reasons to avoid participating in the practice of conspiracy theorizing, in which we explain events using the resources and assumptions of the conspiracy theory tradition. (2018a, p. 29)

However, there is a problem with Stokes's generalist way of framing this. As Pigden incisively points out, conspiracy theorizing is not *a* practice, it is "a *multitude* of different practices. . . . And many of these ways of arriving at conspiracy theories are intellectually kosher" (2018, p. 205). So it seems that we should not disparage conspiracy theorizing generally, but rather evaluate each particular example of conspiracy theorizing and consider, in each case, whether it involves problematic "resources and assumptions."

There is one aspect of conspiracy theorizing that Stokes particularly emphasizes, namely, the "problem that conspiracy theory requires accusation" (2018a, p. 35). Stokes suggests that "interpreting others in the most charitable way possible" is important (p. 36), and that one ought to "refuse to think badly of people until compelled by evidence to do so" (p. 35). But generally speaking, acceptance of official stories often also requires accusation. (The death of Princess Diana, which Stokes sometimes cites in this context, is quite exceptional in this regard.) And many conspiracy theorists think that it is the evidence that compels them to think *less* badly about Sirhan, Oswald, and James Earl Ray, for example. The difference between the conventionalist and the conspiracy theorist is not between those who accuse people and those who do not. Rather, it is a disagreement about *who* is taken to be more likely guilty.

Further, Stokes does his own accusing. He accuses conspiracy theorists of causing unnecessary deaths. Should Stokes have been more reluctant to make such a serious accusation against so many people? If there is a fault with the accusation, it is not that it *is an accusation*, it is with the accuracy or fairness of the accusation. And that has to be adjudicated on a particularist basis, considering the evidence. (Regarding this issue, Basham 2018b challenges Stokes on some of the relevant particulars.)[40]

Stokes is also concerned about the harm that may come from conspiracy theorizing, at least when done badly. He writes, "I touch briefly on some of the reasons why conspiracy theory as a social practice can be harmful, particularly in terms of producing or licensing harassment and other forms of abusive behavior (though of course this is neither a necessary consequence of conspiracy theorizing nor unique to it)" (2018b, p. 190). Stokes's parenthetical remark should be emphasized. Conspiracy theorizing sometimes leads to harassment, but particularism doesn't license *that*. Further, conspiracy theorists have also found themselves harassed, sometimes by supporters of official accounts. Harassment, abuse, and intimidation are, of course, inappropriate. But, as Stokes acknowledges, they are neither essential to conspiracy theorizing nor unique to it.

And yet Stokes thinks there is some close relation between conspiracy theorizing and making accusations. He states that "engaging in conspiracy theorizing involves a willingness to entertain and multiply accusations, and that such a willingness violates a moral comportment we owe to others" (2018b, p. 190). It seems to me that Stokes here exaggerates this duty to others. Self-preservation has long depended on, and justifies, the willingness to consider others as potentially threatening—and thus to entertain an accusation of them. Much of the time this takes the innocuous form of a passing thought, dismissed almost as soon as it occurs. Sometimes, however, such thoughts are not immediately dismissed. Should they be? How are we to protect ourselves from *any* threats if we adopt Stokes's unwillingness to perceive anyone as potentially threatening? We lock our doors, pay for military defense and police protection, and allow the TSA to search us before we get on a plane. Stokes doesn't think, surely, that we should stop any of *those* practices, even though they imply that we can't trust others. Is this different because we haven't named the others we distrust. No. Sometimes we do name them: al-Qaeda, for instance. The difference must be that Stokes simply doesn't think conspiracy theories make accusations against the *right* people. But how are we to judge *that*, unless we are willing to do some conspiracy theorizing?

It is only once one has "entertained" a conspiracy theory, giving it due consideration, that one is in a position to make an ethical judgment about who ought to be accused of what. As Basham puts it, "Epistemology determines ethical response in these cases, not the other way around" (2018b, p. 42). Stokes, in contrast, suggests that "conspiracy theory as a practice licenses a distinctive kind of standing indifference to others' legitimate claims not to be impugned without overwhelmingly good cause" (2018b, p. 199). Framing it this way stacks the deck against conspiracy theories by insisting that they overcome an unreasonably high epistemic burden while cutting off their ability to even begin building a case. If an overwhelmingly convincing theory is required before one may engage in conspiracy theorizing, no such theory could get off the ground. One can't expect to have proof before investigating and theorizing takes place.

Still, there is one positive lesson we can take away from Stokes's challenge. When we engage in conspiracy theorizing, as Stokes maintains, we are "responsible for how we do so" (2018a, p. 36). Indeed. But there is nothing about particularism that implies a denial of *that*. As Dentith comments, "This is true: we should be responsible particularists. . . . We should resist being blithe about how we might go around assessing and investigating particular conspiracy theories" (2018d, pp. 64, 66). Stokes, however, points out that "many published conspiracy theories have not met that moral burden" (2018b, p. 196). Although where the line ought to be drawn is debatable, as is the actual strength of the evidence in particular cases, Stokes does nevertheless make an important point. But it is not a point that undermines particularism. It simply suggests that *those particular* conspiracy theorists who have crossed the line, however many, are individually criticizable on that count. (For more on particularism, see section 10.2.)

2.13. Summary

Let's review. The idea that Popper's argument against *the conspiracy theory of society* implies that *conspiracy theories* can be safely dismissed is based on a simple equivocation. Most conspiracy theories do not imply the conspiracy theory of society. Similarly, contrary to Bale's suggestion, most conspiracy theories are not actually on the implausible side of Bale's "clear analytical distinction." As for Keeley, while he set out to provide a generalist argument against conspiracy theories, he himself recognized that it wouldn't work. And Mandik's absurdist argument against conspiracy theories (and all other historical explanations), and his argument based on not being able to trust the testimony of alleged conspirators, are both nonstarters.

Sunstein and Vermeule (implicitly) and Levy (explicitly) argue that conspiracy theories that conflict with the views of epistemic authorities should be rejected, for we should simply trust our epistemic betters. This is the argument that this book as a whole most seeks to address. Much more will be said about Sunstein and Vermeule in Part II. Part III suggests that (would-be) epistemic authorities in the social sciences have proven unreliable in their analysis of conspiracy theorists. And the appendix suggests that ostensible epistemic authorities are disappointingly unreliable about the events of September 11, 2001.

As for Cassam, his claim that conspiracy theorists are more gullible than conventionalists is question-begging, and his negative depiction of a "conspiracy mentality" is partly based on seriously flawed psychological studies and partly baseless. Clarke's arguments also fail. His argument based on the Fundamental Attribution Error was a "big mistake," and his argument based on the idea that conspiracy theories are insufficiently specific fails both because it is invalid and because at least one of its premises is false. Finally, Stokes's "reluctant particularism" is either biased in its reluctance or it is simply particularism.

In sum, a number of scholarly attempts to articulate a justification for a dismissive attitude toward conspiracy theories have been attempted, and all have failed.[41] The battle between the particularists and the generalists has been a rout. Pigden, Coady, Basham, Dentith, Räikkä, and, in the end, Keeley are right to deny that it is reasonable to dismiss a conspiracy theory just because it is a conspiracy theory. One must adjudicate each theory individually, according to its particular merits and problems. That is, particularism is true. The fact that all sides agree that some conspiracy theories have turned out to be true should have made this conclusion obvious.

That's the situation in philosophy. However, many social scientists, in particular, continue to assume that there is something cognitively wrong with conspiracy theorists. Chapters 7, 8, and 10 address the often-problematic social science literature on conspiracy theories. In the next chapter I address a recent article, authored by a social scientist and two philosophers, which deals with the concept of "conspiracist ideation" and whether such thinking has ever been involved in the revelation of real conspiracies.

THREE

Conspiracist Ideation and the Revelation of Real Conspiracies

In 2018, Stephan Lewandowsky, a cognitive scientist, together with Elisabeth A. Lloyd and Scott Brophy, both philosophers, published an article in the journal *Argumenta*, in a special issue focusing on the ethics and epistemology of conspiracy theories. (I'll often refer to Lewandowsky, Lloyd, and Brophy as "LL&B," in the spirit of their own affinity for acronyms.) I include an analysis of LL&B's article here because it was published in a philosophy journal, and two of the authors are philosophers. In the previous chapter I addressed other relevant philosophy articles, focusing on whether a dismissive attitude toward conspiracy theories is justifiable. The reason I dedicate a separate chapter to LL&B's article is partly because "conspiracist ideation" is a slightly different, though closely related, topic, but mostly because there are so many problems and ambiguities in LL&B's article that sorting them all out—which serves as a convenient way to make a number of points—requires considerable space. In addition, LL&B's article combines philosophy and social science, and to some degree foreshadows, in a general way, the analyses in Chapters 7, 8, and 10 of Part III, which addresses a number of articles in the social sciences.

LL&B, it seems, are terribly concerned about what they perceive as the dangers posed by conspiracy theories. Their worries stem mostly from issues having to do with skepticism regarding scientific consensus, and climate change in particular, rather than political conspiracy theories proper. While their work is undoubtedly well motivated, their concern (some might say "moral panic") over climate change and other scientific "controversies" may at least partly explain how their treatment of conspiracy theories more generally ended up so skewed.

LL&B's article, "When THUNCing Trumps Thinking"—which is transparently partisan[1]—has essentially two aspects: First, the authors suggest that conspiracy theories that have been acknowledged to be true were not uncovered by conspiracy theorists or by "conspiracist cognition." Second, they argue that Donald Trump's tweets exemplify "THUNC" ("Thinking in Unreflective Counterfactuals"), one of the aspects of "conspiracist cognition." In its first aspect, the article attempts to provide a way of distinguishing between "actual conspiracies [which] are preferentially identified by conventional cognition, [and] non-existent conspiracies [which] are the domain of conspiracist cognition" (p. 220). In other words, LL&B are offering a strategy for distinguishing between theories that may identify a real conspiracy and those that only connect with "non-existent conspiracies," *without having to consider the relevant evidence*. Rather, all one has to do is look at the cognition of the theorist. As they cryptically put it: "[I]dentification of the cognitive attributes—and hence inference of the epistemic object—is empirically possible" (p. 219). Citing Sunstein and Vermeule (2009), they claim: "[C]onspiracy theorists typically suffer from a 'crippled epistemology' . . . [which] can be identified by independent means" (p. 219).

LL&B invite the reader to think of conspiracy theories in relation to "delusional beliefs," such as believing that one's "partner has been replaced by an impostor." They explain, "It is only by ignoring the rather infinitesimal prior probability of a wife being replaced by an impostor that perceptual evidence can yield the delusion that this has actually happened. Likewise, paranormal and pseudoscientific claims are typically considered to be epistemically unwarranted" (Lewandowsky et al. 2018, p. 218).[2] They make the same claim with respect to Donald Trump's conspiracy theories, writing: "Trump's claim . . . cannot be sustained without a . . . refusal to consider prior probabilities of various competing worlds" (p. 226). In other words, if one does consider the prior probabilities, then Trump's claims cannot be sustained (regardless of what evidence may obtain). Trump's conspiracy theories are used as an example to illustrate how conspiracy theories can be determined to be unwarranted on the basis of prior probability alone.

LL&B do admit that conspiracy theories are a little different from other types of "delusional beliefs." For there is a "puzzle" involved in assessing "the extent to which conspiracy theories are epistemically warranted." For, as LL&B acknowledge:

> Unlike beliefs in the supernatural or other cognitive delusions that have no discernible echo in reality, conspiracies sometimes do exist. Oliver North did sell weapons to Iran from the basement of the White House, and Richard Nixon's White House was involved in the Watergate break-in, and so on. (p. 218)

And yet, they point out, there are also conspiracy theories offered for which there is "no credible [supportive] evidence." I'm a little puzzled about why this is thought to be puzzling. Some theories turn out to be well supported, and even confirmed as true, while others are not well supported at all. What is puzzling about that? Perhaps it is only puzzling to those who wrongly conceive "conspiracy theories" as a type of "delusional belief." After all, while some specific conspiracy theories may indeed be delusional, that is not a special quality of conspiracy theories; it is true of nearly any type of theory or idea. Some ideas about rabbits, for example, may be delusional. But that doesn't make "beliefs about rabbits" a type of delusional belief.

LL&B believe they have a solution to their "puzzle." They write, "We argue that an analysis of the cognition and argumentation involved in the theorizing can be a potential indicator of whether we are dealing with an epistemically defensible analysis of a potential conspiracy, or with conspiracist cognition that is a candidate for rejection" (p. 218). The phrase "candidate for rejection" is imprecise. What ideas are not *candidates* for rejection? What is actually meant, I suppose, is that, in contrast to theories that are potentially true because they are based on "epistemically defensible analysis," ideas based on "conspiracist cognition" are *not* "epistemically defensible" and thus *should be* rejected. The questions that arise, then, are these: (1) What counts as "conspiracist cognition"? (2) Is conspiracist cognition epistemically problematic? (3) What is the relation between "conspiracist cognition," as LL&B define it, and the thinking of conspiracy theorists? I address these questions below, considering each aspect of "conspiracist cognition," which LL&B also refer to by the label "conspiracist ideation."

3.1. The Seven Characteristics of "Conspiracist Ideation"

In the social science literature, "conspiracy theorists" are often presented as engaging in "conspiracist ideation" (as can be seen in Chapter 7). But there is some ambiguity in the use of this term. In a blog post, Lewandowsky equated "conspiracist ideation" with "conspiratorial thinking."[3] In contrast, the criteria that LL&B use to identify "conspiracist ideation" are much more specific than just "conspiratorial thinking," although in some sense it is a rather expansive term. In this context, "conspiracist ideation" is a construct built from seven mostly independent elements that are grouped together because of their purported prevalence in the thinking of people who also have a heightened propensity to believe conspiracy theories. These are based, in part, on social science studies that generally attempt to measure statistically significant *degrees of difference* between groups. Many of these studies are seriously flawed, as Part III of this book documents. In addition, some of the criteria are based on philosophical essays (such as Keeley 1999), which merely assert an association, providing some (often inadequate) reasoning to motivate the plausibility of the idea. These philosophical essays are at best suggestive and often seriously flawed, as the previous chapter showed.

Although LL&B clearly imply that conspiracist ideation is a form of bad reasoning, there is nothing inherently epistemically problematic about several of the elements of conspiracist ideation, as explained below. Other aspects, however, really are epistemically problematic, but their connection to the thinking of conspiracy theorists is tenuous. And even if there is some statistical correlation between people's propensity to believe in conspiracy theories and these problematic attributes, that cannot be a valid reason for rejecting any particular conspiracy theory.

Below are the seven specific criteria for "conspiracist ideation." LL&B do not specify whether all criteria must be met to count as "conspiracist ideation," or whether a single criterion would be sufficient. If all are necessary, then conspiracist ideation is exceedingly rare. But if only one is needed, then it is both ubiquitous and generally epistemically unproblematic—because cases involving the criteria that *are* problematic are uncommon.

3.1.1. Attribution of Nefarious Motivation

LL&B write, "The first criterion is that the presumed motivations behind any assumed conspiracy are invariably assumed to be nefarious" (p. 221). There is nothing epistemically suspect about the attribution of nefarious motives. Detectives and prosecutors, for example, attribute nefarious motives all the time, with good reason. I address the question of whether conspiracy theorists attribute *implausibly* nefarious motives to alleged conspirators at length in Chapter 8 (see especially section 8.5) and also Chapter 9. The short answer is that conspiracy theorists do not *invariably* assume nefarious motivations, and even when they do assume motives that are *to some degree* nefarious, they are often not implausibly so. To give just one example here, philosopher Juha Räikkä points out:

> It may sound unlikely that the president of a democratic country has a motive to authorize the bombing of a ship owned by an environmental organization. But this is exactly what happened when the *Rainbow Warrior I*, a ship operated by Greenpeace, was sunk by the French foreign intelligence agency in Auckland harbor in July of 1985; the operation, as we now know, was authorized by President Mitterrand. (Räikkä 2009a, p. 197)

Even Lewandowsky is troubled by what he seems to regard as a nefarious conspiracy[4] "to prevent meaningful climate policies aimed at preserving civilization for future generations" (2019, p. 168).

3.1.2. Perceiving Oneself as a Victim

The second criterion is supposed to be a corollary of the first, which is presumably why LL&B provide no explicit reference tying this to the thinking of conspiracy theorists. They explain this criterion as follows:

> [T]he person engaging in conspiracist ideation perceives and presents her- or himself as the victim of organized persecution, while at the same time, at least tacitly, also perceiving themselves as brave antagonists of the nefarious intentions of the conspiracy. Conspiracist cognition thus involves a self-perception of being a victim but also a hero. (p. 221)

Neither perceiving oneself to be a victim nor a hero, nor both simultaneously, necessarily signals epistemically problematic thinking. Whistleblowers, for example, may find themselves justifiably thinking of themselves as both, as the case of Thomas Drake, below, exemplifies. Whether or not it is problematic to view oneself as a victim, or as a hero, will depend on the particulars of the situation. (Surprise! It's particularism again.)

Also, these kinds of self-perceptions are matters of degree. If you've made an effort to do what you believe is right, it's not wrong to think you're a bit of a hero. If you have reason to think you, or people you identify with, are being treated in some way unfairly, it's not wrong to think you are a bit of a victim. We are all victims and heroes to some degree. Recognizing this is neither an epistemic pathology nor unique to conspiracy theorists.

Lewandowsky's own case can serve as an example again. He frames the climate scientists that he collaborates with, and appears to identify with, as both heroes and victims of the above-mentioned nefarious conspiracy. Lewandowsky writes, "Climate scientists have done an admirable job pursuing their science under great political pressure, and they have tirelessly rebutted pseudoscientific arguments against their work" (Lewandowsky 2019, p. 168). From this perfectly legitimate perspective, they are justifiably thought to be heroes, if well-funded ones. And the presumed-to-be coordinated[5] campaign, including threats and abuse, which is intended to thwart the efforts of Lewandowsky and his climate science colleagues to "preserv[e] civilization" makes them at the same time victims.[6] There is nothing epistemically problematic about noticing that. Conspiracy theories can be rational, even when invoked hypocritically by critics of conspiracy theories.

3.1.3. An Almost Nihilistic Degree of Skepticism

LL&B write: "Third, conspiracist ideation involves 'an almost nihilistic degree of skepticism'" (p. 221). They also refer to this as an "overriding suspicion."

They cite Keeley 1999, p. 125. However, Keeley does not actually *show* that this accurately describes the thinking of conspiracy theorists. Indeed, it is an extreme exaggeration that describes almost nobody. Let's look at Keeley's statement in context:

> [T]he challenge of conspiracy theory is that it forces us to choose between an almost nihilistic degree of skepticism and absurdism: the conspiracy theorist chooses to embrace the hyperskepticism inherent in supposing dissimulation on a truly massive scale (by distrusting the claims of our institutions) over the absurdism of an irrational and essentially meaningless world. (1999, p. 125)

That sure looks like a false dichotomy. I think it is fair to say that this aspect of "conspiracist ideation" does not fairly describe the thinking of conspiracy theorists, even as a generality.[7]

A central theme of this book is that supposed epistemic authorities are often not as reliable as we might wish they were—at least when it comes to dealing with conspiracy theories. But that does not imply a nihilistic degree of skepticism. After all, distrust of institutions comes in degrees, and with various possible nuances. LL&B can serve as an example. Given the content of their article, I don't imagine that they have much faith in the reliability of President Trump. But that doesn't make them radical skeptics.

Interestingly, Lewandowsky's own research suggests that the kind of skepticism that would likely support conspiracy theories about the Bush administration's case for war with Iraq tended to be associated with more accurate assessment of information. Lewandowsky explains:

> Our research identified skepticism about the motives underlying the war [in Iraq] as a key predictor of accuracy of information processing. People who were skeptical of the U.S. government's official reason for the invasion, namely to rid Iraq of weapons of mass destruction (WMDs), were found to be better able to reject false information and accept true information then others who accepted WMDs as the *casus belli*. (Lewandowsky 2019, p. 149)

3.1.4. Belief That Nothing Occurs by Accident

LL&B write: "Fourth, the overriding suspicion [i.e., criterion 3] frequently entails the belief that nothing occurs by accident" (p. 221). The idea that "nothing occurs by accident" is an element of the "conspiracy theory of society," and of "the paranoid style," that is often inappropriately applied to conspiracy theorists more generally. Almost no one believes in the "preternatural effectiveness" that is here implied. Conspiracy theories generally neither posit nor imply this. (This is discussed in sections 8.1 and 8.4, as well as 1.3.3 and 2.3).

3.1.5. Conviction That "Something Must Be Wrong" with an Official Account

LL&B write:

> Fifth, although conspiracist cognition may entail the abandonment of specific hypotheses when they become unsustainable, those "corrections" do not impinge on the overall abstraction that "something must be wrong" and the "official" account must be based on deception. (p. 221)

However, believing a theory must be wrong, without knowing exactly what is right, is not epistemically problematic. (See section 2.11.) Suffice it to say here that the theory may, for example, seem to imply something impossible.

The authors nevertheless try to suggest that this aspect of "conspiracist cognition" *is* problematic. Citing Wood et al. 2012, they write:

> [Because of] a fixed belief that an official version is wrong . . . it may not even matter if hypotheses are mutually contradictory, and the simultaneous belief in mutually exclusive theories—e.g., that Princess Diana was murdered but also faked her own death—has been identified as an aspect of conspiracist ideation. (p. 221)

Unfortunately for LL&B, as I explain in detail in Chapter 7,[8] the cited study by Wood et al. is seriously flawed in a way that should have been apparent to any careful reader. Contrary to what Wood et al. claim, it has not been successfully shown that conspiracy theorists have a tendency to believe mutually exclusive theories. (Most of Part III of this book focuses on similar problems with the social science literature on conspiracy theories, which LL&B simply accept uncritically.)

3.1.6. Tendency to Treat Evidence against a Theory as Evidence for It

LL&B write: "Sixth, conspiracist cognition is inherently self-sealing: that is, evidence that counters a theory is re-interpreted as evidence for a conspiracy" (p. 222).

To tie this tendency to conspiracy theorists, LL&B cite Bale 2007, Keeley 1999, and Sunstein and Vermeule 2009, all of which are discussed in the previous chapter, and none of which actually shows that conspiracy theorists invariably do this. Further, as I explain in response to Sunstein and Vermeule 2009 (see section 4.1), official accounts can also be self-sealing. In addition, it has not been established that *when* conspiracy theorists reinterpret evidence in this way, they are *wrong* to do so. For sometimes, what seems at first to be evidence for something, may *legitimately* become evidence against it.

Buenting and Taylor (2010) discuss this under the rubric of "fortuitous data." They argue that, in cases in which the evidence for an official story is too "fishy," "it is irrational to believe the official story instead of the conspiracy theory" (Buenting and Taylor 2010, p. 576). One example of "fishy" evidence is this: "[A] passerby discovered the passport of Satam al-Sugami, one of the (alleged) hijackers [of 9/11]. This is fortuitous. That al-Sugami's passport survived the impact, ensuing fireball, and was found eighty floors below in (surprisingly) pristine condition is lucky" (Buenting and Taylor 2010, p. 574)—so lucky that it is "fishy." Readers will have to make their own judgment about *how* fishy this case is. But, in general, once sufficient evidence obtains that a frame-up is occurring, a reinterpretation of evidence that had been taken at face value is sometimes warranted.

So the kind of reasoning in question here is sometimes appropriate. If it did not come into play in the five cases LL&B choose to consider involving the uncovering of real conspiracies (a supposition that is far from established), that would not show that conspiracy theorists are always wrong to reason this way when they do.

3.1.7. Thinking in Unreflective Counterfactuals

The final aspect of conspiracist cognition is "Thinking in Unreflective Counterfactuals" (THUNC), which LL&B suggest is "typical of conspiracy theories" (p. 228). They explain:

> THUNC is characterized by two attributes: First, it involves hypotheses that are built on a non-existent, counterfactual state of the world, even though knowledge about the true state of the world is demonstrably available. Second, in the context we had initially investigated it, we proposed that THUNC additionally involved conclusions that are logically unwarranted even if the counterfactual state of the world were true. (p. 222)

Put more simply, THUNC is unsound reasoning: either (1) reasoning from false premises or (2) drawing invalid inferences.

LL&B admit, "[W]e are unsure whether [THUNC] is a necessary, or even pervasive, attribute of conspiracist ideation" (p. 228). Nevertheless, they cite Lewandowsky et al. 2015 (which includes LL&B as authors) in an apparent attempt to show "by independent means" (p. 219) that conspiracy theorists have some heightened propensity to engage in THUNC. That article addresses challenges to scientific orthodoxy rather than political conspiracy theories, focusing on "climate denial" (which is a bizarre way of referring to skepticism regarding the scientific consensus on climate change). In any case, the most relevant section of that article compares blogosphere critiques with those of young scholars and defines several "conspiracist attributes" and one "scholarly" characteristic. It finds that the blogosphere critiques were less scholarly than those of the scholars (surprise!). It also finds that the scholars scored lower on "conspiracist attributes." As the authors put it: "It is clear that web comments were rated considerably higher than the PhD comments on all 4 measures that tapped conspiracist attributes, whereas the reverse was true for the single item examining the scholarly nature of a content item" (Lewandowsky et al. 2015, p. 165). Even if we assume that the "conspiracist attributes" are signs of relative unreasonableness, this study supports the unsurprising conclusion that scholars are more reasonable (to some measurable degree) than the average internet commenter. If the authors of the study were truly interested in doing a fair study comparing the thinking of conventionalists with those who question orthodox views, they would have pitted scholars against scholars and random commenters against random commenters.

In the end, we are left with no compelling evidence that conspiracy theorists engage in THUNC more than anyone else—even if that were relevant, which it really isn't. For even if there is some statistical difference in this regard, which there may well be, that would not be a valid reason to dismiss conspiracy theories. To argue otherwise would itself be an example of THUNC (in its second aspect).

To sum up, out of the first six criteria for "conspiracist ideation," four are not necessarily epistemically problematic, and the other two (criteria 3 and 4), are not characteristic of the thinking of conspiracy theorists, even as a generality. And the final criterion, THUNC, is not convincingly tied to the thinking of conspiracy theorists.

3.1.8. Is the Construct "Conspiracist Ideation" Helpful or Misleading?

First of all, both "conspiracist" and "ideation" are pathologizing terms, and so their use in an investigation of the potential merits of a set of cognitive tendencies seems to prejudge the issue. As Lee Basham, in particular, stresses, deploying terms like "ideation" serves to decouple conspiracy theorists from any hint they might have good reasons for their views. Relatedly, Juha Räikkä and Basham have argued that various biases "dispose researchers to neglect how rationality and evidence function in the cognitive practices of conspiracy theorists" (Räikkä and Basham 2019, p. 183). They remark:

> The result is a *pathologizing approach* to conspiracy theorizing. Conspiracy theorizing is reduced to a mysterious illness. Social science research often labels conspiracy theories as mere "ideation," "anomalous belief," "conspiracism" and other vague, dismissive terms. Swept up by the pathologizing approach, many social science researchers deployed this de-rationalizing and dehumanizing lan-

guage in their discussions of conspiracy theorizing and conspiracy theorists. (p. 183)

That seems to be the case here. And so, perhaps it shouldn't be too surprising that there are significant problems with LL&B's analysis.

When LL&B refer to "the demonstrable lack of . . . success of conspiracist cognition" (p. 220), they seem to treat conspiracist cognition as an all-or-nothing trait. But it is not. First, it is possible to exhibit some elements of conspiracist ideation and not others. And, second, the manifestation of any particular element is a matter of degree. For example, even assuming an objective line can be drawn in the abstract between conventional "healthy skepticism" and conspiracist "overriding suspicion" (which is doubtful), different people will pass that line to different extents. And, further, these extents will be neither static nor equal in all directions. One may be very suspicious regarding X but not so much regarding Y. These considerations suggest that identifying the presence of "conspiracist cognition" is at best unlikely to be a useful indicator of epistemic warrant, especially since there is no getting around the need to assess the relevant evidence in order to determine, for example, how much suspicion is *justified* in the particular case. Such issues are likely to be contentious.

However, there are some aspects of "conspiracist cognition," criteria 4 and 7 (THUNC), that may still seem to be of some use. Criterion 4 (an aspect of "the conspiracy theory of society" and of "the paranoid style") may be used against cases in which it truly applies, but such cases a rare, and they don't have large followings. So even if valid, it has limited applicability. How about THUNC? If one can establish that an argument for conspiracy involves THUNC, then that argument gives us no reason to believe the theory. However, there is no need to appeal to "conspiracist cognition" to make this observation. As mentioned above, THUNC is essentially unsound reasoning. Making a case that someone has engaged in unsound reasoning has always been recognized as an appropriate cognitive move. And it applies in all directions. Mixing it together with potentially innocuous modes of reasoning associated with conspiracy theorists, and with exaggerated characteristics inappropriately attributed to them, is not useful. Indeed, it is positively misleading.

Further, unsound reasoning does not imply that a conclusion is wrong. So the suggestion that we can use "conspiracist cognition," or even THUNC by itself, as a way of showing that a conspiracy theory may be safely dismissed is itself fallacious. After all, interesting conspiracy theories are typically supported by multiple lines of argument. Showing that fallacious arguments occur here or there doesn't show that the theory is false. As I stress in Chapter 8, all theories ought to be evaluated on their *best* arguments. Identifying a silly argument as silly is not particularly helpful if a better one exists.

3.1.9. Donald Trump's THUNCing

As the article shifts focus to the apparent THUNCing of one Donald J. Trump, LL&B revise their understanding of the term: "In our revised understanding of THUNC, the counterfactual has been imagined precisely to account for a conspiratorial thesis" (p. 222). The discussion of conspiracy theories put forth by Trump seems to function largely as an attempt at guilt by association, reflecting badly on all conspiracy theories. I do concede that association with the ideas expressed in many of Donald Trump's tweets would indeed be a legitimate cause for embarrassment. And, there do seem to be a surprising number of people who accept Trump's apparently false pronouncements.

However, if Trump is either wrong or lying about many things, or if he is irrational, indeed, if he is mentally ill, *and* if he is also a conspiracy theorist with a considerable following, what does that tell us about *other* conspiracy theorists, such as David Ray Griffin, Peter Dale Scott, Jim DiEugenio, and Bill Simpich, and the theories they espouse about 9/11 and the JFK assassination? Nothing conclusive. But, for those eager to justify their reflexive dismissiveness, perhaps it suggests a hypothesis: these people, being conspiracy theorists, may exhibit the same kind of irrationality characteristic of Trump's tweets. If anyone is tempted by that hypothesis, I encourage them to test it directly: *look*, and *see* if they do. I, myself, see no resemblance between Trump's ravings and the careful argumentation of these and many other conspiracy theorists. That observation, of course, is not meant to exonerate *everybody*. Each theorist ought to be judged on the quality of his or her own arguments. (Particularism again.)

Now, having just indicated the *irrelevance* of LL&B's analysis of Trump's theories, I will nevertheless explain why their analysis of Trump is flawed. They focus on two of his theories, neither or which I support: one involves the supposed wiretapping of Trump Tower, the other involves massive voter fraud on the part of undocumented immigrants. I don't put any stock in the theory that President Obama had Trump Tower wiretapped because, as the authors state, there is no credible evidence (that I am aware of) for it. LL&B and I thus seem to agree that lack of evidence is reasonable justification for not believing a theory. But then that is obvious. So, let's put it aside, for the authors seem to think there is a more interesting critique to be made, based on possible worlds.

First, they point out that if Trump Tower was wiretapped, it was either done with a warrant or without one. Both alternatives, they argue, have problems that make them not only wrong, but irrational. Let's consider them (in reverse order). The central problem with the view that the wiretapping was done without a warrant is this: "There must be a technologically-capable agency willing to obey the president in defiance of the law" (p. 225). This isn't strictly true; there could be rogue operators. In any case, many people went along with the Bush administration's "President's Surveillance Program,"[9] as well as with NSA surveillance programs in operation during the Obama administration, despite their questionable legality. And during the years of COINTELPRO many FBI agents were willing to obey J. Edgar Hoover in defiance of the law. Watergate and the Iran-Contra affair also involved willingness to follow someone's illegal directives. So too did the Tuskegee syphilis study (discussed below). Whether those directives stemmed from the president or from someone of lower rank doesn't seem to be a particularly significant difference. If anything, the higher the authority the illegal directive comes from, the more likely it will be followed. So, by itself, this kind of illegality hardly seems out of the question.

LL&B, however, add other supposed implications, which make the situation seem less plausible. But it is not at all clear that Trump's theory implies them. For example, they state, "[T]here must be one or more individuals operating in complete secrecy from this agency[10] who are nevertheless willing to break that secrecy to inform the current president [Trump] about the surveillance" (p. 225). That is one possibility, but it isn't the only one. In failing to notice this, LL&B, it seems, are being unreflective about the supposed counterfactuals.

Regarding the other version of the theory, if there was a warrant for the wiretap, that would mean, LL&B write, that there is "a judge who is a willing puppet of a 'sick' or 'bad' president" (p. 226). This is a bit of a misleading way to put it, since the judge presumably would not share Trump's opinion of President Obama. So let's put it this way: there is a judge who is willing to issue such a warrant for some unknown reason—maybe a good one, maybe not. Let's remember here that there were employees of the Department of Justice, including John Yoo, who were willing to write memos giving legal cover for the Bush administration when they wanted to engage in torture: the so-called torture memos. So this version of the theory doesn't seem out of the question either.

However, without mention of the torture memos, or COINTELPRO, the authors assert:

> [E]ither scenario is so unlikely, that the possible world in which President Obama either received a warrant or wiretapped Trump Towers without a warrant flies in the face of what the current president [i.e., Trump] himself believes about the Department of Justice or the federal courts. (p. 225)

It is not perfectly clear what Trump is supposed to believe about the Department of Justice or the federal courts that flies in the face of either of these scenarios, though LL&B later suggest it relates to "how the president and the Justice Department or intelligence agencies function on a daily basis" (p. 226). In any case, whatever the authors were thinking when they wrote that, nobody really *knows* what Trump believes about the DOJ or the courts.[11] And even if Trump believes they are generally trustworthy, that need not mean that he naively assumes them to be perfectly so, especially if he has any sense of history.

The second example LL&B focus on is the claim that "millions of illegal aliens voted for Hillary Clinton" (p. 223). LL&B's analysis is skewed regarding this as well. They write:

> [Trump] evidently believes that the electoral system and its polling are somewhat accurate, and in fact, represent his fair-and-square victory over Clinton. But this cannot be true on the same possible world in which massive voter fraud occurred, invalidating the results of the election. (p. 227)

But that's wrong; it *can* be true. A massive, largely one-sided illegal voting campaign in which Trump's Electoral College victory is not in doubt, but his popular vote loss is questionable, certainly *is* possible—they can both occur in the same possible world. It is not even that hard to imagine.[12] It seems that, in their failure to recognize this, LL&B once again have failed to be sufficiently reflective.

Regarding the two Trump conspiracy theories under consideration, LL&B claim, "It is not just that they are wrong, it is that they do not make sense. Their beliefs are not compatible with one another in a way that is required for rationality" (p. 228).[13] But that is not true. The theories in question are only wrong;[14] they *do* make sense. And the reason to think they are wrong is that evidence is lacking. So, as usual, assessing the warrant of conspiracy theories comes down to evaluating the evidence (or, in this case, noticing the lack thereof).

Having said all that, let me acknowledge a couple things: (1) *If* Trump had made an argument for a conspiracy theory that truly didn't make sense, that would be a valid reason to reject it. (2) Trump may well have made such arguments. Indeed, I'm a little surprised that LL&B failed to identify any.

Given these acknowledgments, what is the significance of the failure of LL&B's arguments regarding these two cases? First, it further highlights just how completely flawed their article is—almost everything in it is either wrong or lacks adequate support. And that suggests a question that we will ask in various ways throughout this book: How is the publication of such shoddy work even possible? At the same time, there is a sense in which LL&B's article is not anomalous. Indeed, it is in some senses typical of academic attempts to make conspiracy theorists look bad. And so it provides examples of the *kinds* of things that go wrong when academics attempt to use clever (or not so clever) heuristics to assess conspiracy theories. These include asserting dubious implications and ignoring historical analogies.

But we are not finished with the critique yet.

3.2. LL&B's Five Examples, Plus Three Others

Let's now address what LL&B describe as "the demonstrable lack of . . . success of conspiracist cognition" (p. 220) in exposing genuine conspiracies. To make their case, LL&B select five instances of now-acknowledged-to-be-true conspiracy theories and report the (ostensible) "Source of revelation" and the "Type of source," with no further details. The five cases are the Iran-Contra affair, the Gulf of Tonkin episode, the NSA PRISM program, the FBI forensic scandal, and the Tuskegee syphilis study.[15] LL&B conclude, "In no case is there evidence that the 'theory' was first postulated or revealed by conspiracy theorists. Instead, in all cases the revelation involved conventional means of discovery, such as media reports based on leaks or testimony by whistleblowers" (p. 221). The contrast between being "revealed by conspiracy theorists" and "revelation involv[ing] conventional means of discovery" implies that conspiracy theorists don't use conventional means of discovery, "such as media reports based on leaks or testimony by whistleblowers." But that isn't true. Further, what any of this has to do with "conspiracist cognition," as they define it, is unclear. Their claim seems to involve a certain amount of conceptual confusion and equivocation, or at least lack of clarity.

Although I have not done extensive research on any of the five topics LL&B cite, just a little investigation turns up details that problematize the suggestion that neither conspiracist cognition nor conspiracy theorists played any role. And the more one looks, the more complex the situation appears, and the more simplistic and misleading LL&B's "survey" seems to be.

3.2.1. The Iran-Contra Affair

According to LL&B, the Iran-Contra affair was simply revealed by the "Lebanese newspaper *Al-Shiraa*" (p. 220). The only other detail is the source type, listed as "Print media," though it is not clear what the relevance of that is supposed to be. Although no details are given about the *Al-Shiraa* story, one can surmise that this refers to a story published on November 3, 1986. This is quite an abbreviated account of the exposure of the Iran-Contra affair—it could hardly be more abbreviated. But without *any* details, how can we conclude that neither conspiracy theorists nor conspiracist cognition played any role in the revelation?

In attributing the revelation solely to *Al-Shiraa*, LL&B imply, by omission, that the investigation led by (conspiracy theorist) Daniel Sheehan and the Christic Institute, which predated the *Al-Shiraa* story, was not at all relevant. Let's take a moment to consider Sheehan's account.

First an undisputed fact: a C-123 cargo plane carrying weapons and explosives intended for the Contras was shot down in Nicaragua on October 5, 1986 (almost a month before the *Al-Shiraa* story), and Eugege Hasenfus was the sole survivor. According to historian Kathryn Olmsted, "The Iran-Contra conspiracy began to unravel in October 1986 when one of North's planes was shot down by the Nicaraguan army" (Olmsted 2019, p. 294). Sheehan claims that, two days later, Father William J. Davis, who was working with Sheehan, headed to Nicaragua to speak to Hasenfus. Sheehan explains:

> Hasenfus was totally surprised that we at the Jesuit headquarters knew the Contra weapons-supply operation in such detail. Father Davis made Hasenfus a deal: If Hasenfus would come clean and state, on the record, what he knew about this operation (virtually all of which, Father Davis convinced him, we already knew), then Davis would intervene with Father D'Escoto, a fellow Catholic priest [and foreign minister of the Sandinista government], and ask that Hasenfus be set free. That was too good an offer to refuse. So Hasenfus

> went on camera and spilled the beans on the Contra movement and the American network that was supplying the Contras with arms and explosives. (Sheehan 2013a, p. 507)

Sheehan also claims that on October 15, 1986, he appeared before the House Foreign Affairs Committee. He reports having stated:

> [W]e have direct proof that both of you [Congressmen Henry Hyde and Bill McCollum] have been in repeated meetings directly with Rob Owen, the liaison for Lieutenant Colonel Oliver North directly to the Contras, and that, in these meetings, both you and Congressman McCollum were aware of this criminal operation to smuggle weapons and explosives from this administration to the Contras in direct violation of the Boland Amendment passed by this Congress. (Sheehan, 2013a, p. 509)

If this really happened, then conspiracy theorists were on to essential elements of the Iran-Contra affair before the *Al-Shiraa* story. This conclusion holds even if Sheehan and his group are wrong about most of their more explosive allegations (about which I take no position). But *did* this really happen as Sheehan reports it? I don't know.[16] It is possible, for all I know, that Sheehan's autobiographical account is made up—that Sheehan is either crazy or a liar,[17] and, despite his obvious brilliance, he has self-destructively tarnished his outstanding and undeniable legal accomplishments by making up a story with little to gain from it. If so, LL&B are right to give him no credit. But they have done absolutely nothing to *show* that to be the case.

And, further, what should we think about the award-winning work of Robert Parry, which also preceded the *Al-Shiraa* story, and which included, among other things, the drug-smuggling aspect of the Iran-Contra affair? Is that irrelevant too? Or was Parry's thinking judged to be insufficiently "conspiracist," or his conclusions not sufficiently "acknowledged"? If so, on what bases were those determinations made? The reader is not given any clue regarding these questions.

3.2.2. The NSA PRISM Program

LL&B attribute the revelation of the NSA PRISM program to whistleblower Edward Snowden. But what exactly does that prove? By itself (and we aren't given anything else) it does not give any justification for the conclusion that conspiracist ideation played no role in the revelation of that specific program, or in the revelation of the more general fact that sweeping NSA surveillance on Americans was occurring. Did Snowden refrain from any conspiracist ideation when he deliberated about whether or not releasing this information was important enough for him to accept facing life changing adversity? LL&B don't provide any analysis.

But what does Snowden himself say about his own motivations? He says, "As I advanced and learned the dangerous truth behind the U.S. policies that seek to develop secret, irresistible powers and concentrate them in the hands of an unaccountable few, human weakness haunted me" (Gellman and Markon 2013). This seems to suggest that Snowden was operating from a *deep-seated suspicion* that we may all be or become *victims* of people with *questionable motives*. And though Snowden claims not to see himself as a hero, he says things like, "I don't want to live in a world where there's no privacy and therefore no room for intellectual exploration and creativity." And, "My sole motive is to inform the public as to that which is done in their name and that which is done against them" (Greenwald et al. 2013). So he does frame his actions as protecting victims from people who were subverting the public good. And thus, it seems, the revelation of the NSA PRISM program was indeed uncovered thanks, in part, to (some aspects of) conspiracist ideation.

But the story of the revelation of sweeping NSA surveillance is more complex than "Snowden leaked it." Let's fill in just a few details. Snowden was not the first person to reveal that a lot of highly questionable surveillance was going on. For example, there were revelations stemming from the whistleblowing of Bill Binney and Thomas Drake. Although their revelations were touted by conspiracy theorists (from the sober Peter B. Collins to the loudmouth Alex Jones), the mainstream media largely ignored them (as *The Guardian* now essentially admits; see below). And the US government positively persecuted them. The significance of Snowden is that he provided the documents that proved that Binney and Drake, and the conspiracy theorists citing them, had been right all along (see Welna 2014).

Here is how *The Guardian* describes Drake's situation (in retrospect):

> Thomas Drake . . . blew the whistle on the very same NSA activities 10 years before Snowden did. Drake was a much higher-ranking NSA official than Snowden, and he obeyed US whistleblower laws, raising his concerns through official channels. And he got crushed. Drake was fired, arrested at dawn by gun-wielding FBI agents,

> stripped of his security clearance, charged with crimes that could have sent him to prison for the rest of his life, and all but ruined financially and professionally. . . . Adding insult to injury, his warnings about the dangers of the NSA's surveillance programme were largely ignored. (Hertsgaard 2016)

Drake, it should be mentioned, seems (understandably) to see himself as a bit of a victim. He complains, "The only person who [in connection to the NSA surveillance programs] was investigated, prosecuted, charged in secret, then was indicted, then ended up facing trial and [the possibility of] 35 years in prison was myself" (Welna 2014).

And regarding Bill Binney, simply doing a Google search for "bill binney alex jones 2012" (without the quotation marks)[18] quickly reveals that Alex Jones's Infowars website promotes a 2012 article which quotes Bill Binney asserting:

> Domestically, they're pulling together all the data about virtually every U.S. citizen in the country and assembling that information, building communities that you have relationships with, and knowledge about you; what your activities are; what you're doing. So the government is accumulating that kind of information about every individual person and it's a very dangerous process. (Ms. Smith 2012)

That was before Snowden's revelations.

In addition, in 2005, Mark Klein, an AT&T technician, revealed that AT&T was cooperating with the NSA to copy all the internet traffic of AT&T customers. Once again, conspiracy theorists touted this as evidence of sweeping surveillance, but the mainstream largely ignored it. And the history of relevant whistleblowing and associated accusations of sweeping surveillance extends much further back than that, though the story is too complex to sort out here. Yet, somehow, to LL&B, the NSA mass surveillance story is reducible to this: Snowden revealed PRISM, so conspiracy cognition is unsuccessful, and there is no evidence that, or sense in which, "the 'theory' was first postulated or revealed by conspiracy theorists" (p. 221).

3.2.3. The Gulf of Tonkin Incident

Regarding the Gulf of Tonkin, LL&B cite John White's 1967 letter to the editor as the "source of revelation." But what did White's letter reveal?

White reports that "[the sonarman's] evaluation of the sonarscope picture was negative," indicating that there were no torpedoes fired, and expresses the opinion that "a chief sonarman's judgment in such a situation is more reliable than that of anyone else on the ship" (White 1967). That does sound like pretty strong and relevant evidence. But similar evidence can be found in abundance in countless conspiracy theories that have remained controversial.

What exactly did this particular letter accomplish? It didn't settle anything. Whether or not the Gulf of Tonkin incident involved some kind of conspiracy continued to be debated for decades after this letter was published. After all, the letter doesn't even allege a conspiracy, but rather explicitly leaves open the possibility that it was a "misunderstanding" based on a "bad radio report." So the letter has an element of "There is something wrong with the official story, but I don't know exactly what." It was only much later, and gradually, that the incident became widely regarded as involving the *deliberate distortion of the event for the purpose of justifying the widening of the war.*

And how is this conclusion even possible without some degree of conspiracist ideation? There must be some degree of *suspicion* and *questioning of motives* to conclude that there was a deliberate effort to distort an event to widen a war. So, to put it most charitably, it is not at all clear that this example is evidence that no elements of conspiracist ideation play any role in revealing real conspiracies.

3.2.4. The FBI Forensic Scandal

LL&B credit Frederic Whitehurst for the revelations leading to the FBI forensic scandal. The allegations included, in Whitehurst's words, "alterations of reports, alterations of evidence, folks testifying outside their areas of expertise in courts of law" (Nagle 2017). However, Whitehurst's allegations did not quickly settle the matter. According to a 1995 *New York Times* article: "John W. Hicks, the bureau's laboratory director at the time of the transfer, said that when Mr. Whitehurst persisted in his complaints, they were reviewed within the laboratory and by the bureau's disciplinary office. Both reviews, Mr. Hicks said, concluded that the accusations were unfounded" (Johnston 1995). The whistleblower protection blog provides further context:

> Eventually, investigations were launched into Whitehurst's allegations, but failed to lead to any justice. It wasn't until ten years

> later that Whitehurst was finally vindicated, when a scathing 500+ page study of the lab by the Justice Department Inspector General, Michael Bromwich, concluded major reforms were required in the lab. (Nagle 2017)

During the ten years it took for Whitehurst to be vindicated, were conspiracy theorists helping keep up the pressure, with their suspicions regarding nefarious intent, assumptions that something must be wrong in the FBI, and skepticism regarding the FBI's internal reviews? Probably. LL&B provide no reason to think otherwise. Would the DOJ have eventually vindicated Whitehurst even in the absence of conspiracy theorists? We don't know. Certainly LL&B have given us no reason to think that conspiracy theorists had no effect. They haven't provided any analysis at all.

Further, Whitehurst himself exhibited at least some signs of conspiracist cognition. He stated, "[R]eally what was going on was human rights violations. We have a right to fair trials in this country. . . . And that's not what was going on at the FBI lab" (Nagle 2017). That sounds like he thinks that we are all victims of the wrongdoing he (heroically) exposed. Not only that, Whitehurst later charged that he was being retaliated against by the FBI, and by doing so he frames himself personally as a victim. But that's okay, because, as I've indicated, there is nothing epistemically wrong with these kinds of concerns and beliefs—especially if they result in a favorable $1.16 million settlement, as in Whitehurst's case (Friedan 1998). In any case, Whitehurst's "ideation" does seem to fit at least one criterion that LL&B suggest should qualify it as (partially) conspiracist. Without mention of this discrepancy, however, LL&B simply tally Whitehurst on the other side of the ledger, counting his reasoning as unproblematically conventionalist.

3.2.5. The Tuskegee Syphilis Study

The remaining case involved a forty-year (1932–1972) experiment in which almost four hundred African Americans with syphilis were deceived about their illness and not treated for it, even after the effectiveness of treatment with penicillin had been established, so that the progression of the disease in blacks could be scientifically studied.

LL&B answer the question, "Who revealed the study?" Their answer is whistleblower Peter Buxton. Although the full story is more complex, let's accept that. Perhaps the more pertinent question is, "Why wasn't the study revealed any earlier?" After all, the study was so appalling to most ordinary citizens that once it was revealed to the public, it had to be shut down immediately. But lots of people had been informed of the study over the years: especially doctors and medical researchers, as well as members of the draft board. Doctors were contacted to ensure that subjects were not treated for syphilis. And details of the study were published many times in medical journals while the study was continuing. And Buxton wrote two letters of protest (in 1966 and 1968) to the Public Health Service director of the Division of Venereal Diseases, William Brown, to no effect. And a blue-ribbon panel that was convened in 1969 by the Center for Disease Control to discuss the experiment decided to continue it to the "end point" (Thomas and Quinn 1991).

Perhaps social scientists who study conspiracy theorists could give a little attention to the question of why medical professionals were almost uniformly able to overlook obvious moral problems, not to mention Alabama state law.[19] After all, perhaps what was missing was skepticism regarding institutions (in this case the US Public Health Service) and a disposition to suspect impure (in this case racist) motives. It seems the people in the right places may have failed to exhibit sufficient conspiracist ideation of the appropriate kinds.

Interestingly, in his second letter, Buxton writes, "[T]he racial composition of the study [100% Negro][20] supported the thinking of Negro militants that Negroes have long been used for medical experiments and teaching cases in the emergency wards of county hospitals" (Thomas and Quinn 1991). This indicates that relevant conspiracy theories were in existence and, although they may have been wrong in the details, they were right in the essence: whites would callously perform medical experiments on blacks.[21] So, for blacks, suspicion and caution regarding the medical system were justified. Unfortunately, and ironically, the revelation of the study led to worsening life expectancies for blacks, perhaps in part *because* of the heightened suspicion that it justified.

This case is a somewhat unusual conspiracy theory, if it counts as one at all. Once publicly exposed, it couldn't be denied. And so there was no prolonged period of controversy about it in particular (though, as mentioned, a more general conspiracy theory existed). It is during such periods that conspiracy theorists will seize on evidence of conspiracy, and apologist for the accused institutions will deny, dismiss, and rebut the charges. The interesting cases are ones in which this process of debate is long-standing, complex, and not easy to resolve. So, given this difference, the revelation of the Tuskegee syphilis study is not particularly telling regarding the cognition of those on opposite sides of such controversies.

In sum, the five examples provided by LL&B do little, if anything, to support LL&B's suggestion that conspiracy theorists and conspiracist cognition play no role in the revelation of real conspiracies. Indeed, the more one looks into the details, the more problematic their suggestion becomes.[22] Perhaps that is why LL&B do not discuss any details at all. Further, there are other cases that LL&B omitted that further undermine their suggestion that conspiracy theorists and conspiracist ideation play no role in uncovering real conspiracies.

3.2.6. Operation Northwoods

Consider the revelation of Operation Northwoods, which articulated a number of wild-sounding plots to provide pretexts for a US invasion of Cuba (see section 5.2.1 for an extended quotation from the Northwoods document). The plans were rejected by President Kennedy (or perhaps by Secretary of Defense Robert McNamara), but they were real plans. How do we know about these Kennedy-era plots? In 1991, Oliver Stone produced a film, *JFK*, which puts forward a conspiracy theory about President Kennedy's assassination, and ends with mention that records regarding the assassination have been sealed for fifty years. The public outcry about this secrecy led to the JFK Records Act of 1992, which created the Assassination Records Review Board, which, in turn, declassified the Northwoods document (one copy of which had survived). The document was later revealed by James Bamford in his 2001 book *Body of Secrets*. Conspiracy theorists played a role at the beginning of this chain, and arguably at the end. While Bamford, an author and journalist who focuses on the NSA, is not known for propounding particularly controversial conspiracy theories, he may be considered a conspiracy theorist of a relatively mild variety.[23] Further, that it was Bamford rather than someone from, say, the JFK research community that first publicized the existence of the document seems a matter of chance, as JFK researchers are also known for digging through archives in search of just such documents—with some success. In addition, it is conspiracy theorists proper, more than Bamford himself, that have made the existence of this document as widely known as it is (though most people are probably still unaware of it).

Now, did Stone or Bamford (or any of the conspiracy theorists that comb through documents and might have found this one) engage in "conspiracist ideation"? They probably did engage in the relatively innocuous ones, to some degree.

3.2.7. The Killing of Fred Hampton

In judging whether or not conspiracist ideation can have any merit, LL&B seem to assume that only a "now-acknowledged conspiracy" should count. I don't share that assumption, because cases like the killing of Fred Hampton, it seems to me, are also relevant. In this case, the FBI's involvement in what was essentially a political assassination has not been officially acknowledged, though it seems evident enough. Was it unreasonable for those who fought to expose the FBI's involvement—the survivors of the raid, and relatives of Hampton and Mark Clarke (who was also killed in the raid)—to have regarded themselves as victims? Obviously not; they *were* victims. Is there something "epistemically impoverished" about being suspicious and attributing impure motives to those involved in a gratuitously violent assault culminating in the shooting of Hampton in the head as he lay unconscious on his bed, having apparently been drugged the night before? No, especially given the relevant particulars. To suggest otherwise unjustly adds insult to murder.

3.2.8. COINTELPRO

The most egregious omission in LL&B's survey is COINTELPRO. Who exposed the existence of that rather expansive and appalling program? It is true that a media story played an important role in disseminating the information. And it is true that there were official investigations, including the Church Committee. But the initial media story was made possible only because a group of conspiracy theorists—people who were convinced that the FBI was engaged in illegal efforts to quash dissent but didn't know all the details and couldn't prove it—broke into an FBI office and stole the files that did provide the details and did prove it. As Coady puts it:

> [COINTELPRO] was eventually exposed, not by the fearless investigative reporters of the free press or any other institution of the open society, but a group of leftist "conspiracy theorists" called the "Citizens Commission to Investigate the FBI," which burgled the offices of the FBI and stole documents related to the programme. (Coady 2018a, pp. 297–298)

It is worth noting that most of the news outlets that were sent the documents decided not to run the story.[24] And when the first newspaper to decide to publish the story made that decision, it was reversing a previous decision not to (Medsger 2014). This suggests that it is unlikely that any mainstream news agency would have gone out of its way to uncover this story if it was not handed to the agency by conspiracy theorists.

3.2.9. Conspiracy Theorists Are Never Right for Good Reasons

LL&B suggest that neither conspiracy theorists nor conspiracist ideation are involved in the exposure of real conspiracies. A closely related claim is that even if conspiracy theorists happen to be involved in some cases, it is only because they got lucky, not because they had good reasons. Political scientist Joseph Uscinski, in his recent collection of essays on conspiracy theories, makes this bald assertion: "When conspiracy theorists are right, it is by chance" (2019a, p. 110). And yet in the chapter just prior to this, M Dentith explicitly addresses the issue by providing examples that shows Uscinski's claim to be false. Dentith writes:

> [W]ith respect to the possibility conspiracy theorists were only accidentally right, we can show, with respect to the Moscow Show Trials of the 1930s and the Gulf of Tonkin Incident of the 1960s, that said theorists offered plausible arguments and evidence for their conspiracy theories *at the time*, yet had their warranted conclusions dismissed *merely* because they were "conspiracy theorists" peddling "conspiracy theories." (2019a, p. 96)

The Citizens Commission to Investigate the FBI may serve as an example as well, at least prima facie. They were educated people who presumably did not risk life imprisonment without some good reasons to think their suspicions were right. I would think the burden of proof would have to be on those who say they just got lucky.

3.3. Conclusion

There is a lot that is wrong with LL&B's article: it is not clear how many or to what extent each element of "conspiracist ideation" must be exhibited for something to count as "conspiracy ideation." Four of the criteria are common and unproblematic, and two are problematic but extremely rare. The remaining criterion, THUNC (which is really just unsound reasoning), has neither been convincingly tied to the thinking of conspiracy theorists generally, nor even, in the examples given by LL&B, to the thinking of President Trump (which is a pretty low bar). Even if they had been successful in the latter case, it would not tell us anything about the warrant of other specific conspiracy theories or even conspiracy theories generally. To conclude otherwise, which LL&B seem vaguely to encourage, would itself be an example of unsound reasoning. Further, their "survey" of now-acknowledged conspiracies is fraught with ambiguities, and their implied conclusion, that neither conspiracy theorists nor "conspiracist ideation" played any role in the uncovering of these conspiracy theories, is simplistic and at least misleading. This is all the more apparent when one considers other examples, such as the uncovering of COINTELPRO.

PART II

The Pre-eminent Legal Scholar of Our Time

Many intellectuals scoff at what are called "conspiracy theories." But two Harvard law professors go further. They argue that groups that espouse such theories ought to be infiltrated and undermined by government agents and their allies. While some may find this proposal appalling (as indeed we all should), others may find their argument plausible, especially if they have been swayed by the notion that conspiracy theories (or a definable subset thereof), by their nature, somehow do not warrant belief. I will argue that their proposal not only conflicts with the values of an open society but is also epistemically indefensible. In making my case, I will adopt their favored example, counternarratives about 9/11.

The law professors in question are Cass R. Sunstein and Adrian Vermeule. In 2009, they published an article called "Conspiracy Theories: Causes and Cures," in the *Journal of Political Philosophy*. An earlier, longer, version of the article has been available online since 2008, under the title "Conspiracy Theories." And in 2014, Sunstein revised and republished the article as the first chapter of his book *Conspiracy Theories and Other Dangerous Ideas*.

At the time of the *Journal* publication, both authors were Harvard law professors. Sunstein had just moved from the University of Chicago to Harvard, and shortly thereafter he was chosen by President Obama to head the Office of Information and Regulatory Affairs. He is reputed to be "the most cited law professor on any faculty in the United States," according to a White House website, and "one of America's leading constitutional scholars," according to Obama himself (White House 2009).

Indeed, according to Supreme Court justice Elena Kagan, "Cass Sunstein is the pre-eminent legal scholar of our time—the most wide-ranging, the most prolific, the most cited, and the most influential" (Mangan 2008). So, although some people may regard their proposal as too outrageous to merit rebuttal, I agree with Sunstein and Vermeule on one thing: problematic views ought to be confronted, not ignored. And so, particularly given Sunstein's acclaim, it is worth discussing some of the problems with their proposal and with the arguments and assumptions on which it is based.

Chapters 4 and 5 provide a critical evaluation of the article. Chapter 4 focuses mostly on the inappropriateness of their proposal—the "cognitive infiltration" of groups that espouse conspiracy theories. Chapter 5 focuses more on the purported "causes" of conspiracy theories, as well as Sunstein and Vermeule's misleading presentation of significant considerations. Chapter 6 addresses Sunstein's (sole-authored) 2014 revision of the article. That chapter addresses the questions: "Were Sunstein and Vermeule misunderstood?" and "Is Sunstein's revision an improvement?"

FOUR

Is Infiltration of "Extremist Groups" Justified?

Cass Sunstein and Adrian Vermeule support confronting supposedly harmful or pernicious[1] and "demonstrably false" conspiracy theories (or "extreme views"), rather than ignoring them.[2] One strategy, which they emphasize, they describe as follows:

> [W]e suggest a distinctive tactic for breaking up the hard core of extremists who supply conspiracy theories: cognitive infiltration of extremist groups, whereby government agents or their allies (acting either virtually or in real space, and either openly or anonymously) will undermine the crippled epistemology of believers by planting doubts about the theories and stylized facts that circulate within such groups, thereby introducing beneficial cognitive diversity. (2009, p. 219)

In other words, they suggest undermining conspiracy theories *by engaging in a conspiracy* against groups that promote them. I will argue that, beyond the obvious irony,[3] this recommendation is epistemically indefensible, even when considering the rather extreme case of 9/11 conspiracy theories, which is their primary example.

Sunstein and Vermeule's argument in a nutshell is as follows: (1) Lots of people are susceptible to belief in conspiracy theories. (2) Some of these theories are demonstrably false. They are products of a crippled epistemology—that is, they are based on limited and incorrect information, and fueled by informational and reputational cascades.[4] In an "open society," one is not warranted in believing theories that imply that "knowledge-producing institutions" are as unreliable as some "extreme" theories imply. (3) In addition, some of these theories are harmful, as shown by the Oklahoma City bombing—which is blamed, at least in part, on a conspiracy theory regarding federal agents' actions in the Waco siege of 1993. (4) The government should try to counter demonstrably false and harmful theories. (5) However, conspiracy theorists often dismiss evidence that comes from government operatives. The theories have a "self-sealing quality" that "make[s] it more difficult for outsiders to rebut or even to question them" (Sunstein and Vermeule 2009, p. 207). (6) Therefore, the government should infiltrate groups that espouse conspiracy theories in order to undermine those theories by introducing "cognitive diversity" from the inside.

In this chapter, I will focus on (section 4.1) the notion of "demonstrably false" theories, and how that is determined, (section 4.2) the inappropriateness of the recommended infiltrations in an "open society," and (section 4.3) the slippery slope of the practice of deceit.

However, I will first briefly address the relation between epistemic issues and consequential ones. Namely, when weighing the risk of harm associated with radical counternarratives, especially those involving 9/11, we must acknowledge a serious risk in subverting such narratives through epistemically inappropriate means (beyond the purely tactical risks that Sunstein and Vermeule consider). Bogus official stories, after all, can be much more dangerous than bogus counternarratives. And in the case of 9/11 in particular there is much at stake indeed. Not only the war in Afghanistan, but also the war in Iraq, the global war on terror, domestic warrantless wiretapping, secret prisons, torture, extraordinary rendition, and indefinite detention without a trial, have been, and many of these continue to be, justified on the basis of a particular interpretation of what happened on 9/11. If that interpretation is substantially wrong, then it is likely that we have been engaged in, and are continuing, great injustices, and causing incalculable harms based on falsehoods. If so, it is critically important that our understanding of the event moves closer to "the truth." Now, to be clear, I am not here arguing that the "9/11 Truth Movement" is right in its most basic claims. I am merely pointing out that the relation between potential harm or benefit such theories can engender is connected to whether or not they are true, or at least contain valuable truths. And so, by challenging the claim that 9/11 counternarratives are "demonstrably false" I am also thereby challenging the implicit claim that such narratives are, on balance, harmful. If they are not demonstrably false, and therefore not clearly pernicious, the notion that biased[5] and deceitful[6] means ought to be employed to eradicate them is an odious notion indeed.

4.1. Conspiracy Theories and "Crippled Epistemologies"

Sunstein and Vermeule do admit that there are examples of conspiracy theories that turned out to be true—how could they not? They write:

> Of course some conspiracy theories have turned out to be true, and under our definition, they do not cease to be conspiracy theories for that reason. The Watergate hotel room used by [the] Democratic National Committee was, in fact, bugged by Republican officials, operating at the behest of the White House. In the 1950s, the Central Intelligence Agency did, in fact, administer LSD and related drugs under Project MKULTRA, in an effort to investigate the possibility of "mind control." Operation Northwoods, a rumored[7] plan by the Department of Defense to simulate acts of terrorism and to blame them on Cuba, really was proposed by high-level officials (though the plan never went into effect). (2009, p. 206)

They could have gone on and on with other real examples, such as the overthrow of Mosaddeq in Iran in 1953, Operation Gladio, numerous assassinations of foreign political leaders, Iran-Contra, and so on.

Sunstein and Vermeule define "conspiracy theory" as "*an effort to explain some event or practice by reference to the machinations of powerful people, who attempt to conceal their role (at least until their aims are accomplished)*" (2009, p. 205). It should be noted that, according to this definition, the notion that the Nazis were systematically exterminating Jews would have, at some point in time, clearly counted as a conspiracy theory—one that turned out to be true. This is an important example.[8] It shows that one cannot simply reject a conspiracy theory because it seems too extreme in the brutality it attributes to powerful figures, or because of the scale of complicity that would be required, or because of the industrial efficiency with which it is said to be carried out. Shocking though a theory may be, so too are known precedents.

Now, Sunstein and Vermeule claim, "Our focus throughout is on demonstrably false conspiracy theories, such as the various 9/11 conspiracy theories, not ones that are true or whose truth is undetermined. Our ultimate goal is to explore how public officials might undermine such theories, and as a general rule, true accounts should not be undermined" (2009, p. 205). They take 9/11 conspiracies as their stock example. They cannot mean, however, that theories that postulate insider complicity in the events of 9/11 are demonstrably false in a strict sense—that they can be shown to be false with absolute certainty—for that would require a logical contradiction. But there is nothing contradictory about the notion of insider complicity. So they must mean "demonstrably false" in some weaker sense. But they neglect to say clearly what this weaker sense is, though their contrast with theories "whose truth is undetermined" suggests that "demonstrably false" theories *have been determined* (by whom?) to be false. Perhaps we can infer from their discussion of justification that "demonstrably false" theories are those that are not only false but also unjustified, or unwarranted, though the *degree* to which they must be unwarranted is unspecified. So I will assume (charitably, I think) that, by "demonstrably false," they mean something like the following: the evidence is so overwhelmingly against the theory that it is irrational "from the standpoint of the information available in the society as a whole" to believe it (2009, p. 207). Strictly speaking, the totality of such information is something to which *nobody* has access, so they must intend some reasonable approximation. I will argue that, even by this weaker criterion, their claim that 9/11 counternarratives are "demonstrably false" is false. That is, that it is not irrational for a suitably well-informed person to give credence to such counternarratives, or at least it has not convincingly been shown otherwise. Further, on the one hand, the stronger the sense of "demonstrably false" they intend, the more clearly false is their claim that all counternarratives about 9/11 qualify. And, on the other hand, the weaker the sense of "demonstrably false," the more suspect the inference that illiberal infiltration is justified on the basis of a theory meeting that low standard.

When Sunstein and Vermeule assert that the "various 9/11 conspiracy theories" are demonstrably false, presumably they mean to include both the made-it-happen-on-purpose (MIHOP) and the let-it-happen-on-purpose (LIHOP) varieties. That the latter type is false seems less than obvious, and Sunstein and Vermeule provide no evidence that it is. They do provide *some* evidence that a certain part of *some* MIHOP theories are false, but *that* hardly makes their case. To make the necessary case, clearly, would have required a completely different kind of essay, so one cannot blame them for not doing *that*. But one should at least expect a reference to someone who has. They might have been tempted to cite Popular Mechanics' *Debunking 9/11 Myths* (Dunbar and Reagan 2006), but that book has been answered by David Ray Griffin's *Debunking 9/11 Debunking* (2007). Few fair-minded persons familiar with both books would be able to say honestly that *Debunking 9/11 Myths* puts the issue to rest.

Regarding the destruction of the World Trade Center Twin Towers and Building 7, Sunstein and Vermeule might be tempted to point to the official reports of the National Institute of Standards and Technology (NIST). However, there are reasons a rational and well-informed person might be less than fully convinced by those reports as well. While there are many technical issues that have been raised by others, I will just list four that can be stated briefly. First, the Bush administration has been credibly accused of politicizing science—that is, corrupting science for political expediency.[9] And, NIST was, after all, an arm of the Bush administration. Second, representatives of NIST were evasive about key issues, such as checking for explosives.[10] Third, John Gross, a lead engineer involved in the NIST report, denied being aware of any eyewitness accounts of molten steel at the WTC collapse sites, implied that he is unaware of *any* suggestive evidence, and expressed doubt that the necessary temperatures could have been reached.[11] And yet there are many well-documented eyewitness accounts of molten steel (which may have actually been iron), as well as video footage of what looks like molten steel flowing from the South Tower, supported by corroborating evidence of various kinds.[12] (Indeed, the evidence of temperatures sufficient to melt steel is now quite strong.[13] There should be an unbiased scientific discussion regarding how such temperatures were achieved, or at least a reasonable and open discussion of the evidence in question.) Fourth, in the "draft for public comment" version of their report on the collapse of Building 7, NIST asserted that the collapse rate was about 40 percent slower than freefall.[14] David Chandler, a high school physics teacher, demonstrated in an online video that for over one hundred feet the building collapsed at *very* close to freefall acceleration, and he challenged NIST publicly on the matter.[15] In the final version of their report, NIST conceded the point.[16] Having been forced by clear and incontrovertible evidence into this astonishing concession, NIST then simply pretended that it doesn't matter. This shows that there can be a kind of "self-sealing" quality to official stories too. Officials can deny or ignore evidence, just assert that their results are scientific, and then end the discussion, refusing to have open and fair dialogue or debate. And this is precisely what NIST has done.[17] (See the appendix for more on the questionable reliability of NIST.)

Perhaps the real reason Sunstein and Vermeule think counternarratives positing insider complicity in the events of 9/11 must be false is that they just seem too crazy. However, either there is an explicit demonstration of the craziness (which appears to be lacking) or else it is just an intuition. And, clearly, intuitions regarding the craziness of these ideas vary. As Sunstein and Vermeule themselves point out: "Among sober-minded Canadians, a September 2006 poll found that 22 percent believed that 'the attacks on the United States on September 11, 2001 had nothing to do with Osama Bin Laden and were actually a plot by influential Americans'" (2009, p. 202). And roughly *half* of New York City residents accept, at least, the more modest "knew in advance . . . [but] consciously failed to act" counternarrative (2009, p. 202), that is, LIHOP. But perhaps it is only the unwashed half who espouse such beliefs. Sophisticated people, those in the know, or those with relevant expertise, Sunstein and Vermeule may assume, give no credence to such views.

But that isn't true. Intelligence experts, such as former CIA officers Robert Baer[18] and Ray McGovern[19] (who presumably know something about clandestine operations) take counternarratives about 9/11 seriously. And an owner of a Dutch-controlled demolition company, the late Danny Jowenko, was "absolutely" convinced that Building 7 was professionally imploded (Griffin 2010, p. xiv).[20] This is a short list, but there are literally hundreds of accomplished people (see patriotsquestion911.com) who have publicly expressed serious doubts about the official story. According to a petition found at ae911truth.org, three thousand architects and engineers claim that "there is sufficient doubt about the official story" to require a new investigation into the collapse of the Twin Towers and Building 7.[21]

Now, Sunstein and Vermeule could quibble by saying that although these theories have not *been demonstrated* to be false, they are nonetheless *demonstrably* false, that is, *capable of* being demonstrated to be false. But if that is the case, then, especially given their commitment to maintaining an "open society," should not Sunstein and Vermeule be calling for such a public demonstration? Should they not join the chorus of voices calling for a new, more robust, subpoena-empowered, and open inquiry, which, if Sunstein and Vermeule are right, would shed such light on the relevant issues that those espousing crazy and untenable theories would scatter like cockroaches? This is the cure that would be most compatible with democratic values. If they are so sure about where a thorough, open, and unbiased inquiry would lead, why not support that, rather than infiltrations?

Perhaps they believe that a straightforward and open inquiry would not "be effective" because conspiracy theories are "self-sealing," and conspiracy theorists tend to be nearly impervious to reason and evidence.[22] But the empirical evidence for this is thin, and it is a charge that can cut both ways. I have already argued that NIST's account is in some sense self-sealing. In addition, undercover operations, such as those recommended by Sunstein and Vermeule, are likely to involve ideological commitments and specific beliefs that are significantly self-sealing. Infiltrating agents are unlikely to be unbiased. They would be in a difficult position if they changed their assessment of the conspiracy theory that their job required them to undermine.

Now, suppose that a thorough and unbiased investigation into both sides of the issue concluded that counternarratives of 9/11 are false.[23] What assurance do we have that those who would in time administer and participate in the infiltrations that Sunstein and Vermeule recommend would likewise do due diligence? Given the political nature of conspiracy theories, it seems most probable that "demonstrably false" theories will be determined by considerations that go beyond honest and thorough research. And even a thorough and honest study does not guarantee a correct conclusion. Indeed, as long as an array of accomplished scholars support a position with extensive research and careful argument, even if they happen to be wrong, one can hardly say that the issue is settled. So Sunstein and Vermeule's claim, on which the plausibility of their proposal hinges, that "the various 9/11 conspiracy theories" are *demonstrably* false is itself false. And *even if* they had been right about that case, we would still have little reason to believe that the infiltrations that they recommend, were they to become policy, would *only* be carried out against groups that espoused views that were fairly determined to be demonstrably false.

Further, this problem remains *even if* we grant their assumption of pure intentions on the part of the executive—which is dubious indeed.[24] The problem remains because even a well-motivated government (we are really talking about the executive branch) is unlikely to be impartial or to research exhaustively both sides of these issues. And even if they tried, it is not clear that they could always achieve the certainty that Sunstein and Vermeule presume is possible. Can we say with certainty, after all, whether or not there was a conspiracy in the assassinations of JFK, RFK, or MLK? There are some who say, "Yes, we know it was a conspiracy," and there are others who say, "Yes, we know it was not." But many who spent decades researching these issues, even if they have a strong opinion, do not think these matters are solved. Like it or not, this history is rationally contestable, and in a free society the government should not be in the business of undermining private efforts to determine the truth. For peaceful groups or organizations to be infiltrated and undermined because the powerful do not agree with their account of history is fundamentally antidemocratic and epistemically perverse, since such actions subvert the dialectic of competitive inquiry.

4.2. Open and Closed Societies

Sunstein and Vermeule claim:

> Conspiracy theories that posit machinations by government officials typically overestimate the competence and discretion of officials and bureaucracies, who are assumed to be able to make and carry out sophisticated secret plans, despite abundant evidence that in open societies government action does not usually remain secret for very long. (2009, pp. 208–209)

To substantiate this "abundant evidence" claim, they cite one *New York Times* article and one *New Yorker* article, each of which exposes a secret program, the warrantless wiretapping program and the CIA's secret interrogation program. This, by itself, does not seem ample evidence to support such a strong claim. After all, a pair of examples is insufficient to differentiate a rule from an exception. If all one had to do in order to show the ubiquity of a particular kind of phenomenon was to point out two examples, then Watergate and Iran-Contra by themselves would show that purported conspiracy theories are "usually" true. But let's be more generous. Surely, they *could have* cited more examples, *many* more. But how many would they have to cite to make their claim warranted, namely, that there is "abundant evidence that in open societies government action does not usually remain secret for very long"? If they had said that secrets are *often* exposed, then the citation of several exposed secrets would suffice. But that claim, which is indeed true, is significantly weaker than the claim that they are making.[25] For the fact that, in open societies, secrets are often exposed (eventually)[26] does not conflict with the hypothesis that some, perhaps *many* (perhaps even *most*) secrets are held *very* long. Many examples of relatively long-held secrets can be given: the Gulf of Tonkin nonincident, the USS *Liberty* cover-up, Operation Northwoods, MKULTRA, the Tuskegee experiment, and so on. As for secrets that have been kept so well that they have never been revealed in any way (at least not yet), I admit I can give no examples. I hope the reason I cannot do so is obvious, and the implication clear.

In any case, it is hardly disputable that there are agencies within the US government that are veritable "conspiracy factories." And from what little we know about the various "black ops" that take place, they are often of dubious legality, and are morally suspect.[27] So the claim that "in open societies government action does not usually remain secret for very long" is not well established—if the United States counts as an "open society." But even if that claim were true, we still could not rule out the possibility that *some* secrets, or some *kinds* of secrets, are long held.

Further, the idea that a secret is either revealed or it is not is simplistic and inaccurate. Secrets are often *partially* revealed, as in Iran-Contra: there was a conspiracy exposed for sure, but one cannot credibly assert that we got to the bottom of it. Secrets are also sometimes *disputably* revealed. Many so-called conspiracy theories fall into this category. When a jury found that the government was involved in a conspiracy to murder Martin Luther King, *as one did*,[28] was a conspiracy revealed? Well, most people don't even know about that finding. So it wasn't revealed very widely. But some may say that preponderance of evidence (it was a civil case, not a criminal case) is not enough, or they may question the validity of the verdict for one reason or another.[29] So whether this counts as a conspiracy that was revealed is ambiguous on at least two counts. And this kind of ambiguity is the norm. The upshot is that the simplistic notion that "conspiracies are usually revealed" is easily deconstructed, although it is a notion that continues to be spread.

It has long been understood that it is more rational to believe conspiracy theories in totalitarian societies than in open ones. However, in a truly totalitarian society conspiracy theories could not thrive. "Where all children go to school, and all schools are controlled by the government, the authorities can close the minds of the young to everything contrary to official orthodoxy. . . . The only remaining possibility of unauthorized propaganda is by secret whispers from one individual to another. But this, in turn, is rendered appallingly dangerous by improvements in the art of spying," observed Bertrand Russell, more than a half-century ago (1953, p. 46). So the optimal environment for conspiracy theories would be some place in between, where it is plausible for them to both spread and be true. Indeed, the distinction between open and closed societies is misleading. There is a spectrum (or an array of spectrums), and the extremes are merely ideals. People can legitimately disagree about exactly where the United States is, at this time, on that spectrum. If one wants to get rid of conspiracy theories, it seems, one could push in either direction: toward a more open society wherein such conspiracies have little plausibility, or in the direction of a closed society where such theories are quashed by the government and its allies. Although Sunstein and Vermeule make rhetorical gestures to the value of open societies,[30] their recommendations involve moving *in the direction of* a more closed one.

Supposing for a moment that Sunstein and Vermeule's argument is a good one, where do we stop? Consider an analogy: Brian Keeley—whom Sunstein and Vermeule cite favorably on another issue (2009, p. 210)—argues that theories about the existence of God and of powerful conspirators are similar in that they both have something like a self-sealing quality, due to the fact that they involve agents who are able to maneuver to avoid detection. God is even better at this than the most powerful worldly conspirators. This is what makes theism, in Keeley's words, "the ultimate conspiracy theory" (2007, p. 135). Now, if those in power determine that the belief in the existence of God is "demonstrably false" (and certainly some people do find this belief ridiculous), and if, as is surely the case, some religious extremists may be dangerous, should the government infiltrate religious groups and try to undermine their belief in God? (To do so they would have to conceal their identities, because of the potential distrust of nonbelievers—but who cares?) I hope that Sunstein and Vermeule would agree that the government should do no such thing. But why not? What is the difference between this proposal and their own? Of course, there *are* differences. No analogy is perfect. But both insider complicity in the events of 9/11 and the existence of God are highly contentious, and it is not obvious why a government could justify trying to undermine one belief and not the other.

4.3. Deceit as the Cure for a "Crippled Epistemology"?

In describing their proposal, Sunstein and Vermeule repeatedly employ the phrase "introducing cognitive diversity" (2009, pp. 219, 224, 226). But what does this really mean? Make no mistake! Sunstein and Vermeule regard certain conspiracies as diseases that need to be "cured." They are not advocating infiltrating groups that espouse these theories for the purposes of achieving a more informed, open, and fair inquiry. They want to see these groups undermined, at least, and preferably eliminated. And they are willing to sanction deceit to accomplish that goal.

Sunstein and Vermeule downplay the degree to which deceit is integral to their proposal, though they do admit, at minimum, that the identity of agents, and perhaps the source of the "information" (or "countermisinformation") that the agents introduce, should sometimes be concealed. That is, sometimes, the agents would have to lie about these matters.[31]

They don't explicitly condone deceit regarding the "information" that the infiltrators would introduce to produce the desired "cognitive diversity." However, once begun, the practice of deceit is likely to be extended. To borrow a phrase: "There is a sort of virginity about [honesty]; once one has violated it, it is awkward to refuse other invitations by saying, 'But that would be [dishonest]!'"[32] Indeed, there is a history here, and the actual abuses go well beyond mere deceit. As UC Davis history professor Kathryn Olmsted explains:

> During the cold war, the FBI started its domestic covert action programs, known by the acronym COINTELPRO, in which agents infiltrated dissident groups and eventually tried to "expose, disrupt, misdirect, discredit, or otherwise neutralize" them. The FBI did not just monitor these individuals, but tried to break up their marriages, "seed mistrust, sow misinformation," and provoke them to commit crimes so that they could be arrested. (Olmsted 2009, p. 10)

Sunstein and Vermeule assure us that this time would be different:

> By [cognitive infiltration of extremist groups] we do not mean 1960s-style infiltration with a view to surveillance and collecting information, possibly for use in future prosecutions.[33] Rather, we mean that government efforts might succeed in weakening or even breaking up the epistemological complexes that constitute these networks and groups. (2009, p. 224)

Is this supposed to be reassuring? Instead of being officially charged with the mere collecting of information, the infiltrators that Sunstein and Vermeule envision are to have an even more active role in the pursuit of a more questionable goal. Indeed, according to Sunstein and Vermeule's proposal, the stated objective would be closer to where the corrupted COINTELPRO program actually ended up. They explicitly consider the possibility that the infiltrations they recommend might, if partially exposed, "sow uncertainty and distrust within conspiratorial groups," which could "raise the costs of organization and communication." They consider this a jolly good outcome: these "effects are desirable, not perverse" (2009, p. 225).[34] It is a good thing for government agents to disrupt these kinds of peaceful efforts to organize. Why? Because, on Sunstein and Vermeule's view, such groups espouse beliefs that are irrational, given that we live in an "open society."

In a preliminary draft of their paper Sunstein and Vermeule offered a relevant example. As they explained it:

> A mini-scandal erupted in 2006 when U.S. newspapers revealed that the Lincoln Group, an independent contractor of "influence services," had paid Iraqi newspapers to publish hundreds of "news stories" written by U.S. military personnel but not identified as such, most of which portrayed events in Iraq in cheery terms or rebutted circulating conspiracy theories. The stories were factually true, but selective. (2008, p. 27)

The work of the Lincoln Group, while different from what Sunstein and Vermeule propose, is significantly analogous. And, Sunstein and Vermeule have a revealing perspective on it. First, it seems as though their only objection to the work of the Lincoln Group was that they got caught. Second, they frame the "information" provided in the stories in question as "factually true, but selective." This seems to be a euphemism for lying by omission. Indeed, perhaps the work of the Lincoln Group could be fairly characterized as "bullshit," as it seems concerned more with giving a certain impression than with expressing the truth. In any case, Sunstein and Vermeule seem to suggest that so long as one avoids telling straightforward lies (beyond one's identity and the sources of one's "information"), one is free to spin, mislead, and manipulate to one's heart's content[35]—all in the name of the good, and Truth, of course. But even if agents were able to hold the line at avoiding explicit lies about relevant facts, which is doubtful, it is hard to see how this can be viewed as a credible corrective to a "crippled epistemology."

More detail about the work of the Lincoln Group can provide a much-needed reality check. What was the demonstrably false belief that these planted stories were supposed to counter? What were the truths that they were to engender? Well, it would help the Iraqi people, who were no doubt under the sway of some kind of "crippled epistemology," to see truths (in 2006) like "Iraqi troops would soon be able to replace foreign forces"—according to an article entitled "Misinformation Intern: My Summer as a Military Propagandist in Iraq," in *Harper's Magazine* (Marx 2006).[36] The author, Willem Marx, describes the stories in question as follows: "These were far from exemplars of objective journalism, but Jon [another Lincoln Group employee] had said that I should think of the storyboards not so much as news but as messages Iraqis needed to hear." Perhaps even more tellingly, Marx recounts being told, "[I]t's very straightforward. You just have to keep the military happy." Sunstein and Vermeule were wise indeed to remove mention of the Lincoln Group from the journal version of their paper. Real-world examples reflect poorly on their ivory tower proposal, except of course when those stories are whitewashed.

4.4. Conclusion

Sunstein and Vermeule state that their recommendation of infiltration is to apply only to demonstrably false (and potentially harmful) theories. Their chief example of demonstrably false theories is the set of theories that posit insider complicity in the events of 9/11. What is the proof that settles this issue once and for all? My challenge to Sunstein and Vermeule is this: Can you prove, in a fair forum, that the theories in question are false? ("Proof" in an unfair forum, of course, is no proof at all.) The fact of the matter is this: they cannot prove it. So they wish to enforce their *belief* through epistemically illegitimate means. Their proposals exemplify intellectual cowardice. To adapt the bitingly critical remarks of the Chinese sage Laozi: "The man of 'reasonableness' makes his case, but when no one responds, rolls up his sleeves and resorts to persuasion by other means."[37]

Of course, the point transcends the issue of what to do about alternative theories about 9/11. The point is that we cannot engage in the kind of epistemic shenanigans that Sunstein and Vermeule recommend and, *at the same time*, credibly assert that alternative-to-mainstream theories (about whatever) may be dismissed on account of our fair and unbiased structures and organizations that adjudicate truth.[38]

FIVE

Conspiracy Theories and Stylized Facts

As discussed in the previous chapter, Cass Sunstein and Adrian Vermeule argue that the government and its allies ought to counter "conspiracy theories" or "extreme views"[1] by infiltrating and undermining the groups that espouse them. While they specify that this would apply only to theories that are (or are *deemed to be*) "demonstrably false," they give no hint regarding how such a judgment is to be reached. They suggest that infiltrating such groups is the best way to "cure" conspiracy theorists, by treating their "crippled epistemology" with "cognitive diversity." They base their proposal on an analysis of the "causes" of such conspiracy theories, which emphasizes informational and reputational cascades. Many have criticized their proposal for being antidemocratic, as I have in the previous chapter. The current chapter, however, has a more epistemic focus. Here I argue that Sunstein and Vermeule's argument is flawed in at least the following ways: (1) Their account of the "causes," purportedly explaining the popularity of (supposedly) "demonstrably false" conspiracy theories, are implausible. And (2) the air of acceptability that they attempt to evoke regarding their proposed "cures" relies on misleading "stylized facts," including a deceptive depiction of the relevant history, and a caricature of those who doubt official narratives.

5.1. Conspiracy Theories

Sunstein and Vermeule define a conspiracy theory as "*an effort to explain some event or practice by reference to the machinations of powerful people, who attempt to conceal their role (at least until their aims are accomplished)*" (2009, p. 205). It is worth noting that by this definition Saddam Hussein's purported attempt to conceal the weapons of mass destruction he supposedly had counts as a conspiracy theory.[2] (Were he and his supposed coconspirators not powerful people?) But of course, the phrase "conspiracy theory" is not typically employed to describe official Western accusations against foreign adversaries. So Sunstein and Vermeule's definition does not well capture the actual scope of this phrase in ordinary usage. As I explained in Chapter 1, I think a more accurate description of what is generally called a "conspiracy theory" is *an interpretation of a historical event that runs counter to an official account, and suggests that elements within a Western government have behaved in ways that seem particularly egregious*. In any case, my critique of Sunstein and Vermeule's proposal does not depend on any particular or precise definition.

There has been a fair amount of scholarly work on the philosophy of conspiracy theories in the last several years, as documented in Part I of this book.[3] The bottom line of this work is that *all attempts to explain why "conspiracy theories" (or a definable subset thereof) ought to be dismissed have turned out to be failures*. (Sunstein and Vermeule's attempt in this regard is a failure as well, as my discussion of informational and reputational cascades below shows.) This should not really be surprising, since all sides admit that at least some significant conspiracy theories have turned out to be true, Watergate and Iran-Contra being perhaps the most commonly cited.

It should be born in mind that Watergate and Iran-Contra are each *groups* of conspiracies, not individual cases. Kathryn Olmsted describes the various secret crimes associated with Watergate:

> During the subsequent investigation, Americans learned that the Watergate burglary was only one of many committed by President Nixon's aides, and that his reelection campaign has been systematically spying on his rivals for years. They discovered that the Nixon campaign had paid for these surveillance activities with a multimillion-dollar slush fund that it had accumulated through bribery and extortion. Finally, they found out that the president had personally tried to obstruct the FBI's investigation of the crime by paying off the burglars and attempting to blackmail the CIA. (Olmsted 2019, p. 292)

Regarding the Iran-Contra affair, Olmsted notes:

> The scandal had revealed that unelected staff members in the White House could create, as Senator Daniel Inouye said, "a shadowy government with its own air force, its own navy, its own fundraising mechanism, and the ability to pursue its own ideas of the national interest, free from all checks and balances and free from the law itself." (Olmsted 2019, p. 294)

There are many more examples as well.[4] For instance, the Bush administration's coordinated propagandistic efforts to win support for an invasion of Iraq was a conspiracy[5]—to name just one more here. Those who called it what it was early on were promoting a conspiracy theory, by most definitions. (Alas, if only that conspiracy theory had been more successful, much suffering and death might have been avoided.) There are also scores of conspiracy theories that remain plausible, yet unproven—or at least not widely accepted as proven. Many of these may well be true too, for all we know.

5.1.1. Causes: Informational and Reputational Cascades

Sunstein and Vermeule's major innovations are (1) the suggestion that what accounts for the success of conspiracy theories deemed to be "demonstrably false" are informational and reputational cascades, and (2) the proposed "cure," cognitive infiltration, which is designed to disrupt these cascades. But these cascades are implausible explanations for the success of conspiracy theories, especially for the set of conspiracy theories that they take as their "running example," namely conspiracy theories about September 11.

5.1.1.1. Informational Cascades

Sunstein and Vermeule argue that informational cascades, in significant measure, explain the pervasiveness of "demonstrably false" conspiracy theories. I will quote them at length to show how easily such cascades can be applied to explain the success of (dubious) official stories as well. It is not a phenomenon that has any *particular* relation to conspiracy theories. I have simply changed references to conspiracy theories into references to official stories, as indicated by brackets and strikethrough lettering. (Note that so-called informational cascades do not actually involve cascades of *information*, or evidence, but rather cascades of opinion.)

> To see how informational cascades work, imagine a group of people who are trying to assign responsibility for some loss of life. Assume that the group members are announcing their views in sequence. Each member attends, reasonably enough, to the judgments of others. Andrews is the first to speak. He suggests that the event was caused [just how the government said it was] ~~by a conspiracy of powerful people~~. Barnes now knows Andrews's judgment; she should certainly go along with Andrews's account if she agrees independently with him. But if her independent judgment is otherwise, she would—if she trusts Andrews no more and no less than she trusts herself—be indifferent about what to do, and she might simply flip a coin.
>
> Now turn to a third person, Charleton. Suppose that both Andrews and Barnes have endorsed the [official story] ~~conspiracy theory~~, but that Charleton's own view, based on limited information, suggests that they are probably wrong. In that event, Charleton might well ignore what he knows and follow Andrews and Barnes. It is likely, after all, that both Andrews and Barnes had evidence for their conclusion, and unless Charleton thinks that his own information is better than theirs, he should follow their lead. If he does, Charleton is in a cascade. Of course Charleton will resist if he has sufficient grounds to think that Andrews and Barnes are being foolish. But if he lacks those grounds, he is likely to go along with them. This may happen even if Andrews initially speculated in a way that does not fit the facts. That initial speculation, in this example, can start a process by which a number of people are led to participate in a cascade, accepting [an official story] ~~a conspiracy theory~~ whose factual foundations are fragile. (2009, pp. 213–214)[6]

Not only can these cascades work, in hypothetical cases, either for a counternarrative or for an official story, they make more sense as an explanation for the success of dubious official stories, since official stories tend to have the crucial advantage of gaining early traction.

Sunstein and Vermeule are not actually the first to suggest that informational cascades are relevant to conspiracy theories. In a paper entitled "Are Conspiracy Theorists Irrational?" David Coady describes informational cascades in a context wherein it is the conspiracy theorists who are accused of (perhaps irrationally) exercising "intellectual autonomy" by *refusing* to go along with informational cascades. Keeping that context in mind, consider Coady's rather neutral description:

> [W]hat economists call "information cascades" . . . can occur when people express their opinions about the answer to a certain question in a publicly observable sequence. If the early answers exhibit a clear pattern, people later in the sequence may decide to ignore their own epistemic resources and follow the crowd. This belief forming strategy can be entirely rational from an individual perspective, especially if expertise on the question at issue is reasonably evenly spread amongst the group. The epistemic danger of this strategy, however, is that it can lead to relevant evidence being hidden from those later in the sequence. Thus the epistemic authority of thousands of people can be largely illusory, because most of them have had their beliefs determined by a handful of people at the beginning of the sequence. (Coady 2007, pp. 201–202)

Coady concludes that while it may be "individually rational" to go with the flow of an informational cascade, "those who refuse to follow the crowd, even when the crowd is more likely to be right than they are, are doing the crowd an epistemic favour by making it more likely that the crowd itself (or at least most of its members) gets the right answer in the end" (Coady 2007, p. 202). It is worth noticing, in this context, that doubters of the official narrative of September 11 often point out how quickly an official narrative took form. Even if not explicitly mentioning "informational cascades" by name, they clearly imply that setting up such cascades is a propagandistic device that was employed very early on.[7]

The point is this: while the dynamic that Sunstein and Vermeule describe is undoubtedly real, it cuts both ways. Indeed, it works better as an explanation for the success of questionable official stories. Regarding September 11, some rather strong informational cascades (whether based on accurate information or not) affirming the official story began flowing within the first couple days and have continued unabated. Countercurrents, on the other hand, didn't start flowing with any strength for several years. And, as we will see at the end of this chapter, many of those skeptical of the official story of 9/11 cannot plausibly be regarded as uncritically following an informational cascade. Further, regardless of what informational cascades might flow through a particular group or segment within society, it is a rare individual indeed that would have escaped the mainstream media and their relentless support of the official story. At most, a countercascade could have emboldened some to question the official story, and perhaps to begin to investigate the issue. But it is hardly plausible that a counternarrative informational cascade would overwhelm the official/mainstream informational flood—unless it drew strength in some other way, perhaps from empirical evidence.[8] (Whether such evidence is truly substantial cannot be adjudicated a priori but must be carefully examined.)

5.1.1.2. Reputational Cascades

Before addressing Sunstein and Vermeule's analysis of reputational cascades, I'd like to make an observation about a related phenomenon, "pragmatic hypothesis testing," which philosophers Juha Räikkä and Lee Basham include as one element of what they call "conspiracy theory phobia."

First of all, the basic idea behind "reputational cascades" is also implicit in Nicholas DiFonzo's recent analysis of conspiracy theory transmission: When a person makes a conspiratorial claim within a conspiracy-minded group, "[She] get[s] a nice boost to [her] popularity. . . . And [she is] easily forgiven if it turns out to be false or exaggerated" (DiFonzo 2019, p. 263). While this may well explain the transmission of conspiracy theories in some circles, a very different calculus is to be expected in other relevant circles and circumstances. Räikkä and Basham explain:

> When a person hears a disturbing conspiracy theory, she considers it, in her mind, as a hypothesis. Likely, she tests the hypothesis from a *pragmatic point of view*. If the theory in question is one that asserts things people do not want to believe, the person wants plenty of evidence in support of the theory before she's ready to seriously suspect or believe in it. For her, it is important not to believe in a conspiracy theory if there is no conspiracy—much more important than to believe in a conspiracy theory when there *is* a conspiracy. (2019, p. 182)

For academics and those who work for the mainstream media, in particular, the costs involved in endorsing a conspiracy theory too readily can be high—so much so that in these cases I would put Räikkä and Basham's claim more strongly: regardless of the truth of the conspiracy theory, it is important, pragmatically, for most ostensible epistemic authorities not to believe it or espouse it, perhaps even to maintain a dismissive attitude, unless it is *likely to be quickly confirmed*. That is a high bar. It goes some way toward explaining how epistemic authorities can be wrong as a group: reputational dams hold back would-be informational cascades, preventing some ideas from being expressed and explored. As Räikkä and Basham put it, "[W]hen a person evaluates a conspiracy theory, she accepts one conclusion [namely, that it is untrue] much more easily than another. Her reasoning maybe rational from a pragmatic point of view, but there is a considerable risk that she makes false conclusions" (2019, p. 182). This is because the social costs of being wrong are asymmetrical. And this is especially so for those looked upon as epistemic authorities.

For related reasons, Sunstein and Vermeule's theory regarding reputational cascades is even less plausible than their theory involving informational cascades. They describe reputational cascades as follows:

> Conspiracy theories do not take hold only because of information. Sometimes people profess belief in a conspiracy theory, or at least suppress their doubts, because they seek to curry favor. Reputational pressures help account for conspiracy theories, and they feed conspiracy cascades. In a reputational cascade, people think that they know what is right, or what is likely to be right, but they nonetheless go along with the crowd in order to maintain the good opinion of others.
>
> Suppose that Albert suggests that the Central Intelligence Agency was responsible for the assassination of President Kennedy, and that Barbara concurs with Albert, not because she actually thinks that Albert is right, but because she does not wish to seem, to Albert, to be some kind of dupe. It should be easy to see how this process might generate a cascade. Once Albert, Barbara, and Cynthia offer a united front on the issue, their friend David might be reluctant to contradict them even if he believes that they are wrong. In real-world conspiracy theories, reputational pressures often play a large role, leading people to squelch their own doubts in order to avoid social sanctions. (2009, pp. 214–215)

While their example is hypothetical, Sunstein and Vermeule assert that such reputational cascades "often play a large role" in "real-world conspiracy theories." So let's look at the real world.

The *New York Times* published a book review of Vincent Bugliosi's *Reclaiming History*, which purports to disprove conspiracy theories about the assassination of President Kennedy once and for all. The author opines:

> What Bugliosi has done [by ridiculing JFK conspiracy theorists] is a public service; these people *should* be ridiculed, even shunned. It's time we marginalized Kennedy conspiracy theorists the way we've marginalized smokers; next time one of your co-workers starts in

> about Oswald and the C.I.A., make him stand in the rain with the other outcasts. (Burrough 2007)

Or, consider the case of Professor William Woodward of the University of New Hampshire. According to an article in the *Boston Globe*: "Woodward, a professor of the history of psychology, is a member of Scholars For 9/11 Truth. . . . When news of Woodward's association with the group was published in a local newspaper last month, it sparked a hail of criticism from New Hampshire politicians" (see Long 2006). In another article, James Joyner describes the situation as follows: "A student activist group has joined New Hampshire Governor John Lynch in trying to fire a University of New Hampshire professor for his rather bizarre views on the 9/11 attacks. . . . Gov. John Lynch called Woodward's beliefs 'completely crazy and offensive' and asked the trustees to investigate" (Joyner 2006).[9] In an update to that article, Joyner writes: "[A reader] comments, 'I don't think they should fire him. I think they should ridicule him. Publicly. Relentlessly.' Agreed. That is much more in the spirit of higher education than censorship." Now does this sound like an environment wherein a reputational cascade can plausibly account for the spread of the theory in question? I don't think so. Further, I can testify that if I were basing my decision on enhancing, or at least not tarnishing, my reputation with academic colleagues, advocating "9/11 Truth" would be just about the last thing I would do.[10]

Although some people doubted the official story from the beginning, it seems that, at least for a while, they mostly kept it to themselves. In any case, the 9/11 truth movement didn't really start to take off until around 2005 or 2006. By then, informational and reputational cascades were flowing powerfully in support of the official story. In this context, appeal to such cascades as the explanation for the pervasiveness of 9/11 conspiracy theories is unpersuasive.

In the 2014 revised version of Sunstein and Vermeule's article, Sunstein (now writing as the sole author) claims, "It seems clear that reputational pressures play a large role in the adoption and dissemination of conspiracy theories" (2014a, p. 17). Sunstein provides no evidence that this is true, much less *clearly* true. As I have just argued, in influential circles quite the opposite seems true. Here is how Sunstein now ends his section on reputational cascades:

> In some times and places, doubting a conspiracy theory, or even failing to endorse it, can be literally dangerous, in the sense that doubt-

> ers are ostracized or worse. Some conspiracy theories are able to persist only because people silence themselves. Dissent is an indispensable corrective, but it does not occur. (p. 17)

As usual, Sunstein gives no real-world examples, appealing, apparently, to paranoid imaginations. In the real world, as my example above illustrated, it is those who go against an official story who face potential ostracism. At least for those in the academy, the media, or government, espousing conspiracy theories may be "career limiting," if not career suicide. And so we should consider the possibility that some academics, mainstream journalists, and politicians—precisely the groups most likely to disparage conspiracy theories—may be silencing themselves for a reason quite the opposite of the one Sunstein suggests. This may go some way toward explaining how dubious official stories can persist without the need of a massive conspiratorial cover-up.

5.1.2. The Roles of Availability and of Emotions

Sunstein and Vermeule argue that certain kinds of events, such as terrorist attacks, being highly "available"—that is, being on everyone's mind—can trigger cascades in which "conspiracy theories are invoked both in explaining [the event] and using it as a symbol for broader social forces and large narratives about political life" (2009, p. 215). Fair enough. But this seems to apply as much to official accounts as it does to counternarratives. A good example of this is the theory (whether true or not) that September 11 was simply a product of an Islamic conspiracy to attack the United States, which was (in fact) used as a symbol for the need for a war on terror, as well as for other social changes. Now, Sunstein and Vermeule do make a perfectly valid point when they state, in this context, "Within certain nations and groups, the claim that the United States or Israel was responsible for the attacks of 9/11 fits well within a general narrative about who is the aggressor, and the liar" (p. 18). However, within other groups, closer to home, the claim that Muslims were solely responsible "fits well within a general narrative about who is the aggressor, and the liar." *That* can hardly be doubted. So it is not clear that cognitive availability influences the relative appeal of conspiratorial counternarratives compared to official accounts (which are sometimes also conspiratorial, as in the 9/11 case). What *is* clear is that Sunstein and Vermeule consistently have trouble noticing that their analysis can be easily flipped to work the other way.

In this context, Consider Sunstein's final remarks regarding the "role of emotions":

> Of course, evidence matters, and so long as there is some kind of process for meeting falsehoods with truth, mistaken beliefs can be corrected in principle. But sometimes the conditions for corrections are not present; or, even if they are present, people are strongly motivated to disregard them. (Sunstein 2014a, p. 19)

Indeed. As Lee Basham is sure to remind us, when truths are highly toxic to the establishment, such that all those beholden to the establishment perceive a clear disincentive to articulating them, or even investigating or trying to think clearly about them, then "the conditions for correction" will not be present.[11] Momentous conspiracy theories, such as those involving 9/11 or the assassination of President Kennedy, often involve this kind of toxicity.

Looked at this way, it almost seems as though Sunstein and Vermeule were cleverly hinting to careful readers that they should draw conclusions that are roughly the opposite of what they seem on the surface to be asserting, as David Ray Griffin (2011)[12] has teasingly argued at great length.

5.1.3. Cure: Cognitive Infiltration

For whatever reason, some conspiracy theories, such as those that posit insider complicity in 9/11, persist. So what should be done about this? Sunstein and Vermeule think that we can separate plausible conspiracy theories from "demonstrably false" ones. They imagine an alarming range of possible government responses to those conspiracy theories deemed (by someone) to be "demonstrably false." They write:

> What can the government do about conspiracy theories, and what should it do? (1) Government might ban "conspiracy theories," somehow defined. (2) Government might impose some kind of tax, financial or otherwise, on those who disseminate such theories. (3) Government might itself engage in counterspeech, marshaling arguments to discredit conspiracy theories.[13] (4) Government might formally hire credible private parties to engage in counterspeech. (5) Government might engage in informal communication with such parties, encouraging them to help. Each instrument has a distinctive set of potential effects, or costs and benefits, and each will have a place under imaginable conditions. Our main policy claim here is that government should engage in cognitive infiltration of the groups that produce conspiracy theories, which involves a mix of (3), (4), and (5). (2009, 218)

As an example of a set of theories that are "demonstrably false," Sunstein and Vermeule single out counternarratives regarding the events of September 11, 2001. However, they neither provide a comprehensive proof of this falsity (granted, that would be unreasonable to expect of them) nor do they *point* to such a demonstration (a more reasonable expectation). After all, Sunstein claims in his revised version, "In just a few seconds, people can find a credible rebuttal of a conspiracy theory or produce a rebuttal themselves" (Sunstein 2014a, p. 23), as if finding what seems at a glance to be a "credible" rebuttal would be enough to show that a conspiracy theory was demonstrably false. Sunstein and Vermeule do, in fact, provide a limited critique of their own, but not one that inspires confidence in their conclusion, or in their thoroughness or impartiality. Their characterization of the significance of frames of video footage released by the Department of Defense, which I will discuss below, is an example.

In addition to the problem of misdiagnosis, their proposed cure has potentially dangerous side effects. By suggesting that groups who promote views they deem to be demonstrably false ought to be infiltrated, they are implicitly suggesting that members of those groups, or other "kooks"[14] who hold similar views, are not fully persons—in the Kantian sense of being autonomous rational agents who are thus ends-in-themselves. There is an analogy between Sunstein and Vermeule's proposed infiltrations and the policy toward detainees instituted during George W. Bush's administration. By condoning harsh interrogations, stress positions, and such, the Bush administration set the conditions in which abuse was a predictable result.[15] This is one of the many reasons that these policies were ill-advised, and that the administration bears considerable responsibility for the abuses that occurred—even though the official policy did not, of course, explicitly authorize turning detainees into naked human pyramids, or torturing them to death, or sodomizing them with broomsticks, and so on. How did these outrages happen? Dehumanization. Once someone is regarded as less than fully human, it is hard to avoid a feeling of contempt. And, contempt plus power, or at least the sense that one is working in the service of authority, leads quickly to abuse, as the Stanford Prison Experiment so clearly showed.

Now, how does Sunstein and Vermeule's proposal compare? As suggested above, Sunstein and Vermeule's proposal treats "conspiracy theorists" as less than fully human in the sense that they are regarded as irrational[16]—rationality being the hallmark of humanity. Although this is based on a caricature of conspiracy theorists (which will be shown below), it is a necessary assumption for two reasons. First, the supposition that they believe "demonstrably false" theories suggests contemptible ignorance or stupidity. Second, positing irrationality justifies the notion that one must lie about one's identity, since conspiracy theorists are viewed as incapable of evaluating evidence that does not come from their own kind. In addition, implicit in the proposal to actively undermine selected groups based on the beliefs they hold and promote is the notion that those groups do not really have a right to free speech and assembly. To suggest that they are without such a basic right is to suggest that they are less than human. But once individuals are viewed in this way, it is hard to restrain agents from exceeding their explicit mandates in their effort to please their superiors. Thus, here too, abuse is to be expected. And, as discussed below, infiltration has a history, which isn't pretty. This analysis, that abuse can reasonably be expected to follow if such infiltrations are permitted, holds regardless of whether we grant Sunstein and Vermeule's (unsafe) assumption of a well-motivated executive.

In addition to having dangerous implications, the "conspiracy theorists" that their proposal targets would naturally regard the proposal as straightforwardly insulting. But one ought not object because of the insult. After all, sometimes the truth hurts. The objection that I press in the following section is precisely that they have not expressed the truth. Their proposal relies on clearly false premises and misleadingly stylized facts.

5.2. Stylized Facts

Sunstein and Vermeule argue that conspiracy theorists suffer from a "crippled epistemology" as a result of their informational isolation, and thus they need "cognitive diversity" introduced by infiltrating agents able to reframe their "stylized facts."[17]

Let's discuss stylized facts. Not always negative in connotation, a "stylized fact" can mean a general claim that is widely accepted as true as a result of its (supposed) instantiation in a wide variety of contexts. Its presumed truth, then, serves to limit interpretations of phenomena. For example, the idea that conspiracy theories are unwarranted is a stylized fact in this sense. The common refrain, "I don't subscribe to conspiracy theories," suggests, as a general "fact," that conspiracy theories are always unwarranted, and that assumption (inappropriately) closes off the possibility of serious consideration of certain interpretations of events. Relatedly, "stylized fact" can refer to a simplified expression, or summary, of an empirical reality, which, being simplified, misses some (possibly significant) nuances. For example,

Sunstein and Vermeule's presumption of a "well motivated" government, which they characterize as a "standard" assumption, may count as a stylized fact in both of the above senses. Is it *true* that the government is well motivated? Well, there may be *some* truth in the claim that it is, but that generalization glosses over some rather rough spots that may well be very significant indeed.[18] (Was the Tuskegee experiment "well motivated"?) And the assumption closes off certain perfectly reasonable lines of inquiry.

Sunstein and Vermeule offer no explicit example of conspiracy theorists relying on specific stylized facts, so it is hard to know exactly what they are thinking of. Nevertheless, since circulating these (unstated) stylized facts is apparently taken to be an epistemic sin sufficient to justify government infiltration, I take the phrase to be intended in a negative sense. I will, rather loosely, treat it as meaning simply "a misleading characterization of reality." This is in keeping with Sunstein and Vermeule's emphasis on conspiracy theories being grounded in misinformation or misleadingly incomplete information.

While it is no doubt true that so-called conspiracy theorists sometimes do stylize facts, Sunstein and Vermeule neither provide evidence that the conspiracy theories that they are most concerned about critically depend upon stylized facts, nor even that conspiracy theorists employ stylized facts any more than supporters of official theories do. But it is clear that Sunstein and Vermeule themselves rely significantly on misleadingly stylized facts. I will discuss four. (In this part of my discussion I will not restrict myself to the version of Sunstein and Vermeule's article that was published in the *Journal of Political Philosophy* but will include examples from an earlier version of their paper published online. I think this is fair because the issue is their own tendency to stylize facts, not whether the peer review process ferreted out all such significantly misleading "facts"—which we will see it did not.)

5.2.1. Stylized Fact 1: Conspiracy Theories Are the Stuff of Rumor

Operation Northwoods was a Kennedy-era plan that was brought to light by NSA expert James Bamford in 2001, in his book *Body of Secrets*. Bamford explains:

> [T]he plan, which had the written approval of the Chairman and every member of the Joint Chiefs of Staff, called for innocent people to be shot on American streets; for boats carrying refugees fleeing

> Cuba to be sunk on the high seas; for a wave of violent terrorism to be launched in Washington, D.C., Miami, and elsewhere. People would be framed for bombings they did not commit; planes would be hijacked. Using phony evidence, all of it would be blamed on Castro, thus giving Lemnitzer and his cabal the excuse, as well as the public and international backing, they needed to launch their war. (Bamford 2001, p. 82)

Sunstein and Vermeule characterize Operation Northwoods as "a *rumored* plan by the Department of Defense to *simulate* acts of terrorism and to blame them on Cuba" (2009, p. 206, emphasis added). But there is nothing "rumored" about the document uncovered by Bamford detailing a variety of plans approved by the Joint Chiefs of Staff, which, Sunstein and Vermeule admit, "really was proposed by high-level officials." So why do they style it as "rumored"? Clearly the intent was to downplay the significance of this shocking and relevant example. Further, the plan proposed not only to "*simulate* acts of terrorism," but even to *actually carry out* acts of terrorism and blame them on Cuba. It was a plan that included false flag terrorism, not unlike the acts of terrorism carried out in Italy, from the late 1960s to the early 1980s, often referred to under the heading Operation Gladio (see Ganser 2005).

Further, the analogy to what some so-called conspiracy theorists allege about 9/11 is striking. The Northwoods document even details a plan to blow up an empty plane flown by remote control over Cuba. To give a sense of the level of complexity that the planners apparently considered unproblematic, I offer an extended quotation from the document:

> 8. It is possible to create an incident which will demonstrate convincingly that Cuban aircraft has attacked and shot down a chartered civil airliner enroute from the United States to Jamaica, Guatemala, Panama or Venezuela. The destination would be chosen only to cause the flight plan route to cross Cuba. The passengers could be a group of college students off on a holiday or any grouping of persons with a common interest to support chartering a non-scheduled flight.
>
> a. An aircraft at Eglin AFB [Air Force Base] would be painted and numbered as an exact duplicate for a civil registered aircraft belonging to a CIA proprietary organization in the Miami area. At a designated time the duplicate would be substituted for the actual civil aircraft and would be loaded with

the selected passengers, all boarded under carefully prepared aliases. The actual registered aircraft would be converted to a drone.

b. Take off times of the drone aircraft and the actual aircraft will be scheduled to allow a rendezvous south of Florida. From the rendezvous point the passenger-carrying aircraft will descend to minimum altitude and go directly into an auxiliary field at Eglin AFB where arrangements will have to be made to evacuate the passengers and return the aircraft to its original status. The drone aircraft meanwhile will continue to fly the filed flight plan. When over Cuba the drone will be transmitting on the international distress frequency a "MAY DAY" message stating he is under attack by Cuban MIG aircraft. The transmission will be interrupted by destruction of the aircraft which will be triggered by radio signal. This will allow ICAO radio stations in the Western Hemisphere to tell the US what has happened to the aircraft instead of the US trying to "sell" the incident.[19]

Now back to Sunstein and Vermeule's dismissive language. Their choice of words cannot be written off as a mere oversight on their part. For when we read either the 2008 draft version or Sunstein's 2014 version of this paragraph. the deliberate intent to be dismissive becomes unambiguously apparent. Immediately after the mention of Operation Northwoods, we find: "In 1947, space aliens did, in fact, land in Roswell, New Mexico, and the government covered it all up. (Well, maybe not)" (2008, p. 4; 2014a, p. 4). This trivializes a whole list of significant conspiracies that they could not but admit were real, though the list could have been much longer.

5.2.2. Stylized Fact 2: Clear Evidence Proves Conspiracy Theories False

The online draft for their essay, as well as Sunstein's revised 2014 version, states: "Some theorists claimed that no plane had hit the Pentagon. *Even after the Department of Defense released video frames showing Flight 77 approaching the building* . . ." (Sunstein 2014a, p. 27, emphasis added; cf. Sunstein and Vermeule 2008, p. 20). If Sunstein and Vermeule had bothered to actually look at the video frames in question, they would have seen that they do not in fact show anything recognizable as Flight 77 approaching the Pentagon. Indeed, it is not at all clear what these pictures show. Sunstein and Vermeule presumably knew this because they cited *Debunking 9/11 Myths* pages 60 and 61 as their source. Turning to those pages one reads: "[T]he images [originally leaked in 2002] show little more than a blurry white object approaching the Pentagon and a fireball subsequently erupting from the building" (Dunbar and Reagan 2006, p. 60). And then, regarding the "videos" released in 2006, one reads: "[T]o call it a video is a stretch: It's actually more like a slide show of the images previously released—a blur followed by a massive explosion" (pp. 60–61). This is what Sunstein and Vermeule turn into "[T]he Department of Defense released video frames showing Flight 77 approaching the building." Apparently, either Sunstein and Vermeule were just too busy advocating infiltration to objectively scrutinize the evidence (even the two pages they cite) or else they were "stylizing" their facts.

I don't mean to imply that anything is amiss with this part of the official story of 9/11. But it is noteworthy, nonetheless, that an acclaimed scholar would appeal to "evidence" as useless as the supposed pictures of Flight 77 approaching the Pentagon to justify a deceit-countenancing, antidemocratic, and epistemically suspect proposal.[20]

5.2.3. Stylized Fact 3: Infiltration Is Benign

As mentioned in the previous chapter, Sunstein and Vermeule write:

> By [cognitive infiltration of extremist groups] we do not mean 1960s-style infiltration with a view to surveillance and collecting information, possibly for use in future prosecutions. Rather, we mean that government efforts might succeed in weakening or even breaking up the epistemological complexes that constitute these networks and groups. (2009, p. 224)

This gives the impression that the COINTELPRO operations of the fifties and sixties were benign and passive. But this is far from accurate. Kathryn Olmsted gives a much more honest account:

> During the cold war, the FBI started its domestic covert action programs, known by the acronym COINTELPRO, in which agents infiltrated dissident groups and eventually tried to "expose, disrupt, misdirect, discredit, or otherwise neutralize" them. The FBI did not

> just monitor these individuals, but tried to break up their marriages, "seed mistrust, sow misinformation," and provoke them to commit crimes so that they could be arrested. (Olmsted 2009, p. 10)[21]

Further, Olmsted adds:

> Government officials tried to control how the public interpreted events, sometimes lied about these events, and spied on and harassed those citizens who suggested different interpretations. (2009, p. 10)

Had Sunstein and Vermeule given a fuller and more accurate account of the true history of past practices, it would have aroused a sense that great caution is warranted here. So instead they stylized.

5.2.4. Stylized Fact 4: Conspiracy Theorists Are Ignorant Extremists

Perhaps the most significant stylized fact involves the caricature of so-called conspiracy theorists. Sunstein and Vermeule charge that conspiracy theorists generally have "little (relevant) information" (2009, p. 211) or "skewed information" (2009, p. 210). But these claims are unsubstantiated. Indeed, many people that would count as "conspiracy theorists" by Sunstein and Vermeule's lights are very informed people. Indeed, many have specialized knowledge of one relevant kind or another. But Sunstein and Vermeule ignore them.

For example, if all those who take the possibility of insider complicity in 9/11 seriously count, then that list includes established scholars that have employed their considerable research talents to understanding the dynamics surrounding 9/11, such as David Ray Griffin, Peter Dale Scott, Nefeez Ahmed, Graeme MacQueen, and Lance deHaven-Smith, to name just a few;[22] it includes established scientists such as Steven Jones, Jeffrey Farrer, Niels Harrit, and many others;[23] it includes professional engineers and architects—more than three thousand have joined Richard Gage in calling for a new investigation into the collapse of the Twin Towers and Building 7;[24] and it includes intelligence professionals such as Ray McGovern,[25] Robert Steele, and (with some vacillation) Robert Baer. That is still a short list, but the complete list of highly accomplished people that have publicly questioned the official account is at least in the hundreds.[26] By caricaturing conspiracy theorists Sunstein and Vermeule are able to pretend that informed and sophisticated "conspiracy theorists" do not exist. But these people do exist. And Sunstein and Vermeule's theory of the "causes" of conspiracy theories does not account for them.[27] And the inappropriateness of their proposed "cure" is most clear with regard to them.

Thus, in order to make their proposal palatable, Sunstein and Vermeule needed to stylize their depiction of those who question official stories. Indeed, they go to absurd lengths, worrying that their proposed infiltrators might be asked by conspiracy theorists to commit crimes. As suggested by the discussion of the previous stylized fact, it is more likely that the infiltrators will be the ones that end up proposing criminal activity—even if Sunstein and Vermeule don't explicitly advocate this.

One might counter that some conspiracy theorists, surely, really are dangerous, and may indeed engage in and encourage criminal activity. But this would be true of virtually *any* large category of people. If there are valid reasons to think that a *particular* individual, or a small group of individuals, is dangerous, that is another matter. But the fact that a group thinks that, say, 9/11 was an inside job is not, *by itself*, a valid reason to believe that they are prone to criminality or violence. Sunstein and Vermeule's caricature of conspiracy theorists conflates those who hold views that they regard as false with the much smaller group of those who *in addition* have violent or otherwise criminal proclivities.

5.3. Conclusion

The stylization of the above "facts" is important for the plausibility of Sunstein and Vermeule's argument. (1) If they fully acknowledged the history of real conspiracies and of theories that remain plausible if unproven, that would undermine the efficacy of their dismissive rhetorical posture regarding the ill-defined subset of those theories that they believe should be undermined by covert operations. (2) By whitewashing the history of infiltration, they make their proposal seem less obviously problematic. (3) By presenting a caricature of people who espouse so-called conspiracy theories, they treat them as "other"—something less than human, beings not fully capable of reason. Otherwise, a more honest, straightforward, and respectful response would seem more appropriate than infiltration. And, finally, (4) the bogus claim that there are pictures clearly identifiable as Flight 77 approaching the Pentagon made it possible for them to ridicule conspiracy theorists who continue to believe otherwise. Without recourse to ridicule, Sunstein and Vermeule's responsibility to deal with the relevant evidence in a more sophisticated way would have been more evident. And addressing the evidence in this way would have made establishing the falsity of all theories that suggest insider complicity in 9/11 hopelessly complex. But without establishing the clear falsity of those theories, they could not reasonably frame the members of the so-called 9/11 truth movement as irrational, and thus appropriate targets for cognitive infiltration. In the journal version of their paper, Sunstein and Vermeule drop the reference to Flight 77. I thought they may have done so because they came to realize that it is easily exposed as false.[28] But Sunstein reintroduced the assertion in his 2014 revised version. In any case, it seems they didn't really need to resort to ridicule based on false evidence. The strong bias against conspiracy theories,[29] especially in the academy, evidently makes such ridicule unnecessary.

Peaceful, law-abiding people ought to be allowed to freely assemble and pursue their inquiries without infiltration. And this applies even to those who promote theories that posit state crimes against democracy (SCADs)[30]—which is what the most "dangerous" so-called conspiracy theories typically allege. In the interest of peace and justice, all people ought to be allowed to freely assemble and pursue their inquiries without infiltration—even those, or perhaps especially those, who dare to question official narratives.

SIX

Were Sunstein and Vermeule Misunderstood?

An Examination of Sunstein's Revision

> I've written hundreds of articles, and I remember some and not others. That one [about conspiracy theories] I don't remember very well. . . . There are a lot of things I have written, I guess. There are even more things I'm said to have written. I may agree with some of the things I have written, but I'm not exactly sure.
>
> —Cass Sunstein, "the pre-eminent legal scholar of our time"

6.1. Introduction

The epigraph to this chapter is Cass Sunstein's answer to a confrontational question about why he thinks the government should be involved in the infiltration of groups that espouse 9/11 conspiracy theories. The questioner, Luke Rudkowski, framed the question this way:

> [Y]ou openly advocated government agents infiltrate activist groups of 9/11 truth, and also stifle dissent online. I was wondering, why do you think it is the government's job, or why do you think the government *should*, go after family members who have questions about 9/11, responders who were lied to about the air [they were breathing], survivors whose testimony [inaudible], also government whistleblowers that were gagged because they have released information that contradicts the official story?[1]

The way the question is asked might seem unfair. Indeed, the sudden flurry of unexpected examples might reasonably cause some befuddlement. And so part of Sunstein's answer is understandable: "I hope I didn't say that." Of course he didn't, not explicitly. But Rudkowski is pointing to the *implications* of what Sunstein wrote; he is not suggesting that Sunstein had himself singled out those cases. Rather, Rudkowski is pointing out that some of the people who espouse 9/11 conspiracy theories fall into the categories he lists: family members of those who died on 9/11, first responders (many of whom became sick from inhaling the air they were told was safe), survivors, and gagged whistleblowers with relevant inside information—perhaps Rudkowski is thinking of Sibel Edmonds in the whistleblower category.[2] However unexpected and confrontational the question may have been, it seems that Rudkowski is making a fair point: if government agents and their allies infiltrate and attempt to undermine networks of people who espouse 9/11 conspiracy theories, they would, in all likelihood, be infiltrating groups to which some family members, first responders, survivors, and whistleblowers belong. So what? Well, there just seems to be something *particularly* odious about *that*. And so it is too bad that Sunstein couldn't allow himself to consider this question and provide a thoughtful answer. That would not have required actually remembering the details of what he wrote.

As it turns out, Sunstein did eventually decide to refresh his memory of the article, and he went so far as to revise and republish it as the first chapter of his 2014 book, *Conspiracy Theories and Other Dangerous Ideas*. (Only the first chapter addresses conspiracy theories.) Unsurprisingly, the revision does not mention Rudkowski's question/critique. Likewise, Sunstein did not respond in any substantial way to David Ray Griffin's book-length critique, other than insinuating that there is something amiss about the fact that "as of this writing, the book [*Cognitive Infiltration*] has forty-one reviews on Amazon.com—all of them with perfect five-star ratings" (2014a, p. xi). Indeed, though Sunstein mentions the existence of this book, and also an article entitled "Obama Confidant's Spine-Chilling Proposal" by Glenn Greenwald, and a blog post called "Got Fascism?" by Marc Estrin,[3] none of the critiques within them are addressed with any specificity. Sunstein does not even refer to any of these authors, or any other critic, by name. As for the only two critical articles focused on Sunstein and Vermeule's article that were published years before Sunstein's revision[4] in peer-reviewed scholarly journals—namely, earlier versions of the two previous chapters of this book—it is as though they did not exist at all.

Instead of engaging with the arguments of critics, Sunstein simply writes, in the preface to his book:

> The essay has been seen as dangerous partly because it was read to suggest I had a "plan" to infiltrate not only foreign organizations that believe the United States was responsible for the 9/11 attacks, and that seek to threaten our security, but also conservative organizations as a whole. (Of course, nothing could be further from my mind). (2014a, pp. xi–xii)

While the word "plan" may seem to suggest something vaguely conspiratorial, Sunstein and Vermeule were explicitly making a *proposal*, which is pretty close to having a plan. And that proposal did not seem to be focused exclusively on organizations "that seek to threaten our security" but rather seemed to apply to conspiracy theory groups more generally on the theory that some member *might* be motivated to violence. Those seeking to threaten our security, and those having the potential to be motivated to violence, are two very different categories—the latter includes nearly everyone. As for the issue of whether or not it was reasonable to read Sunstein and Vermeule's essay as aimed only at *foreign* organizations, I address that below. Regarding the charge that Sunstein was accused of wanting to infiltrate "conservative organizations as a whole," I'm not aware of anyone who made that charge, and Sunstein provides no specifics. It certainly did not come from Griffin, Estrin, Greenwald, or myself. But *that*, sadly, is the beginning and end of Sunstein's engagement with critics.

Still, Sunstein's revisions, along with comments made in the preface to his book, can be seen as an attempt to answer the critiques of people like Rudkowski, Estrin, and Greenwald, who stress ethical and legal problems with Sunstein and Vermeule's proposal. For Sunstein's revisions may seem to make his proposal more ethically and legally acceptable, at least to American audiences. Sunstein offers two main clarifications (if not revisions) to his proposal. First, Sunstein now states that (ordinarily) "officials should not conceal their identity," and suggests more generally (yet somewhat ambiguously) that some ethical restraints ought to be observed, or even taken for granted as "obvious." Second, Sunstein now maintains that the proposed infiltrations were intended only for groups of *foreigners*. Sunstein implies that this was his intention all along, and that critiques that suggested otherwise were products of biased readers. Sunstein writes, in the preface, "The irony is that the essay itself has been subject to the very mechanisms it explores about the spread of false information" (2014a, p.

xi). Sunstein's defense comes down to this: isn't it a pity, isn't it an "irony," how badly I've been misread!

Below I will consider whether Sunstein's revision is a significant improvement over the previous version (by which I mean the 2009 *Journal of Political Philosophy* version, not the 2008 version available online). At the same time, I will consider whether or not Sunstein's claim that he has been misread is convincing. I first discuss ethical and legal concerns, focusing on the issue of officials concealing their identity and whether the "focus" had been on *foreign* conspiracy theorists. Then I address epistemic concerns, including "biased assimilation."

6.2. Ethical and Legal Concerns

In the revised version of his essay, Sunstein writes, "Outside of unusual circumstances (above all, genuine national security threats), public officials should not conceal their identity" (2014a, p. 30). Unfortunately, Sunstein does not explicitly restrict the "allies" of government agents from concealing that relationship or their mission. Further, it is not clear what constitutes "unusual circumstances," since Sunstein's choice of words avoids restricting it to the "genuine national security threats" that he mentions. So while this seems like movement in the right direction, this assurance is less than fully comforting.

Further, if this is meant to be a meaningful restriction, it seems to represent not just a clarification but a significant change in Sunstein's position—though Sunstein does not acknowledge this. It is true that both versions mention two variants of cognitive infiltration, one in which "government agents would openly proclaim, or at least make no effort to conceal, their institutional affiliations" (2009, p. 225; cf. 2014a, p. 29) and another in which, "government officials would participate anonymously or even with false identities" (2009, p. 225; cf. 2014a, p. 30). And both versions then add, "Each approach has distinct costs and benefits." But the new version then introduces an ethical concern that is lacking in the original. The revised version says, "[T]he second [approach] raises ethical concerns and is riskier, but it might bring higher returns" (2014a, p. 30). The original version had addressed only utilitarian concerns: "[T]he second [approach] risks perverse results but potentially brings higher returns" (2009, p. 225). The original paper also included mention of "desirable" results potentially associated with discovery or disclosure of infiltration:

> [D]isclosure of the government's tactics will sow uncertainty and distrust within conspiratorial groups and among their members; new recruits will be suspect and participants in the group's virtual networks will doubt each other's bona fides. To the extent that these effects raise the costs of organization and communication for, and within, conspiratorial groups, the effects are desirable, not perverse. (2009, p. 225)

Roughly in the place where Sunstein now appeals to ethical concerns to suggest that government officials should only very rarely conceal their identities, Sunstein and Vermeule had concluded simply that "both [honest and dishonest forms of cognitive infiltration] are potentially useful instruments" (p. 226). Then, following a similar consideration of trade-offs between real-world and online infiltrations, they summed up as follows: "Our main suggestion is just that, whatever the tactical details, there would seem to be ample reason for government efforts to introduce some cognitive diversity into the groups that generate conspiracy theories" (2009, p. 226). There is no hint in the original essay of any ethical restriction weighing against the dishonest variation of infiltration. And even in the revised version, in which Sunstein stipulates that "Outside of unusual circumstances . . . public officials should not conceal their identity" (2014a, p. 30), he still seems to be encouraging, under the right conditions, government officials to "participate anonymously or even with false identities" (2014a, p. 30). If Sunstein is not encouraging this, then why introduce it at all? He puts the idea on the table and then seems to say, ambiguously, "*Ordinarily* it shouldn't be done, but . . ." So while the original paper seemed clearly to encourage the deception in question, the revision provides mixed messages involving ambiguously qualified assurances. I'm not sure that is a substantial improvement.

Perhaps more significantly, the 2014 version suggests that Sunstein had intended the proposed cognitive infiltrations to apply only to *foreign* conspiracy theorists. Sunstein writes, "The focus of the discussion is on situations involving serious security risks, above all risks of terrorism, that arise from conspiracy theories in foreign countries" (2014a, p. 29).[5] But even in the revision, it is not until the twenty-ninth page of the chapter that we are alerted to this supposed "focus" (while there is no such signal at all in the original). Sunstein does mention the purported focus in the preface to his book, where he also writes:

> If I were writing the essays [in this book] from scratch, they would not, of course, turn out the same. With respect to conspiracy theories, for example, I would now work a lot harder to clarify the prescription, so as to avoid the *false impression that the idea of cognitive infiltration is meant to support efforts to spy on Americans who distrust their government*. Civil liberties are a foundation of free societies, and while the essay was meant to take that point as an *obvious background fact*, it should have put the point firmly in the foreground. The focus of the essay is on serious terrorist threats abroad. (Sunstein 2014a, p. xvi, emphasis added)[6]

Although not writing from scratch, Sunstein still could have done better to *focus* on the foreign threat in his revised version, rather than waiting until deep into the chapter to mention that that was his "focus." Instead, he continues to begin the chapter with the results of a poll of New York City residents, followed by a more general poll of Americans, regarding their belief about 9/11 conspiracy theories. That seems to suggest that American acceptance of 9/11 conspiracy theories are in focus.

Then, on the next page, Sunstein writes,

> The main (though far from exclusive) focus involves conspiracy theories relating to (and helping to inspire) terrorism, including theories that are connected with or postdate the 9/11 attacks. These theories exist within the United States and, even more virulently, in foreign countries, especially Muslim nations. The existence of both domestic and foreign conspiracy theories is no trivial matter; they can help give rise to serious risks, including the risks of violence. (2014a, p. 2)

That hardly clarifies that the "focus" is on *foreign* conspiracy theorists. And let's keep in mind that we are not just concerned about *emphasis*. To be assured that his proposal is not to apply domestically, it would have to be made clear that it was *exclusively* intended for foreign threats. However, Sunstein seems, even in his revision, to signal otherwise.

Further, Sunstein adds mention of a 2013 poll of Americans (2014a, p. 2), as well as a number of examples that seem primarily American. Although Sunstein does not specify who was spreading the rumors about FBI involvement in the Boston Marathon bombing, to which he alludes, one can hardly be faulted for assuming that his reference to the "false rumors" that spurred race-related violence "throughout American history" applies to Americans.

Perhaps the reader was supposed to understand that, even though the analysis of the "causes" of conspiracy theories included the spread among Americans, the proposed "cure" was meant to apply only to foreigners. It seems we were supposed to assume, as an "obvious background fact," that "Civil liberties are a foundation of free societies" (p. xvi), and thus that Sunstein and Vermeule could not have intended their proposal to apply domestically. I'm not sure readers can reasonably be expected to assume this, especially given the actual history of interventions—a history that, as I mentioned in the previous chapter, Sunstein and Vermeule whitewash. In general, I think it is reasonable for readers to make judgments about what authors mean by what they actually say. So what hints were readers given?

First, Sunstein's list of possible "government responses" to conspiracy theories seem, at least partly, if not mostly, to involve *domestic* strategies. For example, "ban[ning] conspiracy theories" (response 1) in foreign countries is not feasible, nor is taxing foreign conspiracy theorists (response 2). Admittedly, Sunstein does not go as far as endorsing either of these strategies. Indeed, in his revised version, he writes, "Under current free-speech law, of course, responses 1 and 2 are likely to be unconstitutional, and in a free society, censorship of any kind should be used, if at all, only under the most extraordinarily unusual conditions (for example, to prevent imminent violence)" (2014a, pp. 22–23). Yes, *of course*. And yet, in the original version, referring to their list of possible governmental responses (which include these two), Sunstein and Vermeule write, "Each instrument has a distinctive set of potential effects, or costs and benefits, and each will have a place under imaginable conditions" (Sunstein and Vermeule 2009, p. 218). That seems to leave the door fairly wide open to what Sunstein now acknowledges is "likely to be unconstitutional."

In any case, the mere discussion of these possible responses, since they seem at least partially, if not mostly, domestic, sets the reader up to think that any alternative proposal that Sunstein will endorse will also be a domestic one, or at least applicable domestically. And this impression is only strengthened when Sunstein warns about the dangers of conspiracy theories, pointing out that the Oklahoma City bombing was motivated by conspiratorial beliefs, as he does in both versions (2014a, p. 24; 2009, p. 220). In the revised version he concludes that section writing, "Within a network of members who believe that the federal government is a hostile and morally repellent organization that is taking over the country, akin to a foreign invader, armed resistance might seem a sensible course, at least to some" (2014a, p. 25). From this it is clear that Sunstein is suggesting people in the United States (for that is undoubtedly "the country" referred to)[7] should be worried about the threat of *domestic* terrorism. Surely his readers are thereby prompted to expect that his "cure" will be related to *that* threat. Further, regarding the proposed infiltration specifically, Sunstein compares his proposal with "1960s-style infiltration" (2014a, p. 29; cf. 2009, p. 224), which involved *domestic* infiltrations. Indeed, it is worth remembering what was involved in the COINTELPRO program, which both utilized infiltrations and spanned the 1960s. The Church Committee reports:

> [The FBI] conducted a sophisticated vigilante operation *aimed squarely at preventing the exercise of First Amendment rights of speech and association*, on the theory that preventing the growth of dangerous groups and the propagation of dangerous ideas would protect the national security and deter violence. Many of the techniques used would be intolerable in a democratic society even if all of the targets had been involved in violent activity, but COINTELPRO went far beyond that. The unexpressed major premise of the programs was that a law enforcement agency has the duty to do whatever is necessary to combat perceived threats to the existing social and political order. (Church Committee 1976, section I, emphasis added)

The parallels with Sunstein's justifications are apparent: "preventing the growth of dangerous groups and the propagation of dangerous ideas would protect the national security and deter violence." Similarly, Sunstein's proposal involves "weakening or even breaking up the ideological and epistemological complexes that constitute these networks and groups" (2014a, p. 29). Were Sunstein's readers not likewise expected to conclude that, public officials had "the duty to do whatever is necessary to combat perceived threats to the existing social and political order" even if it meant "preventing the exercise of First Amendment rights of speech and association, on the theory that preventing the growth of dangerous groups and the propagation of dangerous ideas would protect the national security and deter violence"? Were readers really supposed to assume that his proposal, based on this same reasoning and involving similar mechanisms, was supposed to be held in check by democratic ideals that did not suffice in the past? We should also bear in mind the distinction between the contours of *the proposal itself*, and *what was likely to actually occur* were his suggested infiltrations implemented (as emphasized in Chapter 4).

Given all this, what should we say about the "false impression that the idea of cognitive infiltration is meant to support efforts to spy on Americans who distrust their government" (2014a, p. xvi)? If this is a genuine misunderstanding, the fault seems to lie squarely with the author. And further, the addition of a couple sentences of assurance in the revision has not fully corrected the problem.

6.3. Epistemic Concerns

Though Sunstein's revised proposal may be less objectionable ethically and legally than the original seemed to be, I have been concerned at least as much with *epistemic* issues as with legal and moral ones. In particular, I find the quality of Sunstein and Vermeule's analysis to be troubling. It seems that Sunstein's revisions have resulted in no improvements on that score, and some slippage—such as the reintroduction of reference to "video frames showing Flight 77 approaching the building" (2014a, p. 27), and the attempt to associate acknowledged conspiracies with the idea that space aliens landed in Roswell (2014a, p. 4).[8] And the mixed messages discussed above do little to restore confidence.

In addition to all the problems described in the previous chapters, let's consider a couple more reasons to worry about Sunstein's reliability. First, Sunstein updates his essay by making note of what is supposedly a new finding about conspiracy theorists. Alluding to a 2012 study (namely, Wood et al. 2012), Sunstein writes, "[P]eople who tend to believe one [conspiracy theory] are likely to believe another, even if the two are in direct contradiction" (2014a, p. 31). More specifically, he writes:

> Some people do show an unusually strong inclination to accept conspiracy theories. Peculiar as it is, the willingness to accept contradictory conspiracy theories is especially suggestive of this inclination, as reflected in the remarkable finding that those people who believe that Osama bin Laden was already dead at the time when US special forces raided his compound in Pakistan are also more likely to believe he is still alive. (2014a, p. 11)

We've seen essentially this same claim before (section 3.1.5, above), and we'll see it again. However, as will be shown in the following chapter, this claim is based on a rather egregious mistake on the part of the authors of the study—conflating believing something with merely giving it some degree of credence. This rather obvious mistake is overlooked by a number of commentators, including Sunstein, who has repeated his remarks in various venues.

Second, Sunstein's treatment of biased assimilation is now augmented with reference to a study that tends to undermine rather than support Sunstein's contention about conspiracy theorists. In both versions the following claim is made: "Those with strong commitments often engage in 'biased assimilation' of evidence, and conspiracy theorists are likely to be especially biased assimilators" (2014a, p. 28; cf. 2009, p. 224). However, Sunstein provides evidence only for the first part of his claim, "Those with strong commitments often engage in 'biased assimilation' of evidence." This statement, on the face of it, would seem to apply equally to conspiracy theorists and staunch conventionalists. However, ignoring the implication for staunch conventionalists, Sunstein suggests that "conspiracy theorists are likely to be especially biased assimilators." In the revised version this is followed by the following remark: "In the context of the assassination of President Kennedy, for example, biased assimilation has been explicitly observed with balanced information leading believers to be more, not less, committed to their conspiracy theory" (2014a, p. 28, citing McHoskey 1995). That is true; but it is less than half of the truth. Does the study cited actually support Sunstein's claim that "conspiracy theorists are likely to be *especially* biased assimilators"? No. It only supports the conclusion that conspiracy theorists are biased assimilators *like everybody else*. The article cited explicitly maintains that biased assimilation applies to people on *both* sides of conspiracy theory debates. In fact, the article states, "In general, subjects with no prior attitude felt that the proconspiracy evidence was superior, and the majority expressed advocacy of a conspiracy theory at the conclusion of the study" (McHoskey 1995, p. 407). Might that not suggest that the proconspiracy evidence really *is* better, since those who were apparently most neutral thought so? If so, those who made the opposite judgment, apparently due to prior inclinations in that direction, might plausibly be regarded as exhibiting *stronger* assimilation bias. And thus this paper may seem to suggest that it is *staunch conventionalists*, rather than conspiracy theorists, who are "likely to be especially biased assimilators." But I don't want to make *too* much of this, for it is difficult to objectively assess the quality of the evidence given on both sides. The point is that Sunstein has no qualms about making his claim despite reasons offered in the cited article that should have induced caution.[9] The inappropriateness of *that* is what I want to highlight, for Sunstein's one-sided depiction of studies like this suggest a strong tendency toward biased assimilation on his part.

Now, there is little doubt that I am a biased assimilator too, as it seems a ubiquitous psychological tendency. But we should be mindful of what this ubiquity suggests about most "epistemic authorities." If they regard themselves as part of the epistemic mainstream, they will likely have a tendency to fall in line with the reigning paradigm and to support the "consensus" with which they identify, the "official story." Knowing about this ubiquitous psychological tendency gives us a general reason to doubt the objectivity of *everyone* involved in debates over conspiracy theories. But it is the defenders of the official story that make the argument that we should trust epistemic authorities, *as if they were not systematically biased*. Yet, unwittingly, Sunstein has directed us to reasons to think that this is wrong, and to believe instead that "epistemic authorities" with official status will be biased in favor of official stories.

We may now wonder: How *significant* is this bias? That can't be answered a priori. We have to look critically at epistemic authorities and see just how skewed their analysis can be. Examining Sunstein and Vermeule's deeply flawed and misleading, one-sided treatment of conspiracy theories has given us a start. And I hope it has raised some concern. For this problem may be worse than we would have expected.

6.4. Conclusion

Let's sum up. Despite the superficial appearance of sober scholarship, Cass Sunstein and Adrian Vermeule's academic writings on conspiracy theories turn out to be deeply flawed. They offer implausible explanations for the spread of conspiracy theories; they misrepresent scientific studies; they abuse history; they cite misleading evidence; and they rely significantly on dubious bald assertions. They also offer a proposal (cognitive infiltration) that is in tension with their assumptions (that the US government doesn't often conspire against its citizens) and with their espoused values (those associated with open societies), not to mention basic rights (such as freedom of speech and assembly). And when challenged, Sunstein responds with ambiguous assurances and mixed messages. How could these scholars do such a shoddy job? I don't know. But I suspect it has something to do with the same "biased assimilation" that they attribute to conspiracy theorists (in an unfairly one-sided way).

But I'm not really concerned about *why* they got so much so wrong. And I'm not really concerned that *they* got so much so wrong. I'm concerned that *people like them could* get so much so wrong. While the other scholars that I criticize in the following chapters, mostly social scientists, are not nearly as famous as Sunstein or even Vermeule, they are nevertheless scholars who publish peer-reviewed articles, and they also get a lot wrong when it comes to the study of conspiracy theories and conspiracy theorists. It raises the question, Are scholars in general unreliable on this issue? If so, are they unreliable on other (perhaps related or somehow analogous) issues? How does this potential unreliability of scholars, at least when it comes to certain kinds of issues, impact how we ought to think about conspiracy theories? Put another way, how does it impact our judgment of the prior probability of official accounts of controversial events? It should, it seems, tend to lower it, and thus raise the prior probability of conspiracy theories and heighten the uncertainty that comes with accepting that much of what ostensible epistemic authorities tell us we "know" may be wrong.

Some people may find this conclusion scary. They may point out that undermining "epistemic authorities" will have negative consequences. But this merely points out a practical inconvenience. It does nothing to show that the conclusion is wrong. It may be that our epistemic circumstances are simply more precarious than we would like.

Let's find out.

PART III

The Social Scientists

Part III addresses some recent work in the social sciences, particularly psychology and political science, on "conspiracy theories and the people who believe them," which is the title of a recent collection of essays on that topic, edited by political scientist Joseph Uscinski (2019). That collection contains a number of observations that are relevant here. For example, historian Andrew McKenzie-McHarg notes:

> [T]he concept of the conspiracy theory emerged in the social sciences in the 20th-century—and carried with it from the very beginning a set of negative connotations simply because the social sciences defined themselves in opposition to the kind of explanation associated with the concept of conspiracy theory. (2019, p. 78)

In other words, the very nature of social science biases it against conspiracy theories. So it shouldn't be too surprising that we find, as religious studies scholar David Robertson and historian Asbjørn Dyrendal observe, that "conspiracy theories are treated [by scholars of conspiracy theories] along the lines of 'cults' and 'extremism'—that is, as social problems" (2019, p. 419), and that, according to Uscinski,

> While most of the authors working on the topic [of conspiracy theories] do not directly refer to the conspiracy theories they study as pathologies, much of the work could be read that way. Many scholars refer to conspiracy theories as "myths," "false beliefs," "misin-

> formation," and "rumors." . . . The current research agenda appears to be most interested in learning how to dissuade individuals of their conspiracy theories. (2018, p. 234)

Or, as cognitive psychologist Preston Bost puts it, "Much of the published [research on conspiracy theories] . . . explicitly treat[s] conspiracy theories as a blight on public discourse, a problem that persistent research might someday solve" (2019, p. 278).

However, the very fact that this is being recognized and pointed out, often with mild or qualified disapproval, suggests that things are beginning to change. Bost continues:

> The dawning realization that conspiracy belief is not a clinical disorder, and that believers are not sharply distinguishable from nonbelievers in their psychological profile, has reined in some of the more strident language. . . . The deeper we look, the more difficult it becomes to treat conspiracy thinking has alien to rationality, or even alien to ourselves. (2019, p. 278)

Indeed, once we admit that some conspiracy theories turn out to be true, the pathologizing of people who believe them becomes problematic. This point is made well by Bost's rhetorical question: "If the Watergate conspiracy theorist[s] merited psychological study before the Oval Office tapes emerged but not after, even as their beliefs stayed constant, what phenomenon are researchers studying exactly?" (2019, p. 272)

Three of the four chapters in this part address problems in the social science research on conspiracy theorists. Specifically, Chapter 7, "Conspiracy Theorists and Monological Belief Systems," demonstrates that some social science research has been faulty and has been used to unfairly denigrate conspiracy theorists. But it also points to signs that some degree of self-correction is beginning to take place. Chapter 8, "Conspiracy Theories and the Paranoid Style," argues that many social scientists, by assuming that conspiracy theorists exemplify "the paranoid style," have prejudicially mischaracterized conspiracy theorists. And Chapter 10, "Conspiracy Theorists and Social Scientists," focuses on one particular group of social scientists who called for more funding for scientific endeavors to "fight conspiracy theories effectively" (see Basham and Dentith 2016, p. 17). When criticized for this, these scientists responded in a way that, I argue, further undermines the credibility of their project. The remaining chapter, Chapter 9, is not focused on the work of social scientists, but rather delves into the details of some of the examples that were alluded to in Chapter 8, as well as other relevant examples that further support a major theme of that chapter, namely, that conspiracy theories often do not posit implausibly malevolent conspiracies.

The social scientists discussed in these chapters frequently fail to make crucial distinctions. Indeed, some of this work is so blind to such distinctions, and so prejudicially misleading as a result, that it is hard to avoid the conclusion that there is a pervasive bias at work against conspiracy theories and conspiracy theorists (just as McKenzie-McHarg's above-noted observation would lead us to expect). For such a bias would tend to lead to distorted and prejudicial conclusions such as those observed. There is irony in this. As Uscinski points out, "That so many dubious claims [by both scholars and journalists] about conspiracy theories are reported is particularly troubling given that much of the fixation on conspiracy theories stems from concerns about people believing dubious things" (2019c, p. 443).

Recognizing that there is a not entirely rational response to conspiracy theories prevalent in the academy, philosophers Juha Räikkä and Lee Basham turn the tables on those who treat conspiracy theories dismissively and pathologize conspiracy theorizing. They do so by coining the phrase "conspiracy theory phobia" to describe irrational impediments to taking conspiracy theories seriously. They explain:

> People who suffer conspiracy [theory] phobia reject conspiracy theories without an appropriate evaluation of the evidence presented, and the reaction toward any particular conspiracy theory tends to be mockery, contempt, hostility, or straw-person characterization of the arguments presented. (Räikkä and Basham 2019, p. 183)

This is worth keeping in mind as one reads the following chapters.

SEVEN

Conspiracy Theorists and Monological Belief Systems

Recent scholarship has claimed to show that conspiracy theorists are prone to simultaneously believe mutually contradictory conspiracy theories, as well as believe entirely made-up conspiracy theories. The authors of those studies suggest that this supports the notion that conspiracy theories operate within "monological belief systems," in which conspiracy theorists find support for conspiratorial beliefs in other conspiratorial beliefs, or in related generalizations, rather than in evidence directly relevant to the conspiracy in question. In this chapter, I argue that all of that is either wrong or at least misleading.

Focusing primarily on three articles (Wood et al. 2012; Swami et al. 2011; and Goertzel 1994), but addressing others as well, I argue that there is a trend in the social science literature on conspiracy theories that is significantly misleading, presumably due to bias against conspiracy theories and conspiracy theorists. When a field of scholarship contains a number of authors (and, apparently, referees and editors as well) who reason in a rather obviously fallacious manner (or fail to notice obvious fallacies) so as to unfairly malign a class of people, and there does not seem to be an analogous situation with opposite implications, it is reasonable to suspect bias may be involved. Of course, this does not by itself conclusively establish that there is bias (a very difficult thing to do in any circumstance), but the more obvious, egregious, widespread, and unidirectional the phenomenon is, the more likely there is bias. Readers can make their own judgments regarding degrees, but I hope to provide good reasons to think that this phenomenon is rather egregious and widespread. (That it is unidirectional I have taken for granted; I await citation of any examples of peer-reviewed published scholarship that unfairly stigmatizes those who accept "conventional wisdom" or official accounts regarding conspiracy theories.)

Specifically, I show the following: (1) The often-cited claim that conspiracy theorists tend to simultaneously believe contradictory conspiracy theories (based on Wood et al. 2012) is unfounded. (2) A study that purports to show that conspiracy theorists are more prone than others to believe entirely fictitious conspiracy theories (Swami et al. 2011) is one-sided and misleading. In addition, the authors make an error about belief that is analogous to the one made by Wood et al. Further, there is nothing unusual or problematic about the reasoning process that presumably underlies the phenomenon they document. (3) Both of the above studies claim to provide evidence that conspiracy theorists tend to operate within a "monological belief system," an idea first put forward by Ted Goertzel (1994). This label, as described by Goertzel, implies that there is something epistemically problematic about the reasoning of conspiracy theorists. However, in all three studies, the evidence produced supports only a connection with its *unproblematic* aspects. There is no documented evidence that the *problematic* aspects actually apply to conspiracy theorists.

And yet these articles, along with similar others, have been used to imply that something epistemically problematic has been discovered about the reasoning of conspiracy theorists in particular. For example, Martin Bruder and his coauthors write:

> There is increasing evidence that there are stable individual differences in people's tendency to believe in [both new and "classic"] conspiracy theories; if a person believes in one conspiracy theory, he or she will also be more likely to believe in other conspiracy theories. In fact, this tendency even extends to beliefs in mutually contradictory conspiracy theories, and to beliefs in fully fictitious conspiracy theories. Thus, those who believe that Princess Diana faked her own death are also more likely to believe that she was murdered; those who believe in "real-world conspiracy theories" (i.e., that John F. Kennedy fell victim to an organized conspiracy) are more likely to believe that there was a conspiracy behind the success of the Red Bull energy drink—a conspiracy theory that was purposely developed for a social psychology study. (Bruder et al. 2013, p. 1)

A large chunk of the above is quoted by Cass Sunstein in the most recent revision of his article "Conspiracy Theories," in *Conspiracy Theories and*

Other Dangerous Ideas. And the claims included in it are repeated, citing largely the same evidence, by a number of scholars as well as journalists. My central thesis is that all of these assertions, both separately, and especially when taken together to insinuate that there is something wrong with conspiracy theorists, are either wrong or at least misleading.

In the above quotation, Bruder et al. make three related claims about conspiracy theorists. Claim 1: "[I]f a person believes in one conspiracy theory, he or she will also be more likely to believe in other conspiracy theories." Here these authors cite Swami et al. 2010 and could have cited others as well. I do not contest the truth of this point but maintain that the probable underlying thought process is normal and reasonable, not distinctive or problematic. Claim 2: "[T]his tendency [to be more likely to believe in other conspiracy theories] even extends to beliefs in mutually contradictory conspiracy theories." Here Wood et al. 2012 is cited. However, as I will argue below, Wood and his colleagues have not provided any evidence for *this*, though they do mistakenly assert that they have. Thus, the claim is unsubstantiated and presumably false as a generalization, being implausible as well as unsupported. Claim 3: "[This tendency even extends] to beliefs in fully fictitious conspiracy theories." This claim is misleading. Bruder seems to imply that there is something epistemically problematic, or at least unusual, going on. But a close examination of the relevant study (Swami et al. 2011) reveals that that is not the case.

Each of the following three sections addresses one of the above claims. I start with Claim 2 in section 7.1, followed by claim 3 in section 7.2. In section 7.3 I address claim 1 in the context of a larger discussion of so-called monological belief systems, as described by Goertzel (1994). Then, in section 7.4, I discuss some critical points made in an article by two coauthors of Wood et al. 2012 (namely, Sutton and Douglas 2014). They too have come to recognize problems with Goertzel's attempt to attribute "monological belief systems" to conspiracy theorists, though they do not acknowledge the problem with their own earlier study (i.e., Wood et al. 2012).

7.1. Conspiracy Theorists Believe Contradictory Conspiracy Theories!

In a recent review of the literature, Jan-Willem van Prooijen and Paul A. M. van Lange write:

> One of the main research findings on this phenomenon [i.e., belief in conspiracy theories] is that conspiracy beliefs are monological in nature: One conspiracy theory reinforces other conspirational ideas,

making individuals who believe in one conspiracy theory more likely to also believe in other conspiracy theories. (2014, p. 237)

This finding, as described here, is neither surprising nor epistemically problematic (for reasons to be elaborated below). Indeed, the idea that this is "the main research finding" regarding belief in conspiracy theories may cause one to wonder, "Why is such a commonsensical and innocuous finding even regarded as interesting?" The answer is that the idea of being "monological" is thought to have other, more problematic, implications. Conspiracy theorists are thought to take this normal and unproblematic reasoning process—using one belief as evidence for another—too far. Indeed, allegedly, they take it to the point of self-contradiction. Continuing, van Prooijen and van Lange write, "A recent study reveals that this monological belief system even applies to conspiracy theories that are mutually exclusive" (van Prooijen and van Lange 2014, p. 237, citing Wood et al. 2012).

Postponing the discussion of "monological belief systems," let us here focus on the claim, based on a 2012 study (Wood et al.), that conspiracy theorists have a tendency to simultaneously believe two mutually inconsistent conspiracy theories. This finding is also cited prominently by other scholars writing on this topic, including Cass Sunstein, Joseph Uscinski and Joseph Parent, Viren Swami, and Christopher French. Uscinski describes Wood's publication as a "watershed moment" in the social science research on conspiracy theories. And Alfred Moore describes it as "One of the most widely publicized papers on conspiracy theory" (Moore 2019, p. 112).[1] And it is a powerful meme for anyone who wants to encourage a dismissive attitude toward conspiracy theories, because it suggests that the fantastical thinking of conspiracy theorists is not even held in check by the most obvious of contradictions. Below are just a few examples of depictions of this finding that reveal the flavor it is given.

Philosopher Keith Harris writes, "Belief in a given conspiracy theory strongly predicts belief in other conspiracy theories, even in cases where the conspiracy theories are incompatible" (2019, p. 240; citing Goertzel 1994 and Wood et al. 2012).

In *American Conspiracy Theories*, political scientists Uscinski and Parent write: "A predisposition toward conspiratorial thinking . . . explains why people believe theories that are logically contradictory (e.g., many of the people believing Osama Bin Laden is still alive also believe he was dead long before the raid on his compound)" (2014, p. 14; cf. Uscinski 2019, p. 20).

Christopher French, professor of psychology and head of the Anomalistic Psychology Research Unit at Goldsmiths College, University of London, writes in *Scientific American*: "[P]eople who are strongly inclined toward conspiratorial thinking will be more likely to endorse mutually contradictory theories. For example, if you believe that Osama bin Laden was killed many years before the American government officially announced his death, you are also more likely to believe that he is still alive" (2015).

In an article titled "Analytic Thinking Reduces Belief in Conspiracy Theories," Swami et al. write: "[A] growing body of evidence suggests that belief in conspiracy theories is associated with . . . belief in contradictory statements" (2014, p. 573, citing Wood et al. 2012).

The science news website, *LiveScience*, under the headline "Contradictions Don't Deter Conspiracy Theorists," also makes the same claim: "Did Princess Diana fake her own death to escape the public eye? Or was she killed by a rogue element of the British secret service? If you agree with one of these theories, there's a good chance you'll subscribe to both even though one suggests Princess Diana is alive, the other dead, a new study indicates. . . . [And] people who believed bin Laden was already dead before the raid were more likely to believe he is still alive" (Parry 2012).

In an op-ed for *Newsday*, Cass Sunstein writes: "Remarkably, people who accept one conspiracy theory tend to accept another conspiracy theory that is logically inconsistent with it. People who believe that Princess Diana faked her own death are more likely to think that she was murdered. People who believe that Osama bin Laden was already dead when U.S. forces invaded his compound are more likely to believe that he is still alive" (2014b).[2]

Sunstein is suggesting that such people are reasoning in a "remarkably" irrational way.[3] Skepticism toward "authorities," Sunstein elsewhere writes, "even lead[s] suspicious individuals to disregard contradictions between them [i.e., between alternative conspiracy theories]" (2015). Sunstein suggests that conspiracy theorists are so "suspicious" that they are driven to "embrace" mutually contradictory positions.

However, there is a serious problem with all this: the research referred to does not actually show the self-contradictory thinking that is alleged.

This misreading of the study is not entirely the fault of Sunstein and these others (although they do share in the responsibility), for the authors of the study themselves suggest the same interpretation of their findings. The very title of the article in question, "Dead and Alive: Beliefs in Contradictory Conspiracy Theories," suggests the above interpretation. And so there can be no misunderstanding, the article concludes by clearly stating that "Believing that Osama bin Laden is still alive is apparently no obstacle to believing that he has been dead for years" (Wood et al. 2012, p. 772).[4]

Yet this is all unfounded. The article did not document individuals who simultaneously believed two mutually exclusive scenarios (though it would not be too surprising to find some isolated occurrences). The authors of the article in question, Michael Wood, Karen Douglas, and Robbie Sutton, describe two studies, which I examine in turn.

Study 1 tells us that there is a correlation between the *degree* of "agreement" with several statements about the death of Princess Diana. What is regarded as interesting in this study is that some positively correlated statements about the death of Princess Diana seem to be contradictory. The authors write, "Either she was killed by a rogue cell of British secret service (#1) or by business rivals of the Fayeds (#4), or she faked her own death (#3)" (Wood et al. 2012, p. 769).[5] These are described as "mutually incompatible" and "unambiguous contradictions." Clearly, the first two (#1 and #4) are not "unambiguous contradictions." British intelligence and the Fayeds' business enemies could have colluded, as many people believe about the Mob and the CIA with regard to the JFK assassination.[6] However, both propositions #1 and #4 really do (nearly enough) contradict the notion that Diana faked her own death (#3). Now, what does a correlation of this kind really mean, anyway? It means that if a subject rated her *level of* "agreement" to one of these statements as relatively high, on a scale of 1 to 7, she was likely to rate the other relatively high too, as compared to those who rated the first relatively low. It is important to notice that *nobody* would have to *believe* any of these statements to produce this result. And it is also worth noting that "endorsement of the faked-own-death theory was extremely low . . . with a mean of only 1.52 on a 7-point scale" (Wood et al. 2012, p. 770). So, in this context, even a rating of "2" is relatively high! Can that be reasonably interpreted as "belief"? More generally, what does it mean to "agree" to a statement at a level of, say 2, 3, 4, or even 5 or 6 on a scale of 1 to 7? It is not clear what it means. But the obviousness of the incongruence of the genuinely incompatible statements suggests that it does *not* amount to belief. It is more plausibly interpreted as something resembling what is more explicitly measured in Study 2, to which we now turn.

In the abstract to their paper, the authors write, "In Study 2 (n = 102), the more participants believed that Osama Bin Laden was already dead when U.S. special forces raided his compound in Pakistan, the more they believed he is still alive" (Wood et al. 2012, p. 767). But that is not quite accurate. Again, the researchers did not measure "belief." What they measured in this study can more accurately be characterized as the *degree of credence* given to mutually contradictory conspiracy theories. In their own words, they found "significant positive correlation between composite endorsement ratings" (Wood et al. 2012, p. 767). If person A found the theory that Osama bin Laden is still alive relatively more "plausible, convincing, worth considering, and coherent" (Wood et al. 2012, p. 771) than person B did, then person A was likely to also find the theory that Osama was already dead at the time of the raid more plausible (etc.) than person B did. And this is actually *perfectly reasonable*; there is no contradiction here. As the researchers quite reasonably suggest, what seems to be at work here is a mediating belief that authorities are untrustworthy. Indeed, that is not *merely* an obvious and plausible idea; it is also supported by their statistical analysis.[7] If a particular individual is less trusting of the government than someone else (regardless of what level of trust is warranted), he or she is more likely to give greater credence to alternative accounts of contested events. There is simply nothing epistemically dubious about, say, rating both the notion that Osama was already dead *and* that he is still alive as more "plausible, convincing, worth considering and coherent" than someone else with more faith in official stories rated both theories.

Philosophers Juha Räikkä and Lee Basham put the same objection this way, "An examination of the paper [i.e., Wood et al. 2012] reveals that its conclusion is based on a conflation of *suspicion* with *settled belief*. Suspicions do not generate contradictions" (Räikkä and Basham 2019, p. 186 n16; citing the original version of this chapter, Hagen 2018a). Basham had earlier provided the following illustration:

> Imagine you have misplaced your key ring. You suspect you left it in the front door lock. You also suspect you left it in the kitchen. Given your previous behavior, you rate as quite probable, "agree" that it is in the front door and equally as probable, "agree," the keys are in the kitchen. This is an entirely rational cognitive practice. But according to the interpretation of Wood et al., you *believe* your keys are located, at the very same moment, in both your front door and in your kitchen. For those with lost-key beliefs, believing one has left

> the keys in the front door is apparently no obstacle to believing the keys are simultaneously in the kitchen. (Basham 2017, pp. 64–65)[8]

The authors of "Dead and Alive" should have noticed this,[9] since they did notice an analogous issue regarding "interestingness," which they decided to exclude in their analysis because "there is no contradiction in finding two rival theories equally interesting" (Wood et al. 2012, p. 771). Yet they failed to recognize that there is also *no contradiction in finding two rival theories equally plausible, convincing, worth considering, and coherent.*

Elsewhere, Robbie Sutton and Karen Douglas (2014), the other two authors of "Dead and Alive," consider several possible theories that could potentially explain how contradictory beliefs could be "held together." They write, for example, "Beliefs may not support each other, but instead may be held together by believers' perception of their own moral tendencies" (Sutton and Douglas 2014, p. 262). Alternatively, along with other views,[10] they consider that "incompatible conspiracy beliefs may be held concurrently because they are explained by their coherence with 'nuclear ideas' that pull ideas together to form an ideological system" (Sutton and Douglas 2014, p. 259). The problem is that, because there are no simultaneously believed contradictory beliefs documented, *there is nothing here that needs explaining*—nothing, at least, that is peculiar to conspiracy theorists. So while the general question of how beliefs and attitudes hold together is an interesting one, the premise that conspiracy theorists in particular tend to hold *incompatible* beliefs, which would make them especially in need of explanation, is unfounded.

As far as one can tell from Wood et al. 2012, "suspicious" people reason normally and properly. And so the fact that this paper has been used to disparage and even ridicule conspiracy theorists is unfortunate and inappropriate. And the authors are partially responsible for this. But what is significant here is not that an isolated article happens to have been flawed. It is that this flaw, which should not have been difficult to recognize, not only escaped the *notice* of multiple authors and passed peer review, but appears to have been nearly universally accepted in the psychological research community and widely repeated in both academic and wider venues. Further, as we will see, this is *not* an isolated error. There is an almost systemic distortion in the psychological literature, and to some degree the social science literature more generally, regarding which the problems with "Dead and Alive" are symptomatic. (To some extent this may be in the process of self-correction. Surprisingly, it is the authors of "Dead and Alive" themselves who have begun this process, as we will see.)

7.2. Conspiracy Theorists Even Believe Conspiracy Theories That Are Completely Made Up!

In a study led by Viren Swami (Swami et al. 2011), subjects were asked to rate "the extent to which they agreed" with various statements about a fictitious conspiracy theory. Let us pause and think for a moment. To what extent *should* someone agree with a statement they no nothing directly about, have never heard of, and so could not have had *any* opinion about prior to being asked? It seems they should have responded, "I don't know. I have no opinion." (For all they know about it, which is nothing, the theory could be true.) But "I don't know" was not an option. They had to pick a number between 1 (completely false) and 9 (completely true).[11] The sensible thing to do, it seems, is to answer according to how *likely to be true* they judged the statement to be, or how *plausible* it seemed to them. And if they do this, the subjects can be expected to make their judgments based on their views regarding analogous cases about which they do already have opinions, presumably based on *something*. And so, it seems, they did. Based on this, Swami concludes, as if it tells us something interesting (and unflattering) about conspiracy theorists, "[B]elieving in real-world conspiracy theories appears to make it more likely that an individual will also be more accepting of fictitious conspiracy theories" (Swami et al. 2011, p. 460).

Yes, indeed. *So it should.* The subject does not know that the theory is fictitious, and presumably makes her assessment of the probability/plausibility of the fictitious theory based on the only evidence available to her—her other beliefs. These beliefs, presumably, are in turn supported by other beliefs, which may or may not ultimately have reasonably solid grounding. A person who has beliefs that suggest that "such things" do occur would normally rank the fictitious story as more plausible than people who have beliefs that suggests "such things" do not occur or occur only very rarely. And so, of course, those who believe in real-world conspiracies are "more accepting of fictitious conspiracy theories," that is, they find them more plausible or more likely to be true than do those who do not believe in real-world conspiracy theories.

Here is the trick. The experimenters pick fictitious conspiracy theories, making the conspiracy believer end up looking a little silly for giving some degree of credence to a theory that is made up from nothing. In comparison, those who tend not to believe conspiracy theories (who I'll refer to as "conventionalists")[12] seem like the better thinkers.[13] But that is an illusion. The conspiracy theorist and the conventionalist, as far as we can tell, are both reasoning the same. If the researchers had picked a true but little-known conspiracy theory as the test, the conspiracy theorist would have come out looking better. Indeed, that is not just speculation; such an experiment has now been done. As described by Rob Brotherton (of all people):[14]

> A recent study . . . found that people who reject speculative conspiracy theories (such as "evidence of alien contact is being concealed from the public") are also more likely to reject documented conspiracies (such as "How likely is the idea that the government has performed mind-control experiments on its own citizens without their consent?"—a coy reference to the CIA's very real MKUltra program). (Brotherton 2016)

Brotherton is referring to a study led by none other than Michael Wood (namely, Wood 2016), which is, to Wood's credit, more balanced than his "Dead and Alive" paper, discussed above. Here Wood offers a couple plausible explanations.

> [P]eople who are aware of past malfeasance by powerful actors in society might extrapolate from known abuses of power to more speculative ones. Alternatively, people with more conspiracist world views might be more likely to seek out information on criminal acts carried out by officials in the past, while those with less conspiracist world views might ignore or reject such information. (Wood 2016, p. 698)[15]

The first "alternative" is perfectly reasonable, for awareness of known abuses of power *should* incline a reasonable person to "rate the likelihood" of speculative cases higher than she otherwise would have. The second "alternative" points to confirmation bias at work in seeking out the information. Nevertheless, evidence is still evidence, even if one finds it as a result of bias. And confirmation bias cuts both ways.

It is important to keep in mind that neither conspiracy theorists nor the conventionalists can make an a priori claim that their inference was the better one. Which group is more likely to be right in real cases depends upon how common the conspiratorial behaviors of the types in question *really* are. And, to have a sense of that, one has to do an empirical study, not of conspiracy theorists, but of the history of conspiracies, both of the officially acknowledged and of the controversial varieties.

Another point that needs to be emphasized is that neither side is *reasoning* badly, given their beliefs about conspiracy theories in general. It is entirely reasonable for those who believe in conspiracy theories to be more inclined to believe theories that are relevantly similar to other theories that they think they have reason to believe, especially if those theories imply that authorities cannot be trusted. And, likewise, it is entirely reasonable for people who do not believe in conspiracy theories in general to be less inclined to believe in any particular one. This is because the (perceived) prior probability of a given type of explanation influences, and *ought* to influence, one's assessment of the plausibility of a particular hypothesis of that type. The interesting question in this context is not whether one's perception of the prior probability involved *influences* one's judgment about the plausibility of a specific case. (In general, it clearly does and should.) Rather, the interesting question is whether the degree of prior probability is accurately perceived in the first place.[16] In other words, if we want to make a judgment about who has the most appropriate perspective on the issue, we would have to study whether or not belief in conspiracy theories *in general* is warranted. And to do this, we would have to study the relevant history and the empirical evidence surrounding controversial cases. And this is something that these social scientists do not typically want to bother with very much. Just as "people with more conspiracist world views" might be inclined to "seek out information on criminal acts carried out by officials in the past," as Wood has suggested, people with conventionalist worldviews may be inclined not to. Both are forms of confirmation bias. As for academics who shy away from such studies, one can hardly blame them. After all, the evidence supporting conspiracy theories involving the most paradigmatic conspiracy theories, the JFK assassination and 9/11, is mountainous.[17]

Along with the alleged tendency to believe contradictory theories, the supposed inclination to be "more accepting" of entirely fictitious conspiracy theories is cited as evidence that conspiracy theorists operate within a "monological belief system." For example, Swami writes: "Evidence of an association between belief in this entirely fictitious conspiracy theory and real-world conspiracy theories would provide strong evidence for a monological belief system in relation to conspiracist ideation" (Swami et al. 2011, p. 453, citing Goertzel 1994).[18]

It is this notion of "monological belief systems" to which we now turn.

7.3. Conspiracy Theorists Are Monological Thinkers—They Talk Only to Themselves!

The idea that so-called conspiracy ideation is indicative of a "monological belief system" was first suggested by political scientist Ted Goertzel (1994) and now enjoys experimental support (supposedly) from the work of various psychologists, including Michael Wood and Viren Swami. Swami regards Goertzel's claim that, in Swami's words, "conspiracy beliefs form part of a 'monological belief system'" as "[p]erhaps one of the most important conclusions" of studies addressing influences on belief in conspiracy theories (Swami and Coles 2010, p. 562).[19] Elsewhere, Swami et al. elaborate:

> In a seminal study, Goertzel (1994) argued that conspiracy beliefs form part of a "monological belief system" in which a conspiratorial idea serves as evidence for other conspiracist ideation. Thus, for example, recent work [Swami et al. 2010] has shown that respondents who more strongly endorsed conspiracy theories about the September 11, 2001 (9/11) terrorist attacks were more likely to believe in other, unrelated conspiracy theories. (2011, pp. 444–445)

They also tie in the results of their 2011 study:

> [W]e showed that the strongest predictor of belief in 7/7 [the London bombings] conspiracy theories was belief in other, general conspiracy theories. This is consistent with the suggestion that conspiracist ideation forms part of a monological belief system (Goertzel, 1994), where one conspiratorial idea serves as evidence for other conspiracist ideation. In the case of the present study, it might be suggested that belief in a range of conspiracy theories provides a basis for comprehending and accepting 7/7 conspiracy theories. (Swami et al. 2011, p. 452)

There is something strangely asymmetrical about this study. It purports to tell us something about "conspiracist ideation," presumably as opposed to nonconspiracist, or "conventional" ideation. Namely, "conspiracy beliefs" in particular "form part of a 'monological belief system' in which a conspiratorial idea serves as evidence for other conspiracist ideation" (Swami et al. 2011, p. 445). But is there really anything here that applies uniquely to conspiracy ideation or that, in any case, is epistemically problematic or noteworthy? I do not think so. What Swami et al. describe is a characteristic of all sensible people, including believers in conventional interpretations of events. What follows is Swami's summary of his results modified by simply crossing out and replacing individual words so that it applies to conventionalist ideation instead of conspiracist ideation. It starts with a speculative claim that seems to apply equally in both cases. (Note that the first paragraph just sets the context by stating the findings of other studies. It is interesting here because it invites the reader to view conspiracy ideation as associated with mostly negatively framed qualities—the exception being support for democratic principles, which the authors have camouflaged under the acronym "SDP." I have adjusted these qualities to neutralize, or even reverse, the emotive feel while trying not to express more bias than the original.)

> ~~[C]onspiracist~~ [Conventionalist] ideation may initially begin as an individual process, in which a person tries to make sense of some event perceived as threatening or calamitous. In such a scenario, a tendency towards ~~conspiracist~~ [conventionalist] ideation may tend to be more prevalent among individuals who are politically ~~cynical~~ [trustful/naive], show ~~stronger~~ [weaker] SDP [support for democratic principles], have ~~lower self-esteem~~ [higher self-admiration], are more ~~disagreeable~~ [agreeable/acquiescent], and possibly have ~~lower~~ [higher] crystallised intelligence. Systemic factors, such as discrepancies or ambiguities in mainstream explanations for an event, may ~~also~~ play a role in initially shaping *conspiracist* ideation.[20] [On the other hand, mainstream criticisms of conspiracy theories, and constant repetition of conventional accounts, perhaps even demeaning insinuations regarding "conspiracy theories" emanating from the academy, may play a role in initially shaping *conventionalist* ideation.]
>
> Once this process has been initiated, a confirmation bias and avoidance of cognitive dissonance may further the drive towards ~~conspiracist~~ [conventionalist] ideation.[21] However, our results also suggest that the strongest predictor of whether or not an individual will ultimately accept a ~~conspiracy~~ [conventionalist] theory is the presence of earlier ~~conspiracist~~ [conventionalist] ideation.[22] This is entirely consistent with Goertzel's (1994) suggestion that ~~conspiracy~~ [conventionalist] theories form part of a monological belief system, where ~~conspiracist~~ [conventionalist] ideation increases the chances that an individual will accept evidence of ~~novel conspiracy theories~~ [conventional explanations]. Such a system may allow individuals to easily comprehend new phenomena within existing belief systems, but communal reinforcement may also play a role in embedding ~~conspiracy theories~~ [conventional accounts] within particular social groups. (Swami et al. 2011, pp. 459–460, amended)

Now, what exactly is a monological belief system? So far I have only mentioned, quoting Swami, that it involves "a conspiratorial idea serv[ing] as evidence for other conspiracist ideation." Let us look more closely. Ted Goertzel, the originator of the idea that conspiracy theorists are monological thinkers, puts this in more general terms, "In a monological belief system, each of the beliefs serves as evidence for each of the other beliefs" (Goertzel 1994, p. 740). Goertzel's study provides evidence that this manner of reasoning applies to conspiracy theorists, a finding that is later supported by other studies (Wood et al. 2012; Swami et al. 2011). However, there is something strange about pointing out this fact. For the reasoning it describes is both ubiquitous and epistemically unproblematic. *Of course* one belief serves as evidence for another, and so it *should*. There is hardly an alternative.

If one is convinced that al-Qaeda was behind the embassy bombings in Africa, as well as the attack on the USS *Cole*, should that be taken as evidence, as is often suggested, that al-Qaeda was behind the September 11 attacks? The answer, clearly, is yes. It *is* evidence (though, of course, not proof). And because it is evidence, a rational person should, based on this consideration, increase her assessment of the likelihood that that event was carried out by al-Qaeda. In general, the degree of credence one gives to one event is, *and should be*, influenced by one's beliefs about other events.

In this case, of course, the September 11 attacks and the embassy bombings are thought to be *related*, both perpetrated by al-Qaeda. What is supposed to be interesting about the reasoning of conspiracy theorists is that they make inferences about particular conspiracy theories from other *unrelated* conspiracy theories. As Sutton and Douglas explain,

> [T]he most consistent finding on the psychology of conspiracy theories, often used as an explanation for their popularity and persistence, is that belief in one particular conspiracy theory is predicted by belief in other theories—even when they refer to completely unrelated events and protagonists. (2014, p. 255; citing Douglas and Sutton, 2011; Goertzel 1994; and Swami et al. 2010, 2011)

For example, Swami et al. cite evidence that people who "more strongly endorsed conspiracy theories about the September 11, 2001 (9/11) terrorist attacks were more likely to believe in other, unrelated conspiracy theories" (2011, p. 445, citing Swami et al. 2010).[23] But *are* the conspiracy theories in question really unrelated in the relevant sense? That is, are they *epistemically* unrelated? That is what matters. After all, of what significance is being unrelated in some other sense? So long as belief in one theory gives one a good reason to give increased credence to another, there is nothing distinctive or problematic about doing so. Wood, Douglas, and Sutton themselves provide reasonable considerations. In "Dead and Alive," they mention several plausible mediating beliefs, including "belief in deception by authority" and "belief in the effectiveness of intimidation and bribery" (Wood et al. 2012, p. 772).[24] A person who believes that authorities are more deceptive than does someone else, perhaps based on their knowledge of a particular case, has reason to regard conspiracy theories as more plausible than does the other person. (Which person has more accurately assessed the degrees in question is another matter entirely.) This need not be a complete explanation, but it is enough to conclude that there is no mystery here. To be clear, I do not have a strong objection to social scientists proving the obvious. What I do find objectionable is framing it in a way that suggests that there is something distinctive going on that reflects poorly on conspiracy theorists when, in fact, there is not.

A related notion, also supported by empirical studies, is this: "The more conspiracies a monological thinker believes in, the more likely he or she is to believe in any new conspiracy theory that is proposed" (Goertzel 1994, p. 740). As Sutton and Douglas put it, "[P]eople who already believe in existing conspiracy theories will tend to be receptive to new conspiracy theories, exactly as Goertzel (1994) suggested."[25] This effect may seem epistemically problematic, as Goertzel's way of putting it seems to imply. But that is a mistake. Nothing has gone wrong here.

As an analogy, the more a person believes that scientific discoveries have solved social problems (whether or not they have good reasons for believing this) the more likely he or she is to believe that a scientific discovery will solve other problems. It is not just "monological" thinkers who think this way. Everyone does. And everyone *should*. There is nothing wrong with *the inference*, so long as it is not conceived of as proof, but only a matter of increasing likelihood. If one's reasons for one's more primary beliefs, on which other beliefs are based, are not good ones, then that *is* a problem. (The problem is that they are reasoning based on shaky premises, not that the inference itself is problematic.) But the fact that a person is more likely to accept a new conspiracy theory because they already accept other conspiracy theories in no way implies that the more primary beliefs were not well grounded empirically.

Goertzel himself provides the following example: "African-Americans, who are more likely to be aware of the Tuskegee syphilis conspiracy, are predisposed to believe that AIDS may also be a conspiracy, while this idea may seem absurd to people who are unfamiliar with past medical abuses"

(Goertzel 1994, p. 740). This much is presumably true: people who are aware of the Tuskegee experiment and are motivated to consider it,[26] would likely be more disposed, in some degree, to have unconventional beliefs about AIDS than those who are unaware of such historical outrages. And so they *should* be. It is not that they should *believe* that AIDS is a conspiracy of some sort. But they should be more inclined to think that it might be than they otherwise would have been.

Still, there is something about monological belief systems that *is* epistemically problematic. Goertzel describes it this way, "[B]elief systems can be characterized as *dialogical* or *monological*. Dialogical belief systems engage in a dialogue with their context, while monological systems speak only to themselves, ignoring their context in all but the shallowest respects" (Goertzel 1994, p. 740). Sutton and Douglas characterize this as a "key feature" of monological belief systems, writing, "[F]or Goertzel (1994), a key feature of monological belief systems is what we might call a *closed epistemology*. Beliefs are evaluated according to their coherence with other beliefs in the system, rather than external data" (Sutton and Douglas 2014, p. 256). Clearly, this "closed epistemology," in which context is ignored, is epistemically problematic. Sutton and Douglas see the problem as involving two features. They explain:

> Goertzel (1994) identifies two key features of the monological conspiracist mind-set: [1] a closed epistemology in which adherents are indifferent to evidence, preferring to rely on their previous beliefs, and [2] a nomothetic explanatory style in which each event is explained in terms of general patterns in the world rather than the unique, proximal conditions that might have brought it about. (Sutton and Douglas 2014, p. 257)

These two ideas are clearly related. Conspiracy theorists are charged with disregarding evidence, or, in other words, disregarding the relevant "unique, proximal conditions," and instead reasoning primarily by analogy to beliefs they have about other cases, and thus tend to focus on "general patterns." However, as Sutton and Douglas recognize, there is *no evidence* for this—indeed, it seems to be false (see section 7.4 below).

Goertzel has, in effect, equivocated between two ideas. (1) People who believe in one conspiracy theory are more likely to believe (or give more credence to) another—because one belief serves as evidence for another. And (2) conspiracy theorists ignore empirical evidence—and other people too. They "speak only to themselves" (Goertzel 1994, p. 740). They are closed-minded. The first claim is supported by scientific research but is not epistemically problematic. The second claim is epistemically problematic but is unsupported. Indeed, there is some evidence that the opposite is true—conspiracy theorists tend to be more open-minded (Swami et al. 2010, 2013). Goertzel has tried to pin the problematic aspect of his pejorative-sounding "monological belief systems" on conspiracy theorists, while only actually providing evidence regarding the unproblematic aspects. The further evidence that later accumulated likewise only supported the unproblematic aspects.

In sum, the charge (and it does have a ring of indictment)[27] that conspiracy theorists operate within a "monological belief system" suggests that (1) they attempt to maintain an internal coherence among their beliefs, such that hypotheses that cohere well with settled beliefs are regarded as relatively more plausible, for which studies like Swami et al. 2011 provide support, but which is in no way unusual or problematic. It also suggests that (2) conspiracy theorists live in a particularly "closed-off" world.[28] We can divide the latter into two claims: (2a) they do not engage with contrary viewpoints, and (2b) they make little effort to ground their beliefs in empirical evidence. However, the evidence provided by various studies purporting to link conspiracist ideation with monological belief systems actually only supports the former aspect of monological belief systems, the normal and unproblematic aspect. That is, conspiracy theorists form beliefs based in part on their other beliefs, as we all inevitably do. However, neither of the latter aspects (2a and 2b) has been established. Indeed, 2a is false, at least as it applies to a large segment of conspiracy theorists, who clearly do engage with conventionalists, and often seek out opportunities for debate. The same is true of 2b. Indeed, it is almost ironic. Many conspiracy theorists are, as Sunstein admits, "spectacularly well-informed" (2015) about the pertinent empirical details,[29] and they are constantly calling for further empirical studies and demanding the release of relevant records.[30]

7.4. Sutton and Douglas to the Rescue—Sort Of

Sutton and Douglas do make a number of valid points that militate against making any negative inferences about the merit of conspiracy theories based on research linking conspiracy theorists with monological belief systems. They point out that "there is no empirical evidence for key tenets of the monological position" (Sutton and Douglas 2014, p. 259). More specifically, they write:

> [1] [T]here is no evidence that these people explain events in more abstract, general, less case-specific terms. Indeed, even a cursory examination of websites devoted to conspiracies, such as that of the 9/11 Truth Movement, appears to reveal a deep rhetorical[31] attachment to case-specific facts, true or not, regarding the melting point of steel, the burning temperature of aviation fuel, the way the buildings collapsed, and the collapse of neighboring buildings, *inter alia*. (2014, p. 259)

Similarly, philosophers Lee Basham and M Dentith remark: "[A] quick review of popular conspiracy theory websites, as well as dozens of interviews conducted over the last decade with self-identified conspiracy theorists reveals the evidence attentive adherents of popular conspiracy theories amass is impressive" (2015).

Sutton and Douglas continue:

> [2] [T]here is little evidence that conspiracy theorists have a generally closed epistemology. Indeed, some findings suggest that such individuals may tend to be somewhat more open to experience. (2014, p. 259, citing Swami et al. 2010, 2013)
>
> [3] The correlations among various conspiracy beliefs can be explained in other, simpler ways, without invoking a closed epistemology or a nomothetic explanatory style. (Sutton and Douglas 2014, p. 260)

They conclude:

> In sum, we think it is premature, and unfairly pejorative, to portray conspiracy beliefs as a manifestation of monological thinking. Closed-mindedness and willful ignorance of facts are stigmatized characteristics. It is sobering to reflect that scales measuring agreement with conspiracy theories also, by definition, measure disagreement with them. So *we might equally say that correlations between conspiracy items show that rejection of conspiracy theories comprises part of a monological worldview, in which alternatives to official accounts are dismissed in a closed-minded and irrational fashion!* (Sutton and Douglas 2014, p. 268, emphasis added)

Sutton and Douglas are to be commended for their attempt at evenhandedness, and for urging other scientists to be nonjudgmental. They write,

"We would advocate a dispassionate mode of scientific inquiry which is open to the possibility that these theories have conventionally desirable and undesirable consequences, and which is not unduly influenced by the cultural impulse to label them as simply 'good' or 'bad'" (Sutton and Douglas 2014, p. 268). But their attempt at evenhandedness is spotty. In addition to continuing to maintain that conspiracy theorists tend to simultaneously believe contradictory theories, they also seem to accept some baseless stereotypes, such as the view that conspiracy theorists base their ideas on "limited information."[32] They also state, "[T]he majority of conspiracy theories lack evidential support and are resistant to falsification"[33] (Sutton and Douglas, p. 254). They cite a 2002 article by Steve Clarke. Although often cited by psychologists working on conspiracy theories, that article has serious problems explained in section 2.10. In any case, Clarke does not exactly assert (much less prove) that conspiracy theories lack evidential support. In fact, though Clarke does seem to take a dim view of conspiracy theories, it is not because they simply "lack evidential support." Indeed, Clarke writes, "[Conspiracy theorists] are typically quite dedicated in their search for evidence relevant to their favorite conspiracy theory and are usually able to overwhelm you with a deluge of evidence in favor of that theory" (Clarke 2002). So, as it turns out, not only does the claim that "the majority of conspiracy theories lack evidential support" itself lack evidential support, its falsity is admitted by critics of conspiracy theories.[34] Of course, those critics do not believe the evidence as a whole best supports conspiracy theories, but making an empirical case against these conspiracy theories would require actually dealing with the "deluge of evidence."

Sutton and Douglas do attribute some (weakly) positive aspects to conspiracy theories. They point out that conspiracy theories "can be seen as creative, almost heroic efforts by ordinary people to question and to create alternatives to political orthodoxy" (Sutton and Douglas 2014, p. 268). If that does not strike you as a little patronizing, try this one: "They can be seen as products, and affirmations, of the democratic right to entertain minority views, no matter how absurd or unsettling they seem to the majority" (Sutton and Douglas 2014, p. 268). More significantly, Sutton and Douglas do admit that conspiracy theories "may allow people to question social hierarchies, causing governments to be more transparent and democratic" (2014, p. 268). But they cannot seem to quite acknowledge the most significant potential positive value—that controversial conspiracy theories might sometimes be on to something, and thus help reveal important truths, thwart pernicious conspiracies, and serve as a deterrent to other would-be conspiracies.

7.5. Some Final Considerations

Many scholars writing on conspiracy theories, such as most of those mentioned above, seem to assume that conspiracy theories *are neither true nor warranted*. Consider Cass Sunstein's rhetorical question: "Why, then, do they ["spectacularly well-informed" conspiracy theorists] accept theories that are patently inconsistent with reality?" (2015). Elsewhere, Sunstein insinuates that conspiracy theorists' "information" is wrong. But he does not effectively justify that assertion (see Sunstein 2014a).[35] In general, when conspiracy theories are construed as false or "relatively implausible"[36] in the social science literature, no serious argument about the relative plausibility of conspiracy theories is made; it is merely assumed without argument, or with the briefest of dismissive remarks, that any theory that the researchers regard as a "conspiracy theory" is implausible.[37] This begs a rather important question. Still, the desire to avoid this question is understandable. After all, what would it take to prove that controversial conspiracy theories are implausible? Nothing less than addressing the relevant particulars directly—as well as sizing up all the prior probability factors.[38] For the most part, academics shy away from this. Is it not conventionalist social scientists, then, who exhibit *to a greater degree* a problematic aspect of monological belief systems, in their relative unwillingness to engage in debate, and to investigate the factual details?

But perhaps that does not matter. Let us look at the issue once again, this time framed by Sutton and Douglas: "One of the predictions of the monological position is that adherents (vs. skeptics) of conspiracy theories will invoke fewer concrete facts and more general patterns when explaining major events" (Sutton and Douglas 2014, p. 268). Even if they did, so what? This is not really a problem, so long as there is at least *some* attention to facts of the case in question. After all, if we are only talking about a difference in degree, not about abandoning all consideration of proximate data altogether, would it be better to rely more on proximate evidence or more on analogy to other cases and other general considerations (i.e., prior probability considerations)? We cannot render a general verdict. It depends on all sorts of factors. What can be generally said is that to some degree *both* background considerations relevant to prior probability, *and* direct evidence pertinent to the particular case should be considered. Showing that conspiracy theorists tend to rely more on prior probability than on forensic evidence, dubious though that proposition is, would imply nothing interesting about the psychology, or the quality of the reasoning, of those conspiracy theorists.

Consider a similar issue, for which there is some evidence. Compared to conventionalists, "conspiracist commenters were more likely to argue against the opposing interpretation and less likely to argue in favor of their own interpretation" (Wood 2013). This and other tactical differences in argument style (such as greater focus on issues of prior probability, if that can be shown to apply) can plausibly be attributed to *situational* differences. As an analogy, a defense attorney, in any particular case, may stress the character of his client, calling witness after witness attesting to his integrity, while the prosecution appeals to forensic evidence. We cannot prejudge the verdict based on the difference in strategy, nor should we attribute the strategy to psychological differences between defense attorneys and prosecutors. Most plausibly, they have each simply selected the evidence that is best suited to their case. If the facts of the case were different, they may reverse strategies. If such strategy reversals occur only occasionally, so that we find a statistically significant tendency, we still should not infer anything about the psychological dispositions of defense attorneys. More likely, there is something about these kinds of cases by virtue of which one strategy tends to serve the defense more than the prosecution. In the case of conspiracy theories, one situational consideration in particular stands out. Namely, conspiracy theorists have neither the resources nor legal powers available to the government. Conspiracy theorists are therefore more limited in their ability to acquire information and cannot ordinarily force testimony. Though they do their best to wrest information from the government, through freedom-of-information requests, and through the courts, they are nevertheless forced, to a large degree, to deal with the evidence that is put forward by and through official sources. Is it any wonder, under these circumstances, that they would tend to focus on the problems with the official account more than on developing a detailed positive account?

Now, academics, for the most part, are convinced that controversial conspiracy theories are not true. Even when trying to be polite, "open-minded," and exhibit proper scientific neutrality, they signal in subtle ways that they regard conspiracy theorists as nutty—even if they cannot seem to prove it. The focus on conspiracy theorists, without equal attention to conventionalists, itself signals that one ought to think something strange is going on with conspiracy theorists in particular. Driving the point home, some researchers, such as Brotherton and French, classify the study of conspiracy theorists under the heading of "anomalistic psychology" (2014, p. 246).

In alliance with other social scientists, it seems that psychologists have invented a problem so as to posit a psychological explanation for it, and have done so in a way that seems to reflect poorly on conspiracy theories.

But there is nothing here that needs any kind of special explanation, or that should be understood as problematic. The most straightforward view is that different people come to different conclusions about conspiracy theories for the same kinds of reasons that they come to different conclusions about other matters. They weigh up the evidence they are aware of, factoring in some sense of prior probability. Neither the psychological research by Wood and Swami, nor the arguments of Goertzel, indicate that the explanation for conspiracy theorizing resides in anything more interesting than that.

Rather than focusing on conspiracy theorists, many of these lines of investigation could be turned on people who believe official stories. It would be interesting, and arguably at least as important, and would go some way toward bringing balance to this area of research, if some effort was made to explore why it is that so many people believe false or dubious official stories, or ideas that certain elites clearly try to promote even if they do not explicitly state them. For example, why did so many Americans believe that Saddam Hussein was in cahoots with al-Qaeda? (We know that part of the reason is that officials actively encouraged this belief. The question is: Why did people buy it?) Relatedly, why did soldiers in Iraq believe they were avenging the victims of 9/11? Why did people believe the baby incubator story that was used to market the first Gulf War? Why do some people believe that Syrian president Bashar al-Assad would be so irrational as to use chemical weapons right after Obama declared that to do such a thing would be crossing a line that would ensure US military action against him (and almost did)—especially when the evidence presented was weak? This could be framed as an inquiry into the "good German" syndrome.[39] But those are just suggestions for others; I am not a psychologist. As a philosopher, I'm concerned with whether the definition of conspiracy theory is being used inconsistently in a way that is unfair to serious conspiracy theorists, or whether the implied image of a conspiracy theorist that is found in much academic work amounts to a crude caricature, as the following chapter suggests.

7.6. Conclusion

This chapter has shown, in a narrow sense, that several social science articles focusing on conspiracy theories have put forward very flawed findings, and that these flawed findings were accepted uncritically and repeated by other scholars, and more broadly as well. Indeed, these findings were used to disparage conspiracy theorists unfairly, making them appear intellectually unhinged, while it was these scholars themselves, ironically, who were failing to reason clearly. Further, this chapter has suggested that the flaws in question ought to have been noticed (though exactly how obvious these flaws are is a matter of subjective interpretation). One implication is that scholars, and social scientists in particular, ought to be much more careful in their treatment of conspiracy theorists. Unfairly disparaging a large class of people is no small matter.

I will end with a more general worry. Many philosophers, including David Coady and Steve Clarke, have commented that academics have a "low opinion" of conspiracy theorists (Coady 2006b, p. 1), or that conspiracy theorists are "unpopular amongst intellectuals" (Clarke 2002, p. 131). Indeed, it hardly takes a philosopher to notice *that*. But it is troubling to consider this in connection with the lopsided and unfair treatment of conspiracy theorists in the social science literature, for it suggests that these are not just "innocent mistakes" that could have gone either way. Rather, one must worry that bias against conspiracy theories is influencing the results of social science scholarship, with one biased finding building upon another. And while this chapter has been narrowly focused on the treatment of conspiracy theories in particular, it raises the question of the degree to which the social science literature more generally may be influenced by other widely shared biases.

EIGHT

Conspiracy Theories and the Paranoid Style

Do Conspiracy Theories Posit Implausibly Vast and Evil Conspiracies?

In the social science literature, conspiracy theories are commonly characterized as theories positing a vast network of evil and preternaturally effective conspirators, and they are often treated, either explicitly or implicitly, as dubious on this basis. This characterization is based on Richard Hofstadter's influential account of "the paranoid style." In a recent anthology, *Conspiracy Theories and the People Who Believe Them* (Uscinski 2019), several contributors note the continued significance of Hofstadter's idea. For example: political scientist Ted Goertzel maintains that "the paranoid style is still important. Indeed, it may be more pervasive than ever" (2019, p. 234). Philosopher M Dentith notes, "Numerous scholars have adopted Hofstadter's view explicitly," as we will see (2019a, p. 95). And, making the point stronger, in an essay called, "The History of Conspiracy Theory Research," historian Michael Butter and professor of American studies Peter Knight write, "The impact of Hofstadter's conceptualization of conspiracy theory cannot be overestimated . . . his understanding of the phenomenon influences how considerable parts of the public, media, and academics all over the world have conceived of the topic thus labeled ever since" (2019, p. 35).

And yet many significant conspiracy theories do not have any of the relevant qualities. Thus, the social science literature provides a distorted account of the general category "conspiracy theory," conflating it with a subset of that category that encourages unfairly negative evaluations of conspiracy theories. Generally, when evaluating theories, one should focus on the most plausible versions; the merit of a theory is independent of the existence of less plausible versions of it. By ignoring this and glossing over important distinctions, many academics, especially in the social sciences, have misclassified many conspiracy theories and in doing so have contributed to an epistemically unfair depiction of them. Further, even theories that genuinely fit the description of "the paranoid style" cannot be completely dismissed on that basis. All conspiracy theories ought to be judged on the totality of their individual merits.

8.1. The Paranoid Style

In an essay originally published in 1964, "The Paranoid Style in American Politics," Richard Hofstadter described a style of thinking "systematized in grandiose theories of conspiracy" (1965, p. 4). Hofstadter writes:

> The distinguishing thing about the paranoid style is not that its exponents see conspiracies or plots here and there in history, but that they regard a "vast" or "gigantic" conspiracy as *the motive force* in historical events. History *is* a conspiracy, set in motion by demonic forces of almost transcendent power. (p. 29)[1]

Hofstadter acknowledges that "there *are* conspiratorial acts in history, and there is nothing paranoid about taking note of them" (p. 29).[2] However, expressions in the *paranoid style* are products of "uncommonly angry minds" (p. 3) that "tend to be overheated, oversuspicious, overaggressive, grandiose, and apocalyptic in expression" (p. 4) and exhibit "qualities of heated exaggeration, suspiciousness, and conspiratorial fantasy" (p. 3).

The key features that Hofstadter emphasizes are

(1) an all-encompassing "vast" or "gigantic" conspiracy involving
(2) "preternaturally effective" conspirators of "almost transcendent power"
(3) who are "sinister," "fiendish," and "cruel"—they are "demonic forces" (p. 29) or "demonic agent[s]" (p. 32) engaged in "absolute evil" (p. 31).

Clearly, Hofstadter's characterization is meant to be a criticism of such expression, and of the worldview that underlies it. Hofstadter is explicit about this: "Of course, the term 'paranoid style' is pejorative, and it is meant to be" (p. 5).[3]

Citing Hofstadter's famous essay, many social scientists attribute the characteristics mentioned above to conspiracy theories. For example, quoting Hofstadter (p. 14), Viren Swami and coauthors write, "[C]onspiracist ideation is usually described as a belief in the existence of a 'vast, insidious, preternaturally effective international conspiratorial network designed to perpetrate acts of the most fiendish character'" (Swami et al. 2011, p. 444). Likewise, Oliver and Wood remark that "previous scholarship suggests that most conspiracy theories have three common characteristics" (2014, p. 953, citing Barkun 2003):

> First, [conspiracy theories] locate the source of unusual social and political phenomena in unseen, intentional, and malevolent forces. Second, they typically interpret political events in terms of a Manichean struggle between good and evil. As Hofstadter (1965, 29) famously described, "the distinguishing thing about the paranoid style is [to think] that a vast or gigantic conspiracy is the motive force in historical events. History is a conspiracy, set in motion by demonic forces of almost transcendent power." (Oliver and Wood 2014, p. 953)

Jeffrey Bale covers similar ground.

> "Conspiracy theories" share a number of distinguishing characteristics, but in all of them the essential element is a belief in the existence of a "vast, insidious, preternaturally effective international conspiratorial network designed to perpetrate acts of the most fiendish character," acts that aim to "undermine and destroy a way of life." (2007, p. 50, quoting Hofstadter, pp. 14, 29)[4]

On Bale's account of conspiracy theories, the conspiracy is vast, having a "totalistic, all-encompassing quality" (2007, p. 53). The conspirators are preternaturally effective, "unerring in the pursuit of [their] goals" (p. 51), indeed they are "virtually omnipotent" (p. 52). And they are evil, representing "metaphysical evil" (p. 53). More specifically,

> [C]onspiracy theorists consider the alleged conspirators to be Evil Incarnate. They are . . . inhuman, superhuman and/or anti-human

beings who regularly commit abominable acts and are implacably attempting to subvert and destroy everything that is decent and worth preserving in the existing world. (p. 51)

Robert Brotherton includes two of the three of the characteristics (namely, evil and preternatural effectiveness) in his definition of "conspiracy theory," referring to postulation of "unusually sinister and competent conspirators" (2013, p. 9). He claims, "Conspiracy theories assume unusually malign intent," "Conspiracy theories assume that everything is intended," and "[C]onspiracy theories posit an ordered world in which conspiracies are preternaturally successful. . . . According to conspiracy theories almost nothing happens by accident, only by agency" (Brotherton 2013, p. 11). Elsewhere, Brotherton, writing with Christopher French, covers the third base, vastness: "[C]onspiracy theories usually explain momentous events in terms of a proportionally vast, sinister conspiracy" (2014, p. 238).

Finally, while Michael Barkun separates conspiracy theories into various types with differing scopes, he characterizes them generally as involving supremely powerful groups acting in secret for "some malevolent end" (2013, p. 3). Quoting Daniel Wojcik (1997), who in turn appeals to Hofstadter, Barkun suggests that conspiracy theorists "view history as controlled by massive, demonic forces" (p. 3). "Nothing happens by accident" (p. 3), according to Barkun's account. "[Conspiracy theories] attribute all of the world's evil to the activities of a single plot, or set of plots. . . . Because the conspiracy is so powerful, it controls virtually all of the channels through which information is disseminated—universities, media, and so forth" (p. 7).

While the above-cited scholars do not say that conspiracy theories *always* have the qualities mentioned, they nevertheless leave the impression that these qualities are sufficiently characteristic to adequately serve as the principal properties by which conspiracy theories are to be understood, and thus also evaluated.

Although Hofstadter's depiction of a vast conspiracy of fiendish supermen continues to be cited by social scientists as though this somehow provides an authoritative definition of the phrase "conspiracy theory," it was never intended to be this. Hofstadter was not, strictly speaking, discussing "conspiracist ideation" in general, nor the cluster of ideas we now call "conspiracy theories." He was discussing something narrower.[5] While exhibiting the paranoid style *does* imply belief in a conspiracy, the reverse does not hold: espousing a conspiracy theory, even for paradigmatic cases like 9/11 and JFK conspiracy theories, *does not* imply exhibiting the paranoid style. In other words, people who exemplify the paranoid style are merely a *subset* of conspiracy theorists.[6] And the qualities that characterize a subset cannot be taken as a definition of the larger group without risk of serious distortion. And further, the epistemic merits or demerits particular to a certain subgroup should not be uncritically attributed to other subgroups.[7] In this chapter I argue that contemporary social scientists make a serious mistake when they conflate "conspiracy theories" (or "conspiracist ideation") with the qualities associated with "the paranoid style."

8.2. Representing Conspiracy Theories Fairly

The implication of my argument is that many critics of conspiracy theories, and academics who claim to study conspiracy theories or conspiracy theorists "objectively" but do so with a dismissive air, have engaged in a prejudicial misrepresentation. I aim to explain that the common characterization of conspiracy theories in the social sciences—that of belief in a vast, evil, preternaturally effective conspiracy—is at least misleading, especially when generalizations regarding the plausibility of conspiracy theories are inferred. Though undoubtedly *some* conspiracy theorists have such beliefs, these beliefs have not been shown to be characteristic or typical of conspiracy theories. Indeed, it is unlikely that most people who believe conspiracy theories hold these views.

In the discussions below, when I identify someone as a "conspiracy theorist," I am not disparaging him or her. Rather, I do it to show, as explicitly as possible, an example of a conspiracy theorist who does *not* exhibit "the paranoid style," thereby suggesting that being a conspiracy theorist does not imply the negative qualities often insinuated. The fact is, I am a conspiracy theorist myself. And with fellow conspiracy theorists like Charles de Gaulle, Bobby Kennedy, Daniel Sheehan, and David Talbot, as well as Bertrand Russell and John Dewey (see section 2.2 for the latter two), I consider myself in good company. But my company is much broader than that, for most people believe in at least one controversial conspiracy theory. As Uscinski and Parent surmise, "Given previous polling data, we suspect that everyone believes in at least one conspiracy theory, and most people believe in several. So everyone is a conspiracy theorist at least some of the time" (2014, p. 78).

In any case, characteristics of the paranoid style certainly do not define the boundaries of conspiracy theory. And so, even if these qualities were characteristic or typical (which I will argue they are not), it would still be a mistake to make an inference about the plausibility of any *particular* conspiracy theory based on these generalizations, for a theory ought to be evaluated on its *best*, most plausible version, not on characteristics typical of its category.

Let us dwell on that last point for a moment. When determining the best explanation, it is important to formulate the best version of competing hypotheses, and then evaluate *those*. It is of little relevance that one can find crippling flaws in an unsophisticated version of a conspiracy theory or in a flimsy misrepresentation. According to the MIHOP (Made It Happen on Purpose) versions of 9/11 conspiracy theories, insiders orchestrated the attacks of 9/11. That is the essence of the claim, the core thesis. If we flesh it out with poorly thought out and implausible details and then knock it down on account of those details, all we have done is shown that a sloppy version can be debunked. But that is of little significance. The proper way to evaluate a theory is to give the theory its best, most plausible rendering and then evaluate *that*. This is in line with the principle of charity—that one should be evaluated on the most charitable interpretation of one's position—which is a standard academic expectation. At least, critics should turn their attention to what seem to be the strongest advocates of a position, rather than the weakest.[8] This will often require considerable effort to understand the theory and the evidence for it. This can be a daunting task, as some of the more interesting conspiracy theories are rather complicated. Those who make sweepingly dismissive critiques rarely demonstrate that they have bothered to try to understand the most plausible versions of the theories that they so flippantly reject.

To show that we should adopt a more charitable interpretation of what belief in conspiracy theories entails, I address, in turn, the following three questions: (1) Do conspiracy theories imply implausibly vast conspiracies? (2) Do conspiracy theories imply preternaturally effective conspirators? And, (3) do conspiracy theories imply implausibly evil conspirators? The answer in each case will be negative. I then turn the tables and ask whether the *official account* of 9/11 has the qualities of the paranoid style, and also challenge the view that Western leaders ought to be presumed to be "good" or "benign." These considerations, taken together, suggest that a number of social scientists have treated conspiracy theories unfairly, reinforcing prejudicial stereotypes that tend to encourage the inappropriate dismissal of conspiracy theories based on generalities that often do not apply. In the final section I press even further, suggesting (though reluctantly) that even conspiracy theories that *do* imply a vast network involving unusually powerful malevolent conspirators cannot be *completely* dismissed on that account. Each theory ought to be judged on the totality of its individual merits.

8.3. Do Conspiracy Theories Imply Implausibly Vast Conspiracies?

Here I argue that the common insinuation that conspiracy theories are unwarranted because they imply implausibly vast conspiracies is ill-founded. Many conspiracy theories do not posit particularly vast conspiracies, and the charge that they nevertheless somehow *imply* vast conspiracies is contestable and generally lacks substantiation. Further, one cannot make a valid assessment of the plausibility of a purported conspiracy based on its size without also considering significant nuances, such as organizational structure, and important distinctions, such as between the underlying conspiracy and its cover-up.

Many controversial political conspiracy theories do not hypothesize a *vast* conspiracy (though it is hard to know how big something must be to count as "vast"). Why do we have to choose between a lone nut or small group of terrorists, on the one hand, and a vast conspiracy of insiders on the other? Logically, it is an obvious false dichotomy. And empirically, one can easily find actual conspiracy theories that occupy some middle ground. While examples could have been drawn from any number of conspiracy theories, I focus here on JFK and 9/11 conspiracy theories, those being, in my view, the most paradigmatic conspiracy theories (and I happen to think that some version of each is probably true). Let's start with (conspiracy theorist) Ed Tatro, who makes the following point regarding the assassination of President Kennedy:

> If the head of the FBI and the head of the Secret Service and key lieutenants of those two figures are involved in covering up, plus the President of the United States and all his cronies, you don't need a massive conspiracy to get away with this.[9]

Still, the idea that conspiracy theories must involve vast conspiracies persists. For example, regarding 9/11, it is not uncommon to hear the claim that, for it to have been an inside job, "*thousands* of people would have had to be in on it." As one critic of 9/11 conspiracy theories puts it:

> These conspiracy theories are paradoxical mainly because it would take thousands and thousands of people to cooperate in plotting them and executing them. The cover up should involve half of Washington. The very passengers who, according to some conspiracy theories, never died, should be kept alive somewhere. Their relatives should be part of the conspiracy. (Scaruffi 2002)

None of this is necessarily true. Where has it been authoritatively demonstrated that any of these claims must be so? Many conspiracy theorists have much lower estimates, even for versions that involve controlled demolition.[10] These lower estimates may or may not be justifiable. The point is that many conspiracy theories *do not believe* in a vast conspiracy, and it is not fair to assert that they are nevertheless *committed* to such a belief, given their other beliefs, without adequate argumentation. The charge that conspiracy theorists assume a vast conspiracy, even in cases like MIHOP theories about 9/11, is contestable and is in fact contested by at least some (and perhaps most) of these conspiracy theorists.

As in the example above, the purported vastness of the conspiracy that is supposedly entailed by "conspiracy theories" is generally, at best, a weakly supported and sketchy insinuation made by critics of those theories, who do not always seem to have made much of an effort to understand the theory they so dismissively characterize as "vast." In the social science literature, the claim is often not supported at all. Typically, having claimed that conspiracy theories posit "vast" conspiracies, they proceed to reference stock conspiracy theories, such as those involving 9/11, the London bombings, or JFK, without even attempting to show that these conspiracy theories meet this criterion,[11] or offering only a feeble attempt. Regarding the latter, recall an example given in the discussion of Jeffery Bale in Chapter 2: to exemplify the "totalistic, all-encompassing quality" of "conspiracy theories" (in the pejorative sense in which Bale uses that phrase), Bale cites the theory that "umbrella man," who pumped his umbrella in front of JFK's car as the assassination began, was acting as a signal man (2007, p. 53). Yet, clearly, there is nothing totalistic or all-encompassing about this idea.

Further, such critics generally make no distinction between a conspiracy alleged to have brought about an event and the conspiracies involved in subsequent cover-ups. Nor do they acknowledge that many other factors that are not themselves conspiratorial may come into play that may result in large ripple effects emanating from a conspiracy of a relatively modest size. And, as many conspiracy theorists explicitly acknowledge, individuals participating in cover-ups may well be doing so for either noble reasons, or because of a particular incentive structure, or out of fear, either for their career and reputation or because of larger implications.[12] Theories that posit extensive cover-ups of pernicious conspiracies, at least arguably, are generally more plausible than theories in which the pernicious conspiracy *itself* is vast. As (conspiracy theorist) Peter Dale Scott says in his introduction to (conspiracy theorist) Sylvia Meagher's book, *Accessories after the Fact*:

> [T]he unaccustomed reader may mistakenly think that the only alternative to the Warren Report is a vast conspiracy of thought control and the rewriting of history at every level of society. Here I think the more recent experience of Watergate will help us to dispel such paranoia and to understand the mechanics of the Warren Commission syndrome in a more rational perspective. If Oswald had ever, at any point in his career, been involved with any government agency or in any national security matter, it is clear (we now know) that numbers of relatively innocent public servants would participate in a cover-up where they were told that national security was involved. (Scott 1975, p. xvi)

Indeed, today we can put the point even more sharply. That the FBI, CIA, Secret Service, and others *had actually* engaged in various cover-up activities regarding the JFK assassination—concealing, destroying, or altering evidence—is a fact that is pointed out even by prominent *defenders* of the Warren Commission's findings, such as Philip Shenon.[13] So the postulation of a cover-up is no longer something that differentiates conspiracy theorists from their adversaries.[14]

On the most innocent level, there may be large numbers of people who may participate passively in a cover-up, simply going along to get along. For example, Charles Crenshaw, a doctor who assisted in the attempt to save President Kennedy at Parkland Hospital, and the other doctors at Parkland as well, may be regarded as having, in some sense and to some degree, participated in a cover-up, though not because they were part of a massive evil plot. In a book originally titled *JFK: Conspiracy of Silence*, Crenshaw explains why he did not straightforwardly say, "Hell, no, Oswald didn't shoot him in the head, because the President was shot from the front." Crenshaw writes,

> None of us doctors were willing to do that. We all valued our medical careers too much. . . . I believe there was a common denominator in our silence—a fearful perception that to come forward with what we believed to be the medical truth would be asking for trouble. (Crenshaw 2013)[15]

Further, those participating more actively in a cover-up (of sorts) may not see themselves as covering up *for the real perpetrators*. Rather, they may merely recognize (if they force themselves to attend to it) that they are being "just slightly" dishonest or unfair in support of a view that they believe is, nevertheless, more or less true.[16]

All of this is to suggest that we should not be too surprised if a particular appalling conspiracy of modest size is accompanied by a wide range of activities that could be considered part of a cover-up. Thus, the cover-up may appear "vast" (though not strictly conspiratorial in most of its dimensions) even when the underlying conspiracy being covered up is not.

Making some similar points, Dentith and Martin Orr argue that the "relative ease" with which conspirators might maintain secrecy is a function of organizational structure, not mere size. They write:

> A conspiracy can look big, yet only feature a small number of people who know the full extent or aim of the conspiracy. Some members of the conspiracy will be lackeys, goons, useful idiots, or even unwitting conspirators. It is even possible to be involved in a conspiracy without realizing you are conspiring; not everyone in the NSA need necessarily know that the data they are collecting and processing has been illegally obtained, and FBI agents who are using forensic evidence to secure convictions may not have been informed by senior personnel that the evidence they are using is of dubious merit. (Dentith and Orr 2018, p. 440)

However, all of these nuances are left unanalyzed when one makes vague, but clearly dismissive, gestures to vast evil conspiracy theories.

It should be noted that, in contrast to those who describe conspiracy theories as positing vast conspiracies, some scholars actually define conspiracy theories as involving *small* groups. For example, as mentioned in Chapter 1, Joseph Uscinski and Joseph Parent define conspiracy theories in terms of a "small group of powerful persons" acting as the conspirators (2014, p. 32; cf. p. 146). Brian Keeley likewise writes, "A conspiracy theory is a proposed explanation of some historical event (or events) in terms of the significant causal agency of a relatively *small group of persons*—the conspirators—acting in secret" (1999, p. 116, emphasis added; see also Keeley 2019, p. 423). "Small" and "vast" are hard to reconcile. When it comes to *counting* as a "conspiracy theory," it is not clear why size matters. When it comes to *being plausible*, size does matter, but there are important nuances to consider.[17] Scholars who characterize conspiracy theories as being defined in part by their vastness without due attention to the relevant nuances thereby contribute to an unfair assessment of their general plausibility.

Let's take a moment here to consider an actual conspiracy. The 20 July Plot (aka Operation Valkyrie), one of many conspiratorial attempts to assassinate Adolf Hitler, was no small operation. It involved not just the assassination itself, but an ambitious attempted coup d'état. After it failed, thousands of people were executed for their purported connection to it. The number of genuine conspirators, who had actually conspired in advance, may have been an order of magnitude or more smaller than those held responsible, though quite a few high-ranking officers and other elites seem to be genuinely implicated. Others joined in once the plot was underway, forced to choose sides with limited and perhaps inaccurate information. Many who played some role were just following orders, not really knowing what was going on. Some people were aware of the plot (to varying degrees) in advance, though not strictly in on it, as General Friedrich Fromm seems to have been. Some refused to be involved but did not expose those who tried to recruit them.

Although the 20 July Plot may not count, strictly speaking, as a "conspiracy theory," it certainly was an ambitious conspiracy, and has some relevance for assessing the plausibility of ambitious conspiracy theories. For example, even though many people actually conspired, and still more had some level of awareness of the conspiracy, it didn't fail because "someone talked." With a bit better luck, it would have at least partially succeeded—by killing Hitler. If that had happened, one can only speculate on what the received "history" of the plot would have been. While it is tempting to think that the conspirators would eagerly admit their role, since it is easy for us now to frame them as heroes, as does the movie *Valkyrie*, it is doubtful that high treason would have played so well in the time and context. Assuming the conspirators succeeded in taking power and thus had control of the inquiry into the assassination itself, the bombing may have been blamed on patsies, and their plan to pretend that "A treacherous group of party leaders has attempted to exploit the situation" (Jones 2008, p. 160) may have at least partially succeeded; the suggestion that there was a huge plot worked out in advance may have been passed off as a wild "conspiracy theory." Alternatively, perhaps the plot was "near-hopeless" from the beginning, as some maintain (Jones 2008, p. ix). But that, at least, suggests that elite conspirators will sometimes take extraordinary risks. In any case, because it didn't succeed, we have come to know that this kind of large and ambitious conspiracy may at least be attempted. And further, the motives of the July 20 conspirators (who were not exactly liberal humanitarians, but rather largely conservative nationalist authoritarians) seem not so different from the motives that many prominent conspiracy theorists ascribe to those they suspect are behind the assassination of President Kennedy, as described below.

8.4. Do Conspiracy Theories Imply Preternaturally Effective Conspirators?

For the subset of conspiracy theorists that exhibit the paranoid style, conspirators are thought to be "preternaturally effective" (Hofstadter 1965, p. 14). These theories, Hofstadter explains, leave "no room for mistakes, failures, or ambiguities" (p. 36). However, sophisticated conspiracy theorists do not generally assert or assume this.[18] While they *do* assume technically competent agents will be deployed, for example, to pull the trigger, they also assume that things can and do go wrong for the conspirators. They are not able to control everything. In no way approaching "almost transcendent power" (p. 29), the conspirators are thought to be limited human beings who "put their pants on one leg at a time," as (conspiracy theorist) Daniel Sheehan likes to say while making this point.

Here are just a few brief examples. (JFK conspiracy theorist) Charles de Gaulle, himself an assassination-attempt survivor, and president of France at the time of President Kennedy's murder, said, "They wanted to kill [Oswald] on the spot before he could be grabbed by the judicial system. Unfortunately, it didn't happen exactly the way they had probably planned it would," as reported by (conspiracy theorist) David Talbot (2015, p. 567).[19]

Regarding 9/11, the author of *Disconnecting the Dots: How the CIA and FBI Officials Helped Enable 9/11 and Evaded Government Investigations* (2011),[20] (conspiracy theorist) Kevin Fenton, comments:

> [O]ne of the reasons why we know about the withholding [of information from the FBI by the CIA is] because there were constantly problems and things kept going wrong . . . and they tied themselves up in knots dealing with all of these problems.[21]

Also, in many 9/11 conspiracy theories, the fate of Flight 93 is regarded as an indication that something went wrong—that it was supposed to hit a significant target, perhaps the Capitol Building, but that part of the plan was either aborted or thwarted. Likewise, the late collapse of Building 7, at about 5:20 p.m., is also often regarded as an indication of some kind of bungle. It is presumed that it should have been destroyed at the same time as one of the towers, and thus would have not been such a problem to explain.[22] These are just a couple of the most striking purported foul-ups. The view that seems to be widely shared among the conspiracy theorists who point out these kinds of issues is not that the conspirators are preternaturally effective, but rather that *they don't need to be*, especially when anyone who points to evidence of conspiracy can be simply dismissed as a "conspiracy theorist."

But what about milder versions of the critique? Conspirators may not be assumed to be perfect, but they are nevertheless thought to be too effective to be believable—because everyone knows that government officials are incompetent, so the story goes. For example, Cass Sunstein (who was himself a government official for a time) writes, "Conspiracy theories generally attribute to certain agents extraordinary powers: to plan, to control others, to maintain secrets, and so forth" (2014, p. 5). Even this seems to involve exaggeration. Planning, keeping secrets, and exercising some control over others, especially in hierarchical systems, do not seem like "extraordinary powers." But then these things are, after all, matters of degree. As Sunstein reframes it, "[C]onspiracy theorists typically overestimate the competence and discretion of officials and bureaucracies, which are assumed to be capable of devising and carrying out sophisticated secret plans" (2014, p. 5).[23] But it is not at all clear that conspiracy theorists have overestimated this. After all, people *are* capable of "devising and carrying out sophisticated secret plans." Coups, for example, sometimes succeed in this regard. And it is not obvious that conspiracy theories assume *too much* ability of this kind. But even if we assume that conspiracy theories "generally" or "typically" do so, that tells us nothing about the *best versions* of those theories. Do *those* also posit an implausible degree of competence? It is not clear that they do, at least not always. In any case, we cannot assess the reasonableness of a theory, or whether or not it posits too much competence to key figures, without first understanding the theory and the evidence for it. Until we do that, we neither know how much competence the theory requires, nor what evidence suggests that such competence exists.

So, just as it is misleading for social scientists to dismissively characterize conspiracy theorists as believing in "vast" conspiracies, it is likewise misleading to characterize them as positing or implying preternaturally effective conspirators. In significant controversial cases, it is just not clear that conspiracy theorists imply implausibly vast conspiracies or implausibly effective conspirators.

8.5. Do Conspiracy Theories Imply Implausibly Evil Conspirators?

As shown near the beginning of this chapter, scholars who follow in the footsteps of Hofstadter often suggest that conspiracy theories imply implausibly evil conspirators. However, they often equivocate between different senses of "evil," and sometimes use less dramatic language. Let us start with Uscinski and Parent, who define conspiracy theories as involving conspirators "acting in secret for their own benefit against the common good" (2014, p. 32). The problem is that many significant conspiracy theories could plausibly be understood (by the perpetrators, at least) as being *for* the common good, even if the masses, unable to understand the wisdom of the plan, are appalled by it—perhaps due to moral squeamishness. (Some examples will be given below, and more in the next chapter.)

Uscinski and Parent do seem to recognize this. At one point they qualify their claim, writing: "A critical feature of our definition is that the conspiracy must come at the expense of the common good, *at least in the eyes of the conspiracy theorist*" (2014, p. 34; emphasis added). Nevertheless, they often write in such a way as to invite the inference that conspiracy theories assume a cartoonish motive. For example, they write, "By definition, all conspiracy theories posit a knave out to subvert the common weal" (2014, p. 62), and they characterize the suspects in conspiracy theories as "moustache-twirling mountebanks" (p. 62). Not only do such characterizations suggest that conspiracy theories posit evil conspirators, but such cartoonish depictions seem to imply that conspiracy theories ought to be regarded, on this basis, as unlikely, if not silly.

Further, Uscinski and Parent provide the following argument:

> [1] In all conspiracy theories, the villains are often assumed to be terribly conniving and devilish. But oddly, [2] many of these supposed offenders are otherwise considered to be the good (e.g. the President), the benign (e.g. the Secretary of Commerce), and the boring (e.g. the International Monetary Fund). Hence, [3] we might choose to reject a conspiracy theory if it attributes a depravity that is unlikely given one's track record and institutional incentives. (Parent 2014, p. 41, numeration added)

This is a little sloppy. For one thing, it will not do to conclude with "we *might* choose [whatever]." That would be true regardless. We need something epistemically normative here. In addition, the conjunction of "In *all* conspiracy theories" and "the villains are *often* assumed . . ." is awkward and confusing. Is this supposed to be a strong claim that has significant implications or not? Let's tighten this up so that it explicitly and clearly expresses the idea that the authors seem to imply, given the totality of their disparaging comments regarding conspiracy theories (some of which can be found below):

> [1] In all conspiracy theories, the villains are . . . assumed to be terribly conniving and devilish. But oddly, [2] many of these supposed offenders are [*reasonably* presumed][24] to be the good (e.g. the President), the benign (e.g. the Secretary of Commerce), and the boring (e.g. the International Monetary Fund). Hence, [3] [We have a good reason][25] to reject . . . conspiracy theor[ies] [since they attribute] a depravity that is unlikely given one's track record and institutional incentives.

There. That's better. The problem is that by tightening up the argument we have opened it up to a number of challenges. First of all, even if the argument were sound, it would not show that *all things considered*, conspiracy theories ought to be rejected. It would only show that there is *one good reason* favoring rejection. That may be adequately counterbalanced by other considerations. For even if a conspiracy theory implies that a president has done something morally shocking, and even if presidents are reasonably presumed to be good, we may nevertheless conclude that the conspiracy theory is true *if sufficiently compelling evidence was offered.* Think of "the Nixon tapes" or the president's semen on an intern's dress.

In addition, both premises are problematic. Premise 2, that the accused elites are reasonably considered "good" or at least benign, is questionable. Goodness is a matter of degree, and US presidents, and other high officials, have shown themselves to be morally flawed (to put it mildly)—a point I will amplify below.[26] Hence, whether or not we ought to reject any particular conspiracy theory on account of the degree of depravity that it attributes to people in positions of power requires *conspiracy theorizing*, to clarify the details involved, before it can be properly adjudicated. A sweepingly dismissive attitude is not warranted.

Further, premise 1 is false. Many conspiracy theories do *not* posit evil or "devilish" motives. Indeed, critics of conspiracy theories seem to have some level of awareness of this. For when we look carefully, we find that the accusation that conspiracy theories posit evil conspirators is often strangely ambiguous. Do they mean *truly* "devilish" (to use Uscinski and Parent's word)? That is, do they mean evil in the Augustinian sense of doing wrong *because it is wrong*? Are they talking about *diabolical* evil, as some of the quotations in the section near the beginning of this chapter seem to suggest? Or do they mean evil in the more pedestrian sense of *acting immorally for selfish reasons*? It seems that many critics who imply that conspiracy theories are implausible trade on this ambiguity. When we disambiguate, we notice that the stronger the sense of "evil" they intend, the more obviously false is their claim that conspiracy theories typically posit evil conspirators. But the weaker the sense of "evil" they intend, the weaker the inference to the implausibility of such a motivation.

Let's look at a couple of examples of the ambiguity. First, describing the idea that Monsanto has "captured the FDA" and similar organizations in "many, many countries," Uscinski and Parent remark: "This is not about the prosaic world of science and methods; this is about the diabolical world of power and money. There are no shades of gray. The corporate conspiracy has infiltrated and enslaved the scientific and political establishments alike" (2014, p. 147). Is this genuinely "diabolical," or is it just about "power and money"? The latter merely implies an understandable, selfish motive, not an inexplicable, evil one. Despite resorting to quotations from Facebook comments, with which one can presumably prove just about anything, the examples that Uscinski and Parent provide do not clearly show evil or "diabolical" intent.[27] Similarly, Robert Goldberg seems to conflate self-interest with malevolence in his account of the perspective of those who believe JFK conspiracy theories. He writes, "So steep a fall from grace could not be a matter of circumstance, but could only be the work of the self-interested and malevolent" (Goldberg 2001, p. 149). Which motivation is doing the work? Self-interest or malevolence? Even Hofstadter's characterization of the "demonic" bad guy lapses into ambiguity as "a kind of amoral superman: sinister, ubiquitous, powerful, cruel, sensual, luxury-loving" (1965, pp. 31–32). Leaving the issues of power and ubiquity aside, it is not clear that the moral qualities expressed here go much beyond callous selfishness.

In any case, many, if not most, conspiracy theorists have complex views about the motives of the suspected perpetrators. They often recognize that the conspiracies that they find appalling may be regarded as noble, or at least justified, from another perspective. For example, conspiracies to create a North American Union or a global government have plausible innocent motives. Outrageous experiments on unwitting subjects, such as in the Tuskegee experiment, seem to have been motivated, at least in part, by utilitarian considerations (even if racist assumptions were also significant). Serious consideration of adding sterilants to the water supply seems to have been prompted by genuine concerns about overpopulation (see Ehrlich et al. 1977, pp. 787, 788). Even political assassinations and false flag terrorist attacks could be motivated by "good" intentions. This applies even more clearly to cover-ups of such conspiracies. And conspiracy theorists often acknowledge the possibility of such intentions. For example, regarding the JFK assassination, (conspiracy theorist) Jim Garrison expresses the view that Chief Justice Warren agreed to lead a cover-up "for reasons which I think are probably essentially altruistic" (Barbour 1992). Further, regarding the motive for the assassination itself, the case that Kennedy was viewed as a traitor, or at least a dangerous appeaser, is frequently made by JFK conspiracy theorists—David Talbot (2015) makes it in a chapter entitled "For the Good of the Country" (see especially, pp. 559, 560), and so do James DiEugenio (2012, pp. 66, 67), Daniel Sheehan,[28] and many others. Indeed, this seems the dominant view, at least among the serious JFK assassination researchers. And surely many of their readers have understood the point. So the notion that conspiracy theorists assume evil perpetrators cannot be regarded as true, even as a generality, for even the most paradigmatic conspiracy theory.

Of course, good intentions are not the only plausible ones. People have a tendency to be selfish, though presumably not all to the same degree. We also have the power to overcome our selfish desires, though some of us succeed in this better than others. We should not assume that the problem of selfishness does not exist among prominent people. So it is not implausible that the actions of prominent people might be motivated, at least in part, by selfishness. *How* selfish are such people capable of being? *Where* do they draw the moral lines that they will not cross? Those are empirical questions. And the answers presumably vary considerably from individual to individual.

What motivated members of the George W. Bush administration to conspire to manipulate the American people into supporting, or at least not resisting, a war in Iraq? It could be that they really believed it was the right thing to do. But many critics charge that it was really "about the oil," which can be understood as shorthand for the various ways in which people and companies could profit from the Iraq war. Why would it be implausible that such motives played some role? There is no reason to assume, supposing some selfish motives were involved, that those were the *only* motives. There could well have been a confluence of selfish motives, misguided "noble" motives, and a fair measure of confusion and incompetence as well.

And yet it is not uncommon to find conspiracy theorists lampooned, sometimes subtly, for suggesting that someone may have acted immorally for selfish reasons. For example, contrasting "reasonable people" with conspiracy theorists, Uscinski and Parent write:

> Not everyone who opposes GMOs is a conspiracy theorist; reasonable people can disagree about research and fail to see small groups

> of people covertly working against the common good. But the most visible and vocal members of this movement, however, are conspiracy theorists. They believe that genetically modified foods are a corporate plot, led by the giant multinational Monsanto, to profit off unhealthy food. (2014, p. 146)

Rhetorically, the idea that there is some kind of "corporate plot . . . to profit off unhealthy food" may sound a little unhinged. And the way Uscinski frames it elsewhere—"Monsanto wants to poison the world's population"—seems to posit malevolent intent (2019, p. 50). But knowing the history of tobacco companies, why would the idea of a corporate conspiracy to boost profits be thought to be implausible? Presumably, it is not that anyone *wants* to harm people; the conspirators just want to sell their product, *harmful or not*. It is callous and greedy. But it is not implausibly evil or malevolent. And so the idea that Monsanto is likewise selfishly engaging in dangerous shenanigans of some sort is not implausible. There are clear precedents. In addition to the tobacco fiasco, it has recently been revealed that the scientific consensus regarding sugar and fat has been significantly misdirected for decades by research funded by Big Sugar (Kearns et al. 2016). Should we assume this was just a coincidence, or some strange kind of incompetence, or misguided do-gooding? No. Sometimes things are what they seem. At the very least, when there are prima facie conflicts of interest, suspicions of selfish conspiring should not be dismissed.

In sum, conspiracy theorists do not always posit implausibly evil intentions—that may not even be very common. At least sometimes they understand that the intensions might be thought to be noble from the conspirators' perspective; at other times they are regarded as merely selfish. In either case, this provides little if any reason to regard those conspiracy theories as implausible. And so, on all three counts—vastness, preternatural effectiveness, and evil—the conflation of conspiracy theorists with the paranoid style is inaccurate and misleading.

8.6. The Psychopathic Element

Before turning to examples of elite bad *behavior*, I will here add one final consideration regarding the *mentality* of conspiracy suspects. Psychological research suggests that about 1 percent of the general population may be psychopaths (Neumann and Hare 2008). That means that there may be about three million psychopaths in America alone. Further, "high functioning" psychopaths are overrepresented in positions of power (Boddy 2011, p. 257). And so the possibility that psychopaths may play some role in elite conspiracies should not be ignored.

I'll just offer two very speculative examples. First, there is much in David Talbot's account of the life of Allen Dulles, who Talbot argues was instrumental in the murder of JFK, that smacks of psychopathy. Perhaps the most striking passage is this one:

> As the child began floating away toward the lake's deep, cold waters, her pink dress buoying her like an air balloon, Eleanor began screaming frantically. But Allen, who by then was a strong swimmer, was strangely impassive. The boy just stood on the dock and watched as his little sister drifted away. Finally, as if prompted by Eleanor's cries, he, too, began yelling. (Talbot 2015, p. 39)

By itself, this anecdote means little. But taken in context of his whole life (at least as represented by Talbot),[29] one may reasonably wonder whether Dulles was a genuine psychopath. Of course, that would not mean that all those he may have conspired with were likewise genuine psychopaths, but it nevertheless helps us think concretely about the very real possibility that psychopathy could play *some* role in perverse conspiracies.

Second, it is not hard to imagine psychopaths being drawn into a conspiracy to perform unpleasant functions at the lower end of it, such as the thug, codenamed "WI-ROGUE," recruited by the CIA to kill Patrice Lumumba, the newly elected prime minister of the Congo. As Talbot relates, "ROGUE was the kind of morally unhinged man 'who would try anything once, at least,' said his agency supervisors, untroubled by the 'pangs of conscience'" (2015, p. 380).

And so, while I don't think we should make *too much* of the psychopathy angle, neither should we forget it altogether. It may well be part of a complex mix.

8.7. Should Western Leaders Be Regarded as above Suspicion?

Now I want to make two related points. The first is that there seems to be a double standard at work. Namely, although certain academics are quick to accuse conspiracy theorists of exhibiting the paranoid style, even when it is not clear that they really do, official stories are almost never likewise criticized, even when the shoe seems to fit. It may be thought that there is a relevant difference here, at least regarding the attribution of evil motives: on the one hand, while official stories sometimes impute evil motives to people, they only do so to people who are actually evil, or at least plausibly regarded as evil. On the other hand, conspiracy theorists accuse people who are presumably good. This leads directly to my second point (which I promised above to amplify), that the presumption of goodness on the part of Western leaders and elites is not warranted. It is an example of misplaced exceptionalism.

I will begin by elaborating on the first point. As mentioned earlier, Viren Swami and coauthors claim that "conspiracist ideation is usually described as a belief in the existence of a 'vast, insidious, preternaturally effective international conspiratorial network designed to perpetrate acts of the most fiendish character'" (Swami et al. 2011, p. 444). In the very next sentence, Swami implies that "disbeliev[ing] official accounts of the September 11, 2001 . . . terrorist attacks" counts as conspiracy ideation.

However, it seems that *believing* the official account of September 11 also fits this definition. Is not al-Qaeda a "vast, insidious, preternaturally effective international conspiratorial network designed to perpetrate acts of the most fiendish character"? With this question in mind, consider how well Robert Brotherton's attribution of an "unusually malign intent" fits with the *official account* of 9/11. We can do that with Brotherton's own words, by merely substituting "conspiracy theories" with "official accounts":

> The malevolent intent assumed by ~~most conspiracy theories~~ [the official account of 9/11] goes far beyond everyday plots borne out of self-interest, corruption, cruelty, and criminality. The postulated conspirators are not merely people with selfish agendas or differing values (Bale 2007). Rather, ~~conspiracy theories~~ [official accounts] postulate a black-and-white world in which good is struggling against evil (Bale 2007; Barkun 2003; Oliver and Wood, 2014). The general public is cast as the victim of organized persecution, and the motives of the alleged conspirators often verge on pure maniacal evil (Lewandowsky *et al.* 2013).[30] At the very least, the conspirators are said to have an almost inhuman disregard for the basic liberty and well-being of the general population. (Brotherton 2013, p. 11)

The official story of 9/11 attributes wicked motives to Osama bin Laden and the other "evildoers" in his al-Qaeda network. We were told that we must have a (nonsensical) "war on terror" because these evildoers "hate our freedoms: our freedom of religion, our freedom of speech, our freedom to vote and assemble and disagree with each other" (Bush 2001). And, apparently, there are not just a few such people. There are so many that the United States and its allies will be at war with them for decades, at least. If one wants to quibble over the answer to the question of whether the official story truly reflects the paranoid style, why not do the same regarding alternative narratives? After all, many, if not most, of the more serious conspiracy theorists (those that publish books and peer-reviewed articles and give lectures at universities) would claim that both Hofstadter's and Brotherton's descriptions are exaggerated caricatures of their positions.

The difference between the official story and alternative narratives, it will be pointed out, is that bin Laden and al-Qaeda are not "otherwise considered to be the good (e.g. the President), the benign (e.g. the Secretary of Commerce), and the boring (e.g. the International Monetary Fund)" (Uscinski and Parent 2014, p. 41). But why is, for example, the president (of the United States) considered good? Have all US presidents and high officials been *good*? *Maybe* they all have good intentions, as suggested above. But maybe not. Whatever their inner motivation, the fact is that many presidents and other Western elites have sometimes behaved badly, and sometimes conspired in doing so. This supports the second point, that the presumption of goodness on the part of Western leaders and elites is not warranted. Some elaboration will make this point more vivid and concrete, as well as giving some sense of the scope of the problem, so that its significance is less likely to be lost on those prone to exceptionalist thinking.

While not endorsing controversial conspiracy theories, Kathryn S. Olmsted, a UC Davis history professor, provides a case for the prior probability of conspiracy theories based on "*proven* government conspiracies" (2009, p. 3). Her examples do not reflect well on the moral uprightness of those considered "good, benign, and boring." Olmsted explains:

> [A]s the [US] government grew, it gained the power to conspire against its citizens, and it soon began exercising that power. By the height of the cold war, government agents had consorted with mobsters to kill a foreign leader, dropped hallucinogenic drugs into the drinks of unsuspecting Americans in random bars, and considered launching fake terrorist attacks on Americans in the United States. Public officials had denied potentially life-saving treatment to African American men in medical experiments, sold arms to terrorists in return for American hostages, and faked documents to frame past presidents for crimes they had not committed. (2009, p. 8)

She also cites spying on and imprisoning dissenters during World War I (pp. 4, 6, 9), and Nixon's effort to use state power to punish personal enemies (p. 5).

Let's pursue this line a little further. Consider those who championed the post-9/11 invasions of Afghanistan and Iraq. Whether or not the attacks of September 11 involved insider complicity of some kind, those attacks were seized upon by the Bush administration and by influential neoconservatives to support their preexisting plans, which included the invasions of Afghanistan and Iraq. Hundreds of thousands of people have died as a result of those invasions, and yet none of the publicly proclaimed justifications for the invasions are plausible. (1) There never was any significant collaboration between Saddam and al-Qaeda, and the elites who encouraged the idea that there was such a conspiracy presumably knew that the evidence was weak at best. (2) Despite the continuing insistence otherwise, there never was solid intelligence supporting the claim that Saddam had weapons of mass destruction. (3) The idea that the invasions were substantially motivated by a desire to bring democracy to the Middle East, to liberate women, and such, are not plausible. The history of US intervention shows too much support for tyrannical dictators to take that seriously, especially since there are plenty of unpursued opportunities to do good around the world without inflicting the carnage that war inevitably brings.

In addition, a fair number of high officials in the US government have been credibly accused of war crimes. These include not only President George W. Bush (Bugliosi 2008),[31] Vice President Dick Cheney (Noble 2013), and Secretary of Defense Donald Rumsfeld (Ratner 2008), but also Nobel Peace Prize winner Henry Kissinger (Hitchens 2002), General Curtis LeMay, and Robert McNamara. McNamara actually agreed that such an accusation against him would have merit. He said, "He [LeMay]—and I'd say I—were behaving as war criminals" (Blight and Lang 2005, p. 211).[32] President Obama's drone war and extrajudicial assassinations (which arguably constitute murder) are likewise morally questionable. And President Donald Trump is explicit about his advocacy of a policy of killing whole families based on the presumed sins of the father (LoBianco 2015).

So the question of whether or not influential people in the Bush administration might be willing to sacrifice thousands of innocent lives to achieve their ends in the Middle East must clearly be answered in the affirmative, whether or not their agenda was somehow morally justified in their own minds. And we cannot rule out, a priori, analogous conclusions in analogous cases.

Further, it is not just war crimes that one has to worry about, but a whole raft of antidemocratic practices. As Lance deHaven-Smith points out:

> [A]ntidemocratic conspiracies in high office do, in fact, happen. The congressional hearings on Watergate, the Church Committee's discoveries[33] about illegal domestic surveillance, and the special prosecutors' investigations of Oliver North and Scooter Libby revealed that public officials at the highest levels of American government can and sometimes do engage in conspiracies to manipulate elections, wiretap and smear critics, mislead Congress and the public, and in other ways subvert popular sovereignty. (2010, p. 796)

Indeed, there is a long list of officials who have been convicted of significant crimes, often involving corruption.[34] Further, President Bill Clinton has been accused of rape[35] and a pattern of sexual harassment. Regarding the latter charge, Clinton agreed to pay $850,000 to settle the matter. Whatever the truth about the more serious allegations, the Monica Lewinski affair itself shows that Clinton is not worthy of trust—neither in his words[36] nor in his moral character. And President Trump also has issues regarding his treatment of women, and his honesty is questioned ubiquitously.

And what about others who would ordinarily be presumed good, benign, or boring? Surely a celebrated academic would qualify as beyond suspicion, under at least one of these categories. For example, if a person, such as Ewen Cameron, "had been the president of the American Psychiatric Association, president of the Canadian Psychiatric Association, and president of the World Psychiatric Association," then any theory that suggested that he was up to no good could be summarily dismissed. Right? Certainly such a person could not be involved in anything so wild as developing "psychic driving" techniques as part of mind-control experiments for a mysterious and appalling CIA project. And yet, so he was (see Klein 2007, pp. 35–46).

And what about the CIA? Are its members, who are often implicated in conspiracy theories, also "otherwise considered to be good," or "benign"? (They can hardly be considered "boring," if that word has any relevance.) Acquaintance with the Phoenix Program ought to give us some pause, at the very least. The Phoenix Program was a large-scale kidnapping, torture, and assassination program run by the CIA during the Vietnam War. Although intended to serve strategic war ends, it targeted *civilian* dissidents and often misidentified them. And though ostensibly having an intelligence-gathering mission, it is arguably better classified under the rubric of terrorism and mass murder (see Valentine 1990). The details may be debatable, but it is fairly clear that neither Cameron nor the assassins and torturers involved in the Phoenix Program, nor their superiors, were behaving like choirboys. Indeed, what they were doing, from the point of view of commonsense morality, appears to be so wrong that one may be excused for calling it "evil," even if they thought they were serving a higher purpose.

Let's take one more example. According to Book III of the *Final Report* of the Church Committee, in one "shameful chapter" of US history the FBI made concerted efforts to "neutralize [Martin Luther] King as an effective negro leader" (US Senate 1976b, pp. 86, 134). The FBI, an organization that is presumably "otherwise considered to be good," admitted to being in a "rough, tough, dirty business. . . . No holds were barred"[37] (p. 134). And so perhaps we should not be too surprised to find that, as the *Report* states, "The FBI campaign to discredit and destroy Dr. King was marked by extreme personal vindictiveness" (p. 82). The *Report* continues:

> [A]s the FBI now states, its activities were unjustified and improper. In light of the Bureau's remarks about Dr. King, its reactions to his criticisms, the viciousness of its campaign to destroy him, and its failure to take comparable measures against the Advisers that it believed were communists, it is highly questionable whether the FBI's stated motivation was valid. It was certainly not justification for continuing the investigation of Dr. King for over six years, or for carrying out the attempts to destroy him. (p. 85)

None of this is news to the politically alert. So the assumption of a special level of moral purity routinely granted by academics to presidents and other high officials, and to government agencies, seems at least inappropriate, if not bizarre.[38]

Therefore, when conspiracy theorists attribute seemingly heinous acts to people of influence in Western democracies, they should not be dismissed as positing something wholly unrealistic. At the same time, we must remember that even if *some* conspiracy theorists *do* suppose that the perpetrators are evil in a particularly strong and implausible sense, they generally *need not* do so, and many of their fellow conspiracy theorists *in fact* do not do so. As I have stressed, those seriously interested in assessing the truth or falsity of a conspiracy theory ought to focus on the best versions. At least, one cannot reasonably throw out a more plausible version on account of the existence of a less plausible one.

8.8. On the Possibility of Malevolent Global Conspiracy

I have been arguing that many conspiracy theories do not have the problematic characteristics of the "paranoid style." And so prejudicial assumptions that attend the paranoid style should not be assumed in reference to these other conspiracy theories. I have further argued that since Western elites have behaved badly in the past, and indeed have participated in harmful conspiracies, accusations of other past, present, or future bad behavior on their part should not be regarded as so implausible that it may be safely dismissed on that account.

And now a couple further points must also be made, though somewhat reluctantly. First, even though the various aspects of the paranoid style do make *those types* of conspiracy theories *relatively* implausible, even theories expressed in the paranoid style cannot be *completely* dismissed on that account. As Lee Basham points out in his insightful article "Malevolent Global Conspiracy," "Even if we believe nothing is now afoot (certainly my dominant reaction) it seems reasonable to believe that at some point in the future a serious attempt will be made at malevolent global conspiracy" (Basham 2006, p. 101). And even Hofstadter himself acknowledges:

> Nothing entirely prevents a sound program or a sound issue from being advocated in the paranoid style, and it is admittedly impossible to settle the merits of an argument because we think we hear in its presentation the characteristic paranoid accents. Style has to do with the way in which ideas are believed and advocated rather than with the truth or falsity of their content. (1965, p. 5)

True indeed, as the following case exemplifies.

David Aaronovitch quotes David Ray Griffin, who does not generally typify the "paranoid style," as saying, "We have become entranced by demonic power, so focused on lust for wealth and control that almost anything becomes possible" (Aaronovitch 2010, p. 229). Given Griffin's process theology, being "demonic" can probably be best interpreted as being "directly opposed to divine purposes," which is how he characterizes the policies that 9/11 has made possible in his book *Christian Faith and the Truth behind 9/11* (Griffin 2006, p. viii). Griffin makes clear that he takes 9/11 to have been undertaken by insiders in support of an imperialistic project: "[T]he Bush-Cheney administration had orchestrated 9/11 in order to promote this empire under the pretext of the so-called war on terror" (p. viii). Alluding to a Christmas card sent out by Vice President Cheney that suggests that "an empire can rise" only with God on its side (p. vii), Griffin asserts that "there can be no greater distortion of the original Christian message than the use of Christianity to support imperialism." From this it is clear that Griffin posits empire building as a key motive behind 9/11. Whether or not it is appropriate to characterize actions stemming from such a motive as "demonic," it is hard to deny that empire building, which has generally involved the imposition of much suffering, has been a historically significant motive, and thus not a highly implausible one. And so, Griffin's statement, which seems to exhibit (part of) the "paranoid style," can be unpacked in a way that reveals the underlying idea to be a plausible one, especially if one remembers the brutal history of imperialism/colonialism.

Further, at the deepest level, although my own religious worldview does not include genuinely Sinister Forces, I think we must admit that fundamental metaphysical worldviews are unavoidably controversial. For all anyone really knows, Sinister Forces (whatever that might mean metaphysically) may exist.

And so, we must admit, one must look neither merely at the "style" of a theory, nor at controversial metaphysical presuppositions, which may or may not be genuinely implied by the theory. Rather, to really be justified in confidence regarding the truth or falsity of a particular theory of any historical event, one must carefully investigate the relevant empirical facts and evaluate the quality of the best arguments on a case-by-case basis. There is no shortcut that will help with difficult cases. (Though not all cases will be difficult, such as those seeming to involve fundamental contradictions or clear impossibilities with no compelling supportive evidence or counterargument.)[39] The idea that each conspiracy theory ought to be judged on its particular merits—particularism—applies even to theories positing large-scale, effective, and pernicious conspiracies. Never forget, for example, the Holocaust (about which more will be said near the end of the next chapter).[40]

NINE

For the Greater Good

Conspiracy and Moral Justification

(an interlude)

In the previous chapter I made reference to numerous examples of conspiracy theories that did not imply implausibly evil intentions. In the interest of maintaining the flow of the argument, I did not digress much into details. But not only are such details often fascinating, exploring them is very useful for developing a sense of the subtleties that must be considered when thinking about conspiracy theories. Whole books could be (or in some cases have been) written on each of the topics below, so by comparison even this treatment will seem barely to scratch the surface. It is nonetheless a surface worth scratching.

Focusing mainly on topics that came up in the discussions in the previous chapter, I here discuss some relevant details that further support or extend a major theme of that chapter: that many conspiracy theories do not imply implausibly evil intentions, and so cannot reasonably be dismissed on that basis. I start with examples that are not even prima facie immoral, and then proceed to increasingly morally problematic examples. One purpose of exploring these examples is to develop a more nuanced understanding of the possible motivations involved in the carrying out of, or the covering up for, a conspiracy. These examples also reveal, in concrete ways, that perversity does not imply maliciousness. This understanding can help us see why certain conspiracy theories are not as implausible as they may have at first blush seemed.

According to Brian Keeley, one of the features of "unwarranted conspiracy theories" is that "The true intentions behind the conspiracy are invariably nefarious" (1999, p. 117). He adds, "I am aware of no popular conspiracy theory according to which some group of powerful individuals is secretly doing *good* but desperately hopes its schemes will not be revealed." (He does, however, reconsider this position in Keeley 2007, p. 141.) Similarly, Stephan Lewandowsky, Elisabeth Lloyd, and Scott Brophy write, "[T]he presumed motivations behind any assumed conspiracy are invariably assumed to be nefarious: We know of no conspiracy theory that would acknowledge the benign motivations of the presumed conspirators" (2018, p. 221). Here, strangely, they cite Keeley 1999, as though his claim to ignorance justifies their own. And they fail to recognize that Keeley revised his view. In any case, they are overlooking a lot. They fail to notice that conspirators may believe that they are doing good while recognizing that their actions would nonetheless be very unpopular and possibly illegal. And there are lots of examples, such as those described below, in which at least some conspiracy theorists acknowledge this. (Many more examples could have been given: those who believe in a moon-landing hoax generally do not believe that the moon landing was faked *for nefarious reasons*.)

9.1. The North American Union and the New World Order

While some conspiracy theories involve obvious atrocities, even if they may be thought of as sacrifices for a greater good, not all do. To use an example cited by Uscinski and Parent, consider the idea that "the governments of the United States, Canada, and Mexico have a clandestine plan to establish a new North American currency, the Amero, as part of a supposed New World Order, which would extinguish the sovereignty of the United States" (2014, p. 33). Although many Americans would be appalled at the idea of extinguishing US sovereignty, there is nothing clearly immoral or harmful about this plan, even though there will surely be winners and losers, and even if it is actually a misguided idea, all things considered. The only thing that is prima facie morally problematic about it is the clandestine, and thus undemocratic, process by which it is supposed to be undertaken. In cases like this, there is not even a prima facie reason to regard the conspiracy as "against the common good," let alone "evil."

Now consider the related "New World Order" conspiracy theories. Feeding such theories, David Rockefeller writes in his *Memoirs*:

> Some even believe we [the Rockefeller family] are part of a secret cabal working against the best interests of the United States, characterizing my family and me as "internationalists" and of conspiring with others around the world to build a more integrated global political and economic structure—one world, if you will. If that's the charge, I stand guilty, and I am proud of it. (2002, p. 405)

Why is he proud? Because he believes (or at least claims to believe) that it is a good thing. (It may be against the best interests *of the United States*, but not bad *all things considered*.) Rockefeller was perhaps most explicit in a 1991 Bilderberg meeting, when he reportedly said, "[S]upranational sovereignty of an intellectual elite and world bankers . . . is surely preferable to the national autodetermination practiced in past centuries" (McLaren 2003, p. 163). The reason that I am aware of these quotations is that *conspiracy theorists so frequently point them out*. So those various conspiracy theorists must know that there is a way of looking at the New World Order that is not at all evil, even if they question Rockefeller's sincerity. (It is not hard to see how a world run by bankers would be appealing to Rockefeller. And so he might reasonably be considered likely to be biased in his assessment, though not hell-bent on wrongdoing.)

Similarly, one of the founders of the Bilderberg group, Denis Healey, admits:

> To say we were striving for a one-world government is exaggerated, but not wholly unfair. Those of us in Bilderberg felt we couldn't go on forever fighting one another for nothing and killing people and rendering millions homeless. So we felt that a single community throughout the world would be a good thing. (Ronson 2001)

Indeed, both the idea of a Pax Americana and the vision of some UN-based New World Order could be conceived of, and probably *are* conceived of, at least by some people, as good things, even if they may be unpalatable to many people—thus the need for secrecy in their establishment. And here again there is nothing even prima facie immoral about the fundamental idea. Similarly, Bertrand Russell has been criticized by conspiracy theorists for his support for something like a "New World Order." Russell hoped for the development of a "single government of the whole world, possessing a monopoly of armed force and therefore able to enforce peace" (1953, p. 113). There is little reason to think Russell was being insincere, in part because this vision has such obvious appeal; it is not prima facie evil or against the common good, even if there are some problems with it.[1]

Whether or not Russell, or like-minded others, would support morally dubious *means* to achieve this end is another question entirely.[2] Well-meaning people might be willing to, as they would see it, sacrifice a few for the benefit of the world. Don't people make that judgment whenever they support war, for example? As Henry Kissinger remarked, rationalizing immorality in foreign policy, "[S]ometimes statesmen have to choose among evils" (Roth 2002). The idea of prima facie immoral activities being justified (or perhaps rationalized) by utilitarian reasoning is explored in the examples below, starting with relatively less egregious cases.

9.2. Population Control and Medical "Science"

There are cases that do involve prima facie immoral activities, which may nevertheless be easily imagined to be motivated by good intentions. For example, secretly adding sterilants to the water supply in order to make it more difficult for people to procreate would be considered appalling to many people, and to accuse a modern Western government of doing this would certainly count as a conspiracy theory. Though not carried out (as far as we know), this idea was dispassionately considered by authors that include John P. Holdren, who later became President Obama's science adviser (see Ehrlich et al. 1977, pp. 787–788). Although the authors opposed the strategy, they nevertheless seem to consider it to be potentially warranted in imaginable circumstances given the perceived problem of overpopulation, which they considered to be potentially calamitous.[3]

They start by acknowledging problems with this proposal, writing:

> Adding a sterilant to drinking water or staple foods is a suggestion that seems to horrify people more than most proposals for involuntary fertility control. Indeed this would pose some very difficult political, legal, and social questions, to say nothing of technical problems. (pp. 787–788)

After listing several practical problems with this approach, they articulate a specific version of the proposal, for further consideration:

> Physiologist Melvin Ketchel, of the Tufts University School of Medicine, suggested that a sterilant could be developed that would have a very specific action—for example, preventing implantation of

> the fertilized ovum. He proposed that it be used to reduce fertility levels by adjustable amounts, anywhere from 5 to 75 percent, rather than to sterilize the whole population completely. In this way, fertility could be adjusted from time to time to meet a society's changing needs, and there would be no need to provide an antidote. (p. 788)

The authors conclude, "[T]he risk of serious, unforeseen side effects would, in our opinion, militate against the use of *any* such agent, even though this plan has the advantage of avoiding the need for socioeconomic pressures that might tend to discriminate against particular groups or penalize children" (p. 788). Clearly, the authors believe that *something* must be done about overpopulation, and that a nonproblematic solution may not be found. In this context, none of the proposed tactics are written off completely. The authors, like good utilitarians, simply weigh the costs and benefits. The violation of individual rights is not a major concern here, they reason, because "Where the society has a 'compelling, subordinating interest' in regulating population size, the right of the individual may be curtailed" (p. 838).

One of the several other unpleasant tactics considered is forced abortion, regarding which they write:

> [Based on the obligation to promote] the general welfare . . . or under the equal-protection clause of the Fourteenth Amendment . . . it has been concluded that compulsory population-control laws, even including laws requiring compulsory abortion, could be sustained under the existing Constitution if the population crisis became sufficiently severe to endanger the society. (p. 837)

After all, the authors reason, "If society's survival depended on having more children, women could be required to bear children, just as men can constitutionally be required to serve in the armed forces" (p. 838).

I have chosen to describe this particular case partly because it is not so well known. There are other cases that could be cited as well, including ones in which prima facie immoral actions were actually undertaken by elites, not just talked about, which sacrificed individuals (ostensibly) for the good of the many. Perhaps the best-known example is the Tuskegee experiment. Author and journalist Jesse Walker succinctly describes it as follows:

> In a 40-year ruse, the Public Health Service told hundreds of black sharecroppers that it would give them free health care. Rather than

> inform the patients that they had syphilis, the doctors deliberately left the disease untreated in order to study whether the illness affects blacks and whites in different ways. (2014)[4]

Similarly, the *New York Times* reports, "From 1946 to 1948, American public health doctors deliberately infected nearly 700 Guatemalans—prison inmates, mental patients and soldiers—with venereal diseases in what was meant as an effort to test the effectiveness of penicillin" (McNeil 2010). While these experiments are rightly regarded as deplorable, the perpetrators are not generally regarded has having evil intentions. Indeed, they seem to have thought that they were conducting valid medical experiments that promised to increase scientific knowledge and benefit people (even though it is hard to deny that racist ideology played some role).

These cases are cited to illustrate the plausibility of appalling elite conspiracies undertaken *for the common good*. Or, to put it another way, it shows that characterizing conspiracy theories as being "against the common good" is too simplistic, imprecise, and, indeed, misleading, for it encourages a dismissive attitude that may not be appropriate—even if racism, discrimination, and callous disregard are, in fact, involved to some degree.

It might be objected that conspiracy theorists tend to view people like Holdren and his coauthors as insincere, merely *pretending* that they have good intentions. Indeed, some conspiracy theorists do. But not all. Not only is it *possible* to think that perverse policies might be undertaken for "noble" reasons, it is not uncommon for conspiracy theorists to explicitly suggest this. For example, (conspiracy theorist) Jeremy Day explains a mechanism of pernicious social control, as he sees it:

> Everyone thinks, "I'm going to do this because it is good for the world." . . . Everyone has an agenda. And now you hand someone a billion . . . or fifty billion dollars and suddenly if they think, "You know, there is really too many people on the planet. That's really the big problem. I'm going to put five billion toward influencing congress, and, you know, we're going to have some eugenics here, but we won't call it that"—I mean, why would that not be going on? And so, everyone has their agenda, and some people have a few billion dollars to execute that agenda.[5]

To be clear, I'm not endorsing Day's view. But Day's conspiracy theories cannot be dismissed on the basis that they imply implausible evil, since he does not posit evil intentions at all.

9.3. The "Benign" JFK Assassination Cover-Up

As emphasized in the previous chapter, it is important to make a distinction between a cover-up of a conspiracy and the underlying conspiracy itself. Even if the conspiracy is itself appalling—like the brutal assassination of President Kennedy—the motives for a cover-up may be more easily seen as "benign," or at least less morally problematic. This is the first of three sections addressing different aspects of the assassination of President Kennedy (though not exclusively). This section highlights some considerations suggestive of a "benign" cover-up. The following section is about purported motives behind the assassination itself. And the next section discusses a structure that may enable elite conspiracies of this kind.

Let's look at Chief Justice Earl Warren's motives in agreeing to head up the Warren Commission. First consider Warren's recruitment and the motives of President Johnson (and perhaps those of Deputy US Attorney General Nicholas Katzenbach as well). Johnson once said, "I never believed that Oswald acted alone" (cited in Olmsted 2009, p. 117).[6] Historian Kathryn Olmsted surmises that "he did not really want to know for sure" (p. 117), for he was worried that if there was a conspiracy, Castro might be implicated. According to Olmsted, "If [Johnson] had proof that Castro was behind the murder, then he would be forced to invade Cuba, and the Soviets might respond by launching World War III" (p. 117). As Olmsted also notes:

> In the memo [recommending the creation of the commission,] Katzenbach told President Johnson that the public must be satisfied "that Oswald was the assassin; that he did not have confederates who are still at large; and that the evidence was such that he would have been convicted at trial." In other words, Johnson must convince the public of something he personally did not believe. (p. 118)

To do so, Johnson "invoked the specter of nuclear holocaust to persuade [Warren]" to front a commission that was to reach preestablished conclusions (p. 118). Recordings of President Johnson reveal that, apparently, Warren was convinced by Johnson that it was important for the American people to believe that Lee Harvey Oswald acted alone in assassinating President Kennedy (whether it was true or not), because otherwise there might be a nuclear exchange with the Soviet Union resulting in the deaths of forty million Americans.[7] Now, that might seem ridiculous from our perspective today, if we lack awareness of the particulars. But Warren seems to have really believed it.[8] (Strictly speaking, what is important here is not that Warren believed it, but that at least some serious conspiracy theorists view Warren as being motivated by this belief. And *this* cannot be reasonably doubted.)[9]

Here, very briefly, are some of the particulars. Referring, presumably, to Oswald's (ostensible)[10] visits to the Soviet Embassy and Cuban Consulate in Mexico City, during which he supposedly had contact with a KGB agent apparently involved in assassination plots, Valery Kostikov, President Johnson describes how he got Warren to agree to head the commission: "I just pulled out what Hoover told me about a little incident in Mexico City" (Newman 2013). As a result, Warren, who had earlier refused to be involved with the commission that would later bear his name, agreed to go along with it—and Johnson says that Warren started crying and said, "I'll just do whatever you say."[11] It appears that Warren may not merely have been agreeing to head up the commission, he also knew what the commission, for the good of the country, needed to "prove"—regardless of whether or not it was true. Many, if not most, JFK "conspiracy theorists," especially the serious ones (who have published or given talks on the assassination), understand Warren's role in this light—and a few other members of the Warren Commission are viewed similarly. They are *not* thought to be covering up for the "real assassins," though they *are* thought to be participating in a cover-up nonetheless. As (conspiracy theorist) Jim Garrison said:

> So, after the meeting [with President Johnson], after probably initially saying "no" thirty times, Chief Justice Warren comes out of the oval room with tears running down his face and *for reasons which I think are probably essentially altruistic* has [agreed] to head an investigation in which he knows that the result has to be, in one way or another, has to be false. (Barbour 1992, emphasis added)

A slightly more modest interpretation is also plausible: Warren may well have succeeded in convincing himself that the lone gunman theory was true, but he can hardly be regarded as unbiased, given what he thought was at stake.

If Warren was complicit in a kind of conspiracy, it is a small one (and even this conspiracy is not unambiguously explicit). This little conspiracy is obviously *related* to other, and larger, conspiracies (if the conspiracy theorists are right), but many, if not most, conspiracy theorists do not regard it as part of a single vast conspiracy. They have more complex and nuanced positions. If critics of conspiracy theories are to avoid the charge of attacking a straw man, they must take account of those complexities and nuances.

It should be noted that so-called conspiracy theorists are not alone in positing a well-meaning cover-up. FBI special agent James P. Hosty, who was assigned to investigate Oswald upon Oswald's return from the Soviet Union and was involved in Oswald's interrogation following the assassination, also held this view. Hosty remarks:

> I feel, based upon what I know now, there was what I would call a "benign cover-up." Yes there was—they were concerned about Oswald's connections to the Soviet Union and to Castro. And they were fearful that if the American public were to find this out they could become so incensed that it could have possibly led to an atomic war.[12]

The twist here is that the cover-up that Hosty is referring to, which involves the FBI's concealment of information, is one in which Hosty himself admits involvement.[13] Further, the position that there was such a "benign cover-up" has been used by Philip Shenon to defend the "surviving staff lawyers" of the Warren Commission (2013, p. 13). According to Shenon, Warren Commission members thought that the FBI was engaged in an "orchestrated effort" to, it seems, cover up certain facts related to the assassination. (And Shenon seems to agree.) Shenon writes:

> To several of the commissioners, the articles [stating that the FBI would report that "the lone and unaided assassin of President Kennedy"] amounted to an orchestrated effort by the FBI, and probably by Hoover personally, to cement public opinion around the idea that there had been no conspiracy to kill the president—certainly no conspiracy that the FBI might have been able to foil. The bureau was trying to force them [the commission members] to reach conclusions before they had weighed any of the evidence, it appeared. (2013, p. 72)

At this point, the evidence of an FBI cover-up seems quite strong. Wouldn't it be a little odd to admit that the FBI was engaged in a cover-up of some kind and yet dismiss as implausible the notion that the members of the Warren Commission were doing something analogous? (See Shenon 2013, pp. 145–146, for an example of the Warren Commission contemplating destroying records of their own conversations.)

The CIA, likewise, engaged in a "benign cover-up." As Shenon reports:

> Half a century after JFK's death, in a once-secret report written in 2013 by the CIA's top in-house historian and quietly declassified last fall, the spy agency acknowledges what others were convinced of long ago: that [CIA director John] McCone and other senior CIA officials were "complicit" in keeping "incendiary" information from the Warren Commission. According to the report by CIA historian David Robarge, McCone, who died in 1991, was at the heart of a "benign cover-up" at the spy agency, intended to keep the commission focused on "what the Agency believed at the time was the 'best truth'—that Lee Harvey Oswald, for as yet undetermined motives, had acted alone in killing John Kennedy." (Shenon 2015)

The focus here has been on the "benign" motives of at least some of those engaged in a significant cover-up. But notice also that not everyone involved need have the same or analogous motives. A cover-up that is largely "benign," motivated mostly by either good intentions or to prevent exposure of institutional screw-ups, can occur even if the underlying truth, unbeknownst to most of those involved in the cover-up, actually involves an appalling conspiracy.

9.4. Assassinations and False Flag Terrorism

Now let us turn to cases that are, prima facie, more strongly immoral: those involving assassination or the killing of innocent civilians. Even these cases can be thought of as "good," that is, justified in utilitarian terms. For example, a relatively strong case can be made that the murder of President Kennedy (assuming it was an elite conspiracy) was motivated by the sincere belief, for which significant justification could be offered, that Kennedy was selling out to the communists,[14] and putting America in grave peril, not to mention being an outrageous philanderer and moral degenerate. As mentioned briefly in the previous chapter, not only *could* such a case be made, it *is* made by, among others, David Talbot, James DiEugenio, and Daniel Sheehan—all JFK conspiracy theorists:

Talbot writes: "Over the final months of JFK's presidency, a clear consensus took shape within America's deep state: Kennedy was a national security threat. For the good of the country, he must be removed" (2015, p. 560).[15]

DiEugenio writes:

> [T]he hawks in this administration—[William] Harvey, [Curtis] LeMay, Dulles—had now seen Kennedy bypass two perfect opportunities to rid America of the 'Cuba problem.' By the end of the Missile Crisis, they must have felt that if Kennedy had not acted by then—with Cuba nearly a nuclear base aimed at America—that he was not going to act ever. (As we will see, they were correct.) The Cuban exiles felt the same way. Kennedy's no-invasion pledge was shattering to them. As one investigator discovered, whenever there were gatherings of exiles in safe houses or at training camps, the word 'traitor' was now used in relation to JFK. (2012, pp. 66–67)

Sheehan states: "When one is playing the long game here, one could argue that . . . if in fact Kennedy . . . had been allowed to continue going forward [with his disarmament plans] that we never would have ended up winning the cold war, and that the opportunity that we have to actually export democracy, to export democratic capitalism, to do all of those things—that these have been made possible by the people who had the courage to do what was necessary to do." In other words, "the President was killed because he should have been."[16]

Similarly, many members of the 9/11 Truth Movement either view the suspected perpetrators as misguided neoconservative ideologues or are at least open to that possibility. They may suspect some degree of self-deception or believe that motives were mixed. Some, such as the leadership of Architects and Engineers for 9/11 Truth, try to remain neutral about such questions, sticking to scientific issues as much as possible, at least publicly. The individual members of this organization presumably have varied opinions.

Finally, let me acknowledge that, regarding both the Kennedy assassination and 9/11, conspiracy theorists do tend to view those events as designed, at least in part, to enable or facilitate the escalation of war. But a conspiracy to support war may be thought to be for the good. (I happen to think that war is *almost always* an abominable idea, but most people don't seem to agree with me.) Even conducting a false flag operation on the scale of 9/11 could be conceived of as justified for a greater good, just as allowing the attack on Pearl Harbor, if it was known in advance, could be (more plausibly) justified as necessary in order to get the United States into World War II, and thus save the world from tyranny. In the case of 9/11, the purpose may be to secure a Pax Americana. Indeed, Donald Rumsfeld is reported, by (conspiracy theorist) Peter Dale Scott, to have explicitly framed 9/11 as "the kind of opportunit[y] that World War II offered, to refashion the world" (2007, p. 2).

9.5. National Security and the "Vast" Secret Team

Let's look at one more example relevant to the Kennedy assassination. This one focuses less on motivation and more on size and structure. But it also can help us understand how a "vast" structure can be developed based on prima facie good intentions (national defense) which can be subverted by a few members with their own complex and perhaps diverse intentions.

Fletcher Prouty is the author of *The Secret Team: The CIA and Its Allies in Control of the United States and the World*, and *JFK: The CIA, Vietnam, and the Plot to Assassinate John F. Kennedy*. At the time of the assassination of President Kennedy, Prouty was chief of special operations for the Joint Chiefs of Staff. Prouty *did* use the word "vast"[17] to describe a large complex network of well-placed "team"[18] members who could be called upon to acquire information, borrow resources, exert influence, and initiate courses of action within an organization. It is important to stress that Prouty is not *theorizing* the existence of such a network; he is describing a network that he personally worked with. He is either lying about his firsthand experience, or a network of this type existed (at least during the 1950s and 1960s). There is room, of course, for interpretation regarding degrees—*how* extensive and *how* effective this network was. Whether or not, or to what degree, it was involved in controversial events, such as assassinations, is also up for interpretation, as Prouty admits.[19] And there is also room for differences in moral evaluation, which may be influenced by the answers to various factual questions. Prouty himself is persuaded that elements within this team were involved in assassinations, including the assassination of President Kennedy. But he does not seem to think that the existence of this kind of network is *necessarily* problematic. In Prouty's view, the operations of the "Secret Team" became problematic because of contingent historical circumstances.

As Prouty presents it, having a secret team created certain efficiencies that facilitated various legitimate national defense operations. The main problem, in Prouty's view, was oversight. Oversight can easily be lost at the top if there is a reason to want to turn a blind eye to certain activities, perhaps to enssure "plausible deniability." However, when mechanisms of oversight are lost for this reason, one loses sight of more than just the narrow issue regarding which one seeks deniability. In addition, dispersed and informal mechanisms of oversight at lower levels throughout the network are lost as a matter of course, as Prouty explains in some detail:

> [T]he CIA would politely offer to provide a few men to work with the regular customs personnel to take the burden for such activity from them. This was the way it was put in the first place, and the customs office would gratefully accept the assistance. The CIA would go through all the necessary steps to get authorization for increasing the manpower allocations in the customs service by the number of men it planned to put there and then to make arrangements to reimburse the customs office for the payroll and other costs of the office. . . . In the beginning this would all be done with elaborate open-handedness, even to the point where the new agency men would receive training and other prerequisites of the job. However, as the years passed, most of this procedure would be forgotten, and few would recall that those special assignments had even originated with the Agency. . . . This has happened quite extensively in a great many places all throughout the Government. There are CIA men in the Federal Aviation Administration, in State, all over the DOD, and in most other offices where the CIA has wanted to place them. Few top officials, if any, would ever deny the Agency such a service; and as the appointive official departed, and his staffs came and went, the whole device would be lost with only the CIA remembering that they were still there. . . . On the basis of security [Dulles] would place people in all areas of the Government, and then he would move them up and deeper into their cover jobs, until they began to take a very active part in the role of their own cover organizations. This is how the ST [Secret Team] was born. Today, the role of the CIA is performed by an *ad hoc* organization that is much greater in size, strength, and resources than the CIA has ever been visualized to be. (Prouty 2011, pp. 304–306)

The important point to recognize here is simply that *none of this is implausible on its face*. Indeed, there is little question that the CIA does engage in this *kind* of thing, as revelations about Operation Mockingbird show, to cite just one example.

Since we are exploring details here, allow me to digress for a paragraph to mention some relevant facts. The Church Committee report revealed:

> [T]he CIA maintained covert relationships with about 50 American journalists or employees of U.S. media organizations. They are part of a network of several hundred foreign individuals around the world who provide intelligence for the CIA and at times attempt to influence foreign opinion through the use of covert propaganda. (US Senate 1976a, p. 192; cf. Bernstein 1977)

Relatedly, in 1975, then-CIA director William Colby, testifying before the House Select Committee on Intelligence ("the Pike Committee"), explains:

> CIA personnel, when they go overseas, obviously must go under some other title. This is, in some cases, a title of another agency of the government. In some cases it is some other title outside of the government. For reasons of continuity, CIA personnel sometimes have to retain that identification while they serve a tour inside the United States.[20]

While this is commonsensical, and not inherently pernicious, such explanations help us begin to form an understanding of how and why covert networks exist within the United States. Further, it helps us recognize that it is not implausible that parts of such a network might at times get out of control in the way that Prouty suggests.

Now, the *extent* of the network that Prouty refers to, its *level* of effectiveness, and the *degree* to which its members were willing to engage in criminal or appalling activities need to be determined empirically. Such matters cannot be judged a priori, or adjudicated by means of some simplistic and imprecise heuristic, such as the notion that it amounts to a vast or malevolent conspiracy.

It is also important to recognize that, while the network Prouty describes is "vast" (according to his own choice of words), only a small portion of that network would need to be involved in any particular operation. (Perhaps more of it will be called upon after the fact to participate in the cover-up.) And, further, most actors will not fully understand how their action fits into the whole. So the idea that the whole network, or even the fraction of it that participates in some way, is "in on the conspiracy," in the sense of being aware of and consenting to its full scope, is a mischaracterization of the view. Compartmentalization based on the "need to know" segregates various actors. And various feints of various kinds are, Prouty reports, going on all the time. So few people know whether something is as it appears or not, even if they are participants. Sometimes they just know their own task (but not whether or not it is a mere feint). Other times they are told a story that they may believe, such as seems to be the case with Earl Warren. Some actors are "agents," genuinely part of the "team." Other actors are "assets," whose actions are generally to some greater degree manipulated. Assets (and even agents) may be witting, unwitting, or partially witting. In the latter case, like Warren, they may wittingly participate in a biased investigation with a foregone conclusion; but they *unwittingly* participate in the cover-up of the CIA's involvement in the killing of President Kennedy (if that is what happened), because they do not know or believe that is what happened. They think they are covering up something else, such as institutional or personal incompetence or blunders. Or, as in Warren's case, they are doing it for some larger reason—for the good of the country.

Sophisticated conspiracy theorists, recognizing that a vast conspiracy would be less plausible than a more modest one, try to figure out how the conspiracy could have been pulled off without being vast. That is, they try to think of the most plausible conspiracy theory that they can. This is part of the *theorizing* regarding conspiracies. Opponents of conspiracy theories would naturally not be inclined to engage in such theorizing. But failing to do this results in a misunderstanding of their subject. This theorizing process is *necessary to formulate the best conspiratorial explanation* that can be constructed. Without this, one cannot fairly assess the likelihood of a conspiracy. One would end up comparing a relatively well-articulated hypothesis (the official account) with a sketchy caricature of its rival. And that is largely what happens when mainstream sources, and purported epistemic authorities, pontificate on conspiracy theories.

9.6. Torture and Murder as Policy

President Trump is explicit about his advocacy of a policy of killing whole families based on the presumed sins of the father (LoBianco 2015). Trump's 2016 campaign slogan was, "Make America great again." Trump's popularity, and subsequent electoral victory, suggests that this resonates with many Americans. A big part of being "great," it appears, is being "strong." Marco Rubio's campaign slogan emphasized this explicitly: "The world is a better place when America is the strongest country in the world." What is required to be strong and great? One major issue is the response to ISIS, and radical Islam more generally. Trump's unabashed strategy: "He would kill the families of terrorists in order to win the fight against ISIS" (LoBianco 2015). So, apparently, being "great" is compatible with murder—fighting terrorism with terrorism. The point is this: *willingness, on the part of average Americans as well as elites, to advocate or approve of the commission of atrocities for the purpose of national security, or even for crass national self-aggrandizement, is not even rare.*

Consider also how torture is viewed in America, post-9/11. When I taught a course called Moral Problems, every semester some students expressed support for the use of torture. My students, needless to say, were not evil. They thought it was necessary for defensive purposes, and they suspected that the tortured individuals, who were presumed to be terrorists, probably deserved it anyway. Polls suggest that such support for torture is widespread—59 percent approve, according to a 2014 poll (Goldman and Craighill 2014). So, to suggest that high-level government officials would be willing to utilize torture is not to suggest they are unusually immoral.

Indeed, it is clear that members of the Bush administration regarded waterboarding as justified—Dick Cheney famously said that using it was a "no-brainer" (Eggen 2006). And legal memos were produced by the Office of Legal Counsel justifying so-called enhanced interrogation techniques (see Lewis 2005). One of the authors of these memos, John Yoo, went so far as to suggest that it might be permissible to crush the testicles of an innocent child to get someone else to talk. When challenged on this, Yoo responded, "I think it depends on why the president thinks he needs to do that" (Blumenthal 2006). In addition, President Obama authorized the extrajudicial assassination of American citizens overseas as well as drone strikes that kill innocent bystanders—and publicly defended this. And, making excuses for clear violations of the Geneva Conventions in the formerly secret "harsh interrogations" program, President George W. Bush once remarked, "That's like—it's very vague. What does that mean, 'outrages upon human dignity'? That's a statement that is wide open to interpretation" (Bush 2006). All of this shows that a theory that attributes prima facie immoral actions to prominent Americans (or others), even to the point of perpetrating "outrages upon human dignity," cannot be summarily dismissed as implausible on that account because, in part, *they may well think they are doing the right thing*. (A mix of moral awareness among various actors is also plausible.)

In *Evil: Inside Human Cruelty and Violence*, Roy Baumeister argues that there are four pervasive causes of evil, where "evil" is understood as "crime, violence, oppression, cruelty, and the rest" (1997, p. 375). The causes are (1) "desire for material gain, such as money or power," (2) "threatened egotism" (3) "idealism," and (4) "the pursuit of sadistic pleasure," the fourth being by far the least common (pp. 376–377). It is the third cause that deserves special attention here. Baumeister elaborates:

> When people believe firmly that they are on the side of the good and are working to make the world a better place, they often feel justified in using strong measures against the seemingly evil forces that oppose them. Noble ends are often seen as justifying violent means. In reality, such means often discredit and contaminate the noble goals, but this outcome is rarely anticipated. Human nature inclines people to align themselves in groups that square off against each other, each group seeing itself as good and the other as bad. Group competition can evolve into brutal conflict in which each side sincerely sees itself as the good guys who need to take strong measures to defeat the forces of evil that oppose them. When the perpetrators are driven by idealism, the victims do not get much mercy. (1997, p. 377)

Baumeister describes these causes of evil as "pervasive." Relatedly, a strong case has been made that people who commit actions that may be regarded as "evil" are generally psychologically normal (Bartlett 2005).[21]

Much of the above can be viewed on the model of law enforcement personnel falsifying, or misleadingly characterizing, evidence against someone they believe (or believe they *know*) to be guilty. In such cases of "noble cause corruption" the wrongness of the activity is not obvious to the perpetrators. Indeed, the point is that law enforcement personnel who do this kind of thing, and too often they *do* (see Friederdorf 2015), presumably think they are doing the right thing. Or, if they are doing it for careerist motives, they have probably at least rationalized it to themselves. Though generally regarded as blameworthy, indeed criminal, such activity is not commonly regarded as evil. On the contrary, it is understandable; it is all too human.

9.7. The Holocaust as a Conspiracy

Sometimes, however, intentions can be worse than merely selfish; they can be murderous—even if there is a twisted idealism involved. The Holocaust provides an example of such a murderous conspiracy. Though some have questioned its relevance to reasoning about conspiracy theories, its relevance has been defended in the philosophical literature. David Coady, for example, writes:

> [T]he Holocaust itself was to a large extent brought about by conspiracy. Although here and now belief in the Holocaust does not qualify as a conspiracy theory, because it also has official status, there was a time and place (*i.e.*, Nazi occupied Europe) in which what would now be called "Holocaust Revisionism" was the official story, and belief in the Holocaust was belief in an accurate conspiracy theory. To believe in the Holocaust was to deny the official story that the Jews were merely being resettled. (Coady 2003, p. 205)[22]

Basham and Dentith make the same observation, adding a few details:

> [T]he Holocaust began as a conspiracy. Prepared in secret councils of the Nazi party, the conspiracy culminated at the Wannsee conference of 1942. The contents of this conference were hardly broadcast to the world or its intended victims. The Nazis assured the world it was "relocating" Jews, even forcing family members already in the extermination facilities to write letters to their relatives in "ghettos" (often rural camps) encouraging them to get on the trains, as life, they were forced to write, was so much better at the extermination facilities. When officials were challenged about their intentions and actions they argued anything more sinister than relocation was an outrageous conspiracy theory. (2019, pp. 82–83)

More generally, Basham rhetorically asks, "Would the state-sponsored horrors of the nineteenth and twentieth century—our many Holocausts—have occurred without intentionally hidden, evolving conspiracies?" (2006, p. 133) He asks this in response to Brian Keeley's suggestion that these horrors were largely committed *in the open*. Keeley had written, "[W]hen it comes to committing large-scale evil, secrecy is often largely unnecessary. Six million European Jews, Gypsies, and homosexuals can be rounded up and systematically murdered without a global conspiracy" 2003, p. 109). And he had asked, "[W]hy waste time and energy maintaining a conspiracy when history shows that one can get away with it in the open?" (Keeley 2003, p. 109). Basham answers this directly:

> The answer: So one *can* do it, and get away with it, *in the open*. Which returns us to the critical point: Were such open acts nevertheless

> the expression of a prior conspiracy? In all the obvious cases—the genocides against indigenous North and South Americans, the Jewish holocaust, the Stalinist "wrecker" show trials, and many others, the evolving plans came first, and they were hidden from their intended victims and those who would have successfully intervened on the behalf of the victims. The early stages of putting these plans into public action were disguised, or rationalized as the product of other considerations. Only these plans' end-game moves and ultimate results were publicly manifest, long after there was nothing ordinary people could do to stop them. In every case, it began with a conspiracy. It had to. (2006, p. 134)

Now let's look at this again from another angle. In reference to the Holocaust and "the Great Terror of 1934–1939 in the Soviet Union," philosopher Juha Räikkä suggests that "it is contestable whether these should be called genuine conspiracies" (2009a, p. 193). He explains: "[I]t was generally 'known' what was going on. What was not known was who was responsible, how extensive the action was, and so on. But relatively early, most or at least very many people did have a suspicion that the official stories were not completely correct" (pp. 193–194). However, by *that* vague criterion, if today's most paradigmatic conspiracy theories about JFK and about 9/11 are one day confirmed to be true, we can look back at them and say that it is "contestable" whether they were really "genuine conspiracies" in the first place, because many people "knew" or suspected that "the official stories were not completely correct," even if they didn't know all the details.

These considerations suggest that the Holocaust, and our other "many Holocausts" as well, should count as relevant precedents when considering conspiracy theories. And this suggests that it is important to remain on guard against even monstrous conspiracies.

9.8. How Monstrous Is Too Monstrous?

Genocide—killing people regarded as "other"—is one thing. But would anyone in government or the intelligence services go so far as to directly kill many innocent people of "their own" for some further objective, say, to promote war? To get some perspective on that question it is useful to consider the Russian apartment bombings of 1999. Just two years before 9/11, there were a series of apartment bombings in and near Moscow which killed hundreds of people. The last incident occurred in Ryazan on the night of September 22, but this time the bomb did not go off because it was discovered and defused. For two days the official story was that this was another attack by dastardly Chechen rebels, thwarted by an alert citizen and local authorities. But that all changed when it was discovered that FSB (Russian intelligence) agents had planted the device.[23] The new story was that it was just a drill; there was never any real danger, for the "explosives" were just sacks of sugar. Many of the problems with this story were exposed on Russian TV,[24] and are documented in a book, *Blowing Up Russia*, by Alexander Litvinenko and Yuri Felshtinsky.[25] Litvinenko and Felshtinsky argue that the incident was a "failed attempt by the FSB to blow up an apartment building in Ryazan" (2007, p. 63), as an act of false flag terrorism. They also argue that the bombings in Buinaksk, Moscow, and Volgodonsk were successful versions of the same, both justifying war and propelling the new prime minister, recent FSB director, and former KGB officer Vladimir Putin, to the presidency. (Litvinenko, incidentally, was the former FSB agent who died in London in 2006 after being poisoned with polonium-210. To suggest that he was in this way assassinated by Putin, though not conclusively established, is generally not regarded as a conspiracy theory.)

Putin derides the suggestion of FSB involvement in the bombings as "raving nonsense." He argues, "There are no people in the Russian secret services who would be capable of such a crime against their own people" (Litvinenko and Felshtinsky 2007, p. 146). Foreshadowing an argument currently being made by philosopher Patrick Stokes against conspiracy theorizing in general (see section 2.12), Putin adds: "The very suggestion is immoral." And yet, according to the British journalist Edward Lucas, "The weight of evidence so far supports the grimmest interpretation: that the attacks were a ruthlessly planned stunt to create a climate of panic and fear in which Mr Putin would quickly become the country's indisputable leader, as indeed he did" (Lucas 2014, pp. 37–38). And historian Karen Dawisha, though more cautious in her accusation, remarks: "[T]o blow up your own innocent and sleeping people in your capital city is an action almost unthinkable. Yet the evidence that the FSB was at least involved in planting a bomb in Ryazan is incontrovertible" (Dawisha 2014, p. 222). And Oxford University professor of Russian history Robert Service praises the book *Blowing Up Russia*, calling it "As vivid a condemnation of the Putin regime as has yet been written" (back cover blurb).

The above quotations notwithstanding, what really happened in Ryazan, and the other Russian cities, cannot be established in a couple paragraphs. Even so, this case does provide relevant food for thought.

9.9. Implications for 9/11 Conspiracy Theories

Compared to MIHOP versions of 9/11, the Russian bombings are far less technically complicated, and thus less prima facie implausible. And yet there is a critical similarity: elites are thought to have blown up buildings in their own country, directly killing many innocent people, in order to garner support for war (and ancillary objectives). If one is willing to entertain the idea that the Russians might have done this, on what basis can one refuse to consider whether their American counterparts would not have? At least for those willing to consider that the Russian bombings *might* have been false flag events, even the controlled demolition hypothesis should not be written off *on account of its monstrousness*. The fact that the casualty count was about ten times higher in the case of 9/11 does little to diminish this conclusion. Once this kind of thing is regarded as justifiable, a larger number of casualties can be justified by larger potential gains.

It has already been argued at length that, on the one hand, political elites and intelligence agencies have a history of (prima facie) immoral behavior, sometimes costing hundreds of thousands of lives, if not millions. The Iraq war that followed 9/11, and was in part "justified" by it, was based on a conspiracy of deceit. As the Downing Street Memo relates, "[I]ntelligence and facts were . . . fixed around the policy." Noting this, philosopher Charles Pigden explains:

> [T]here *was* a conspiracy to talk up the threat of weapons of mass destruction and to justify the war, and the conspirators protected themselves from public scrutiny partly by resorting to conspiracy denialism. The upshot was that hundreds of thousands of people (perhaps over a million) died an early death and millions more were reduced to lives of misery, terror, and destitution. Conspiracy denialism was used to help justify a massive war-crime. It doesn't get much worse than that. (2019, p. 207)

Pigden does not mean to imply anything about alternative interpretations of 9/11, but his comment nevertheless highlights the degree of devastation that came from a real conspiracy related to the 9/11 conspiracy theories. As horrific as the devastation on 9/11 was, the devastation resulting from the known conspiracy to foment war (involving, presumably, at least some of the same actors) was much worse. As I have stressed in previous chapters, neither the Iraq war conspiracy nor the 9/11 conspiracy theory imply that the conspirators had evil intentions. The intentions in both cases may have been similar, if not the same—perhaps a largely ideologically mediated desire to reshape the world for the better (in their estimation), mixed at some level with callous self-interest. The fact that "a new Pearl Harbor" would be helpful in forwarding a neoconservative agenda is explicit in the PNAC document *Rebuilding America's Defenses* (Donnelly 2000, p. 51; see section 2.11), and the fact that 9/11 was exploited in the service of these goals is apparent. I must admit that there are significant differences between, on the one hand, starting a war that *ends up* costing the lives of hundreds of thousands of people, and, on the other, *directly killing* nearly three thousand people. Nevertheless, accepting the fact that a group of conspirators would do the former does raise the prior probability that they would do the latter.

One should not forget, in this context, that Western intelligence agencies have been involved in murderous projects, such as the Phoenix Program, and domestic government agencies have been willing to use their own citizens as guinea pigs for medical experimentation (the Tuskegee syphilis study), mind manipulation (MKULTRA), and Cold War radiation experiments. There is also some serious question regarding the possible involvement of Western intelligence in the false flag terrorist attacks associated with Operation Gladio in Italy and other European countries. Considering this history together with both the Russian apartment bombings and the Iraq war conspiracy, 9/11 conspiracy theories should not be dismissed on the assumption of moral self-restraint. They should be evaluated, like other conspiracy theories, on the evidentiary particulars.

TEN

Conspiracy Theorists and Social Scientists

Ostensibly authoritative sources, such as social scientists who study conspiracy theorists, are generally expected to be logically rigorous, intellectually honest, and unbiased. This chapter suggests that this expectation may not always be justified. Specifically, it exposes a number of significant problems in an attempt by a group of social scientists to defend the scientific study of conspiracy theorists.

I focus on five problems. First, these social scientists cite seriously flawed studies as though they offered robust and noteworthy findings. Second, their critique of a central distinction, between "generalist" and "particularist" approaches, is based on a misconstrual. Third, they misrepresent their own previously stated intentions. Fourth, they misrepresent a critique of those intentions. And fifth, they fail completely in their attempt to show that, regarding the inappropriate pathologizing of conspiracy theorists, their critics are as guilty as they are. All this raises a troubling question: How can we have confidence in a program of study undertaken, in part, by scientists who are so prone to error and misrepresentation?

Conspiracy theorists labor under considerable abuse from ostensibly reliable sources in the media, government, and academy. Often judged collectively, conspiracy theorists are commonly dismissed as irrational and deluded ignoramuses. While it is true that *some* conspiracy theorists appear only weakly tethered to reality, and logical and factual mistakes can easily be found in their work, not all conspiracy theorists are equal in these respects. And so grouping all conspiracy theorists together and judging them as if they formed a homogeneous class is both unfair to the better-grounded conspiracy theorists and is seriously misleading. Judgments about Alex Jones (a provocative host of a conspiracy-oriented radio show) may not apply to Peter Dale Scott (a scholar whose books are published by the University of California Press), even though they both find the official narrative of the September 11 attacks, as well as the JFK assassination, to be dubious. Further, and more to the point of this chapter, while it is not surprising to find errors and sloppiness of reasoning in the work of some conspiracy theorists, we should expect more from mainstream and "authoritative" sources, such as the academics who study conspiracy theorists. So it would be particularly troubling to find that some social scientists—often social psychologists[1]—have engaged in sloppy reasoning in what appears to be a self-serving agenda to "scientifically" reinforce the stigmatization of conspiracy theorists. But that is precisely what we find.

Our saga begins with a joint statement published in *Le Monde* by a group of social scientists calling for more research on conspiracy theorists in order to "fight" the "disease" of conspiracy theorizing more effectively.[2] In response, a number of scholars, including myself, signed an open letter criticizing this agenda (Basham and Dentith 2016).[3] In response to us, the authors of the *Le Monde* statement (minus Karen Douglas) published a sprawling rebuttal entitled "'They' Respond" (Dieguez et al. 2016), published in the *Social Epistemology Review and Reply Collective*. Several of us, in our own ways, responded in the same venue. This chapter is an extended version of my initial response (Hagen 2017).[4]

10.1. Uncritical Citation of Seriously Flawed Studies

The *Le Monde* authors claim merely to want more "rigorous empirical research" on conspiracy theorists (Dieguez et al. 2016, p. 25). Some of this supposedly "rigorous empirical research" on conspiracy theorists does, it turns out, just happen to have unflattering conclusions about conspiracy theorists. However, at least some of these conclusions are based on flagrant errors in reasoning[5]—errors that the *Le Monde* authors uncritically overlook, citing these studies as if they were unproblematic and authoritative, and thereby implying that there is some kind of noteworthy problem in the reasoning of conspiracy theorists.

For example, citing Wood et al. 2012 and Swami et al. 2011, respectively, the *Le Monde* authors state that conspiracy theorists "simultaneously endorse flatly contradictory conspiracy theories [and] readily accept experimentally made-up conspiracy theories" (Dieguez et al. 2016, p. 25). That makes it sound like there is some pretty sloppy reasoning going on. And there is. But it turns out that it is the social scientists doing the research that are guilty of it, as are those who uncritically cite this research and exaggerate its findings. Specifically, the claim that conspiracy theorists "simultaneously endorse flatly contradictory conspiracy theories" is not supported by the evidence and is presumably false, and the claim that they "readily accept experimentally made-up conspiracy theories" is a misleading exaggeration. In the former case, as I explain in some detail in Chapter 7, the authors of the study conflate *belief* with various ways of *giving some degree of credence*. For example, the respondents judged various conspiracy theories to be relatively more "plausible, convincing, worth considering, and coherent" than did others (Wood et al. 2012, p. 771). Notice that, while "endorsing flatly contradictory conspiracy theories" sounds irrational, there is nothing irrational about giving two contradictory theories more credence in the relevant respects than someone else does. Indeed, being alive to conflicting theories is not just unobjectionable; it is often an epistemic virtue, emblematic of a good detective or scientist.[6] In addition, the subjects were, in a significant sense, *justified* in giving more credence to these views than did their conventionalist counterparts, as I explain below.

As for the claim that conspiracy theorists "*readily accept* experimentally made-up conspiracy theories" (emphasis added), this is an exaggeration of the finding. The subjects rated "the extent to which they agreed" with statements regarding a fictitious conspiracy theory. The authors of the study find that "believing in real-world conspiracy theories appears to make it more likely that an individual will also be more accepting of fictitious conspiracy theories" (Swami et al. 2011, p. 460). Notice first that to be "more likely" to accept a theory than some others (who are more skeptical about conspiracy theories in general) is not the same as to "readily accept" the theory (that's the exaggeration). Notice too that "believing in real-world conspiracy theories" *ought to have the observed effect*. And thus, the study does not tell us anything interesting about the reasoning of conspiracy theorists. For all we can tell from these studies, conspiracy theorists are simply reasoning the way reasonable people should.[7]

The phenomena stems, in all likelihood, from the participants giving some weight to prior probability considerations—just as one should. Judging the prior probability of a hypothesis is simply a matter of estimating the likelihood of the hypothesis *before considering any direct evidence about the particular case*. In the case of the fictional conspiracy theory, the participants know nothing about the proposed theory (because it is fictional—but they don't know *that*). So, in making a judgment of how likely the proposed conspiracy theory is, they have nothing to go on except prior probability considerations. The most obvious and salient prior probability consideration in a case like this is the prevalence of relevantly similar phenomena. Obviously, people who tend to believe that other conspiracy theories are true will tend to (and indeed *ought* to) believe that a newly proposed conspiracy theory is "more likely" than people who tend not to believe other conspiracy theories.

And this is also why, compared to conventionalists, conspiracy theorists are *justified* in giving more credence to conspiracy theories that happen to be mutually inconsistent. The mutual inconsistency, as I have already mentioned, is a nonissue, since the participants in the study did not indicate that they simultaneously *believe* inconsistent theories. But given their other beliefs, namely, that various other conspiracy theories are true, they are justified in assigning a higher prior probability to various other analogous theories, compared to conventionalists. And thus they are justified, in a significant sense, in giving more credence to those theories, all else being equal. One might be tempted to say they are not *ultimately* justified, because their other beliefs about conspiracy theories are false. But that move would prejudge the most significant issue in question and reveal precisely the bias that I worry is undermining the attempt at science we are discussing.

Now, the reasoning process just described relates to what the *Le Monde* authors proclaim to be "the most robust finding in the rather recent field of social-psychological conspiracy theory research." Namely, "[P]eople who believe in one conspiracy theory tend to believe in other, unrelated, conspiracy theories" (Dieguez et al. 2016, p. 24). They cite ten studies, Wood et al. 2012 and Swami et al. 2011 among them, along with Goertzel 1994.[8] From this documented tendency, which is entirely epistemically unproblematic, as just explained, the *Le Monde* authors posit an inauspicious-sounding "conspiracist mindset." Now note, first of all, that "conspiracist" is even more clearly derogatory than "conspiracy theorist." The word "conspiracist" implies belief in conspiracy theories *where no such belief is warranted.* So a central question has been begged, per usual. And the rhetorical slant of much of the social scientific work suggests that, "of course, being more inclined to believe in one conspiracy theory just because one believes in another is surely irrational." *But it isn't.* It is no more or less rational than not believing in one conspiracy theory because one does not believe in another. Both are products of perfectly normal and reasonable inference making, as explained above.

So, the *Le Monde* authors, like at least some of the social scientist they cite, take a perfectly innocuous type of inference, which is not at all peculiar to conspiracy theorists and, with the help of exaggeration and uncritical appeal to flawed studies, frame it as though it represents some sort of abnormal cognitive error that applies uniquely to conspiracy theorists. They thereby appeal to faulty science to unfairly malign an already marginalized group.[9]

10.2. Overgeneralizing Particularism

In Chapter 2, I discussed the distinction between particularism and generalism. But if the *Le Monde* authors are right, this distinction makes no sense. If their arguments are successful, "particularists" such as myself will have to hang our heads in shame and admit that our whole project is incoherent. But how strong are their arguments? Read on and judge for yourself.

Let's begin by reviewing the definition of the distinction in question, as it was first proposed by Joel Buenting and Jason Taylor:

> According to the *generalist view*, the rationality of conspiracy theories can be assessed without considering particular conspiracy theories. On this view, conspiratorial thinking *qua* conspiracy thinking is itself irrational.[10] The *particularist view* about conspiratorial thinking denies that the rationality of conspiracy theories can be assessed without considering particular conspiracy theories. (2010, pp. 568–569)

That seems like a perfectly reasonable distinction. But the *Le Monde* authors claim that "on closer inspection, this partition turns out to be meaningless, self-serving and self-refuting." Ouch, that's a real zinger. How do they show this? They have several strategies. First, they accuse us self-proclaimed "particularists" of inconsistency. Specifically, they try to show that we are equally generalists. They explain, "Basham *et al.* (2016) essentially claim that conspiracy theorizing is generally warranted because there are conspiracies: that is a generalist view."

Ah . . . well . . . yes . . . one could say that this is a "generalist view" *in some sense* of that phrase. It is, after all, a *generalization*, though we never intended to imply that we reject all generalizations of all kinds. That would be crazy. The important thing to note is that the idea that "conspiracy theorizing is generally warranted" is *not* a generalist view in *the specific sense in question here*. If we had asserted or implied that "conspiracy *theories* are generally warranted," then they would have us. We would have been caught *assessing* conspiracy theories in general rather than evaluating particular conspiracy theories each on their own merits. But we did not do that, for we do not hold that position (if I may presume to speak for the group). Although the two claims *sound* very similar, they *mean* two very different things. To say that conspiracy theorizing is warranted is just to say that, since conspiracy theories are sometimes true, it is reasonable to consider them and to evaluate the evidence for them. This tells us nothing about the warrant of any particular conspiracy theory, which has to be investigated on its own merits. Indeed, conspiracy theorizing is warranted in general *because* particularism (in the specific sense in question) is true.

Now, the *Le Monde* authors should have understood this, for they accurately understood that particularism, in their words, "refers to an approach to conspiracy theories based on the examination of each specific claim of conspiracy and its respective argumentative and evidential merits (or shortcomings)" (Dieguez et al. 2016, p. 23 n. 4). And so, according to this view, we cannot rule out a conspiracy theory because of the mere fact that it is a conspiracy theory. That is, generalism is false. So, in order to evaluate a conspiracy theory, we need to "theorize" (if you will) about it, determining what is the most plausible version of it and exploring how well the evidence seems, on the whole, to support it. To point out the fact that this can be characterized as "a generalist view" in the larger sense of involving one sort of generalization or another is a silly non sequitur. It certainly does not make the distinction between generalist and particularists *as it applies to the evaluation of conspiracy theories* "meaningless, self-serving and self-refuting."

However, that was just their first swing. They follow with the assertion that "conspiracy theorists are generalists, in that they tend to endorse several and varied conspiracy theories." But this is another non sequitur. Whether or not conspiracy theorists tend to be generalists themselves would not be germane, *even if* that meant they tend to be generalists in the *relevant* sense. The debate is about what perspective a person *should* take regarding conspiracy theories, not what conspiracy theorists themselves happen to think. In other words, particularism would still be true even if all conspiracy theorists happened to be generalists in the relevant sense. And further, once again, the *Le Monde* authors are *not* referring to generalism *in the relevant sense*. For endorsing several and varied conspiracy theories does not make one a generalist in the relevant sense. The issue is whether conspiracy theorists consider the particulars of these "several and varied" conspiracy theories, which any acquaintance with the arguments put forward by conspiracy theorists suggests that they typically do.

The *Le Monde* authors do say one thing that is right on this issue. They point out that generalism might not only lead to the inappropriately flippant rejection of conspiracy theories, but a generalist orientation *in favor* of conspiracy theories might equally "lead to their uncritical acceptance." True indeed. However, that just shows that generalism is epistemically problematic regardless of its directional orientation. If their point has any relevance to the merits of particularism at all, it gives one further reason to endorse it. It certainly does nothing to advance the claim that the distinction between generalism and particularism is "meaningless, self-serving and self-refuting."

But the *Le Monde* authors are still not done. They also claim that they themselves are, in some sense, particularists. If they are particularists in the *relevant* sense, as they seem to suggest rather tentatively, that would be good news. They *ought* to be. But if they are merely particularists in some other sense, as their more concrete examples suggest, that wouldn't be germane even if their status as particularists in the relevant sense had some bearing on the issue—which it doesn't. The claim that they happen to be particularists themselves is simply not germane. Even if it is true in the relevant sense, it does nothing to support their bold claim regarding the conceptual incoherence of particularism; if anything, it undermines it.

Now, this distinction between particularists and generalists was first articulated *for a reason*. The distinction was designed to make it easy to point to two contrasting perspectives actually taken by philosophers engaged in the debate over conspiracy theories.[11] There are those who think that the whole class of ideas counting as "conspiracy theories" can be regarded as unwarranted based on something close to a priori reasoning (these are the generalists), and those who think that one must treat each conspiracy theory on its own particular merits (the particularists). It is a clear and useful distinction that makes it easy to refer to two actual and distinctive camps. If making the distinction clear also makes generalists feel a little silly and inclines them to forsake generalism, either by claiming to have been some kind of a particularist all along, or by reframing the distinction as a spectrum along which they can take some middle position (perhaps by pretending that we particularists don't already occupy that position),[12] then the distinction is not just conceptually coherent but is doing some real work. It seems to me, then, that the *Le Monde* authors' assertion that the distinction is meaningless is both poorly supported and wrong.

10.3. The Misrepresentation of Their Own Agenda

Let's now consider the original agenda of these authors. What were they up to? The very title of their *Le Monde* statement makes it clear: "Let's fight conspiracy theories effectively." They worry that the "wrong cure might only serve to spread the disease" (see Basham and Dentith 2016, p. 17). The "disease," of course, is conspiracy theorizing,[13] which they conflate with "conspiracism," expressing their desire to help "fight against this particular form of contemporary misinformation known as 'conspiracism'" (p. 17). In putting it this way, they reveal their bias: the generalist presupposition that conspiracy theories are a form of misinformation. They believe that "the growth of conspiracy theories" is "a major problem" (p. 17). And so, they aim to provide research that will help "remedy the problem" of "adherence to conspiracy theories" (p. 18). This research is necessary, they reason, because "Conspiracism is indeed a problem that must be taken seriously" (p. 17)—again conflating conspiracy theories with conspiracism.

It was this objective with which we took issue. But now, in response to our criticism, they have recast their position. Although they had originally characterized the intentions of governmental initiatives to undermine conspiracy theories as "laudable" (p. 17), they have reframed their original *Le Monde* letter in the following ways:

> [Our commentary] cautioned against governmental initiatives to counter conspiracy theories among youths and advocated for more research on the topic. (p. 20)
>
> [We] took issue with French governmental and local initiatives designed to tackle the apparent proliferation of conspiracy theories among youths. (pp. 20–21)

Both of these statements are technically true, but quite misleading. These ways of putting it make it sound as though they are *against* governmental initiatives to counter conspiracy theories. Reinforcing this impression, they go so far as to suggest that they are, in part, trying to "ascertain whether there is a problem [with conspiracy theories] at all" (p. 21), and that they want to "help everybody become *better* conspiracy theorists" (p. 20). But that is not at all the impression one gets from the *Le Monde* statement, as indicated above.

In reality, the original *Le Monde* statement was *not* cautioning against governmental initiatives to counter conspiracy theories. Rather, they expressed full support for that *objective*. They were merely cautioning against doing it without first funding more research (to be done by themselves), so that, armed with this research, the government could counter conspiracy theories *more effectively*. Here is how they pitch it:

> [The current] more or less random campaigns [to combat belief in conspiracy theories] are *expensive*, and *this investment is automatically taken from more methodical studies of the phenomenon*. It is therefore urgent that we launch widespread research programmes aimed at evaluating present educational initiatives rather than continuing to promote them. (Basham and Dentith 2016, p. 18, emphasis added)

As an aside, it seems a tad hypocritical of the *Le Monde* authors to charge us with a "self-serving" interpretation (see Dieguez et al. 2016, pp. 22, 23 n. 4) while they are calling for more funding for research in which they would like to engage. But for critics of conspiracy theories, double standards are the norm.

In our response we took issue with their objective, namely, to aim scientific research (much of which turns out to be seriously flawed) toward pathologizing and fighting a perfectly legitimate, and indeed important, cognitive activity. But now I am taking issue with something different. I'm taking issue with the way they, in their response to us, have misleadingly characterized their own previously expressed purpose.

Though they attempted to recast their intentions, they have not fully retreated from activism. They say that they "thought . . . that something should be done" (p. 21). About what? Why, about "ideological polarization . . . hate-speech and misinformation" (p. 21). But who said anything about those things? It seems that a number of questions have suddenly been begged. Then, almost admitting what their original position had been all along, they worry that "early and hasty endeavours had the potential to misfire or simply be ineffective" (p. 21). Endeavors to do what? Now they seem to be suggesting that they are for efforts to reduce hate speech and misinformation. But their original statement was about being ineffective in undermining conspiracy theories. Rather than straightforwardly defend *that* position, they equivocate between conspiracy theories and "ideological polarization . . . hate-speech and misinformation."

Regarding conspiracy theories, they now present themselves as perfectly neutral, writing:

> So, what were "they" up to? Quite simply, "they" advocated for more research. "They" figured that, before "fighting" against, or "curing," conspiracy theories, it would be good to know exactly what one is talking about. Are conspiracy theories bad? Are they good? Are they *always* bad, are they *always* good? . . . "They," in fact, are "just asking" some questions. (Dieguez et al. 2016, p. 21)

Once again, this is a clearly misleading representation of what they were up to. They now ask in a neutral voice, "Are conspiracy theories bad?" Yet they had already answered this when they described belief in conspiracy theories as a disease and conflated it with "contemporary misinformation known as 'conspiracism'" (Basham and Dentith 2016, p. 17). Have they suddenly turned over a new leaf? If so, why not be honest about what they had originally said, and admit a change of heart?

Later in their response, under the heading "A Cure?," they once again reframe their purpose in neutral terms. They write, "What 'they' had in mind, as must be clear by now, was to study how people, on their own or under some external influence, *think* and come to endorse some *beliefs* about such things" (Dieguez et al. 2016, p. 32). They maintain that they just want to use objective science to answer questions such as whether a "remedy is not needed after all, as the disease might be transitory, or even not a disease at all" (p. 33). They continue, "Scientific research turns out to be the best currently available tool to answer such questions, and that's where the analogy lies with programs devised to counter conspiracy theories." It's a curious position, if we are to take it seriously. They support "programs devised to *counter* conspiracy theories," wanting to try to make such programs *more effective*, because, they seem to suggest, "Who knows? We might end up finding that there was no problem to begin with!" But how likely is it that biased researchers, funded by grants directed for a purpose that aligns with that bias, are going to produce findings that run directly counter to that purpose and so support the conclusion that no more such funding is warranted? No conspiracy theory is needed to recognize this as a flawed, if not intellectually dishonest, approach.

10.4. The Misrepresentation of Our Critique

Naturally, since the *Le Monde* authors misrepresented their original position, they needed to misrepresent our critique of it as well. And so they did. They did not focus directly on the substance of our actual critique, namely, that seeking to use what passes for science to assist the state in undermining belief in conspiracy theories (without concern for whether or not there is justification for those theories) is a bad idea. Instead, they attributed to us a number of positions that we never asserted. Then they produced a wide variety of points in response to these positions, some of which are unobjectionable, others quite problematic, but none directly germane to our central complaint.

For example, they suggest that our objection to their project involved the idea that "everything there is to know on the matter is in fact already known, and that any further attempt to investigate the topic would be a 'grave intellectual, ethical and prudential error,' or worse, a genocidal crime against the masses, destroying lives 'by the thousands, even millions'" (p. 21). Wow! Did we write anything as crazy as that? Or, more likely, is this an egregious misrepresentation of our critique? Let's find out. While it is true that much of the social science research on conspiracy theorists is deeply flawed (as discussed above), we did not even mention this in our objection to their proposal. We certainly did not claim that "any further attempt to investigate the topic" would be necessarily problematic. After all, we ourselves, in our own ways, investigate the topic. No. *That* was not the problem we were pointing out. Neither did we suggest, needless to say, that *merely investigating the topic* would destroy lives by the thousands or millions. So, what exactly *did* we write? We wrote this:

> Political conspiracy theorizing in Western-style democracies should not be restricted, because to do so is a grave intellectual, ethical, and prudential error. As such, the declaration by respected scholars like these is likewise a grave intellectual, ethical and prudential error. (Basham and Dentith 2016, p. 15)

So, quite plainly, we were not saying that any investigation would be inappropriate. We were saying that there should not be an effort to restrict (it would have been better to have said "undermine") political conspiracy theories. *That* is what would be the "intellectual, ethical, and prudential error." And, remember, that is precisely the goal that the *Le Monde* authors were originally supporting, though they are now, in their response, not straightforwardly admitting.

We continued, writing: "Conspiracy theory saves lives, by the thousands, even millions, if we would let it. Its automatic dismissal leaves blood on our hands" (p. 16). What were we talking about? Certainly *not* that merely investigating the topic would result in untold carnage. Perhaps our explanation bears repeating:

> High-placed political conspiracies of lesser ambition often lie behind the political catastrophes of recent history. Very recent. For example, the catastrophe of the invasion of Iraq comes to mind. There is little doubt in the public or [among] scholars that NATO, and many other governments, were intentionally misled and manipulated into this war, particularly by the U.S. government. This truth, well-

> evidenced at the time of grave decision, was silenced as an "outrageous conspiracy theory" by heads of state, mainstream media and yes, certain members of academia. Thus, a war that ultimately led to the death of hundreds of thousands, and a desperate global refugee crisis, was powerfully enabled by an anti-conspiracy theory panic. One that these scholars would seem to like to embrace and nurture as general policy. (p. 14)

We gave other examples as well. So, quite plainly, we were saying that it is engaging in an effort to disable a mechanism for thwarting potentially disastrous conspiracies that "leaves blood on our hands," not merely investigating the topic. Further, let me be emphatically clear about this: The *Le Monde* authors were *not* originally advocating investigating the topic in a fair and neutral way. They have a clear bias: they assume that conspiracy theories are a disease that needs to be cured. And they have an explicit agenda: to "fight conspiracy theories effectively."

Now, I am not opposed to activism, and there is nothing inherently wrong with having an agenda. Indeed, I have an agenda in writing this. I am making a case for what I believe to be true and defending what I think is important. But here is the crucial difference: I am not pretending to be a neutral scientist, objectively collecting the data and letting it speak for itself. These scholars, on the other hand, do claim to be in precisely that business. Perhaps that is why they have a hard time admitting their agenda. And so, having been called out for their agenda, they are now trying to claim that all they wanted to do was to dispassionately and scientifically investigate the topic. They are "just asking questions" (p. 28; cf. pp. 20, 21) and gathering data, they claim. But they are not convincing. As shown above, that position is refuted by their own words in their original statement.

The *Le Monde* authors also claim that we "call . . . for more conspiracy theories and less 'conspiracy theory panic'" (p. 20). Here they are half right. It seems fair to say that we are against "conspiracy theory panic," but it is silly to say we want "more conspiracy theories." For my part, I would say that I want fairness toward conspiracy theories (a desire also expressed by Basham 2017). I do not want to see the state allied with biased social scientists for the purpose of producing research designed to help the state undermine legitimate conspiracy theorizing. But that is not the same as calling "for more conspiracy theories," as if we think that the more conspiracy theories in circulation the better, regardless of their merits. No. We were calling out those who would employ science to try to undermine a legitimate and important activity.

In addition, they also suggest that we accused them of being part of a conspiracy (p. 30). But we did *not* maintain that they were *secretly* up to something morally dubious. Their morally dubious agenda was *openly* articulated in a public forum. However, given their bizarre response, it now seems that they are retrospectively trying to pretend that they were up to something different from what they clearly and repeatedly stated originally. But I, speaking just for myself, do not maintain that they *plotted* any of this. No, in this case, I favor a cock-up theory.[14]

10.5. Pathologizing Conspiracy Theorists

Another central concern that we raised was their pathologizing of conspiracy theorizing, suggesting that conspiracy theories are a "disease" (Basham and Dentith 2016, p. 17). Basham 2017 addresses this issue more broadly. I've chosen here to focus narrowly on reasoning errors in their attempt to vindicate themselves by suggesting that we are equally guilty of the same offense. They accused us of inconsistency since we oppose the *general* pathologizing of conspiracy theories and yet some of us had, on their reading, pathologized certain *particular* conspiracy theories. Hmmm. Actually, even if they had read us correctly (which in at least one case they have not), there is nothing inconsistent about that.

Since I was one of those accused of this supposed inconsistency, and since they have indeed misread me, I'll use their critique of my work to set both matters straight. Specifically, they accuse me of "delegitimiz[ing]" Roswell conspiracy believers (Dieguez et al. 2016, p. 26). Neither did I intend to do that nor would it have been in any way significant if I had. Here is what I wrote:

> [Sunstein and Vermeule's] deliberate intent to be dismissive becomes unambiguously apparent. Immediately after the mention of Operation Northwoods they write: "In 1947, space aliens did, in fact, land in Roswell, New Mexico, and the government covered it all up. (Well, maybe not)." This trivializes a whole list of significant conspiracies that they could not but admit were real, though the list could have been much longer. (Hagen 2011a, p. 13; see section 5.2.1 above)

I was objecting to an obvious appeal to ridicule and inappropriate trivialization of agreed-upon facts by throwing in a widely disbelieved example, accompanied with a snarky comment. As for my own position on the issue of alien visitations in general, and the Roswell incident in particular, I have no firm opinion, as I have not studied these issues in any depth (interesting though they are).

The point of the claim that I delegitimized Roswell conspiracy believers is that I had thereby, presumably, engaged in the pathologizing of a *particular* group of conspiracy theorists, as others in our group are likewise accused. This is a problem, they think, because we were critical of their attempt to pathologize conspiracy theories *in general.*

There are multiple layers of problems with their analysis. To begin with, as I have just explained, I had not even claimed that Roswell conspiracy believers were wrong, or that their belief is poorly evidenced. I did not take a position on that, and I have none. But even if I had, it would not follow that I pathologized them. Asserting that someone's position is wrong, or is not well evidenced, does not suggest that the person is defective. But that *is* what the *Le Monde* scholars seek to do. They aim to describe a presumed-to-be-defective conspiracist "mindset" (Basham and Dentith 2016, p. 18; Dieguez et al. 2016, pp. 20, 23–25, 29–30, 34). And they advertise that their studies will help make efforts to undermine conspiracy theories more effective. Their project is a delegitimizing one. Ours is not. And further, even if I had pathologized *a particular group* of conspiracy theorists, that would not mean I had acted hypocritically in criticizing the *Le Monde* scholars for pathologizing conspiracy theorists *in general.* (After all, while it is wrong to generically pathologize atheists, Republicans, or Norwegians, that does not mean there are no individuals in those groups who may legitimately be regarded as, in some sense, pathological.) At minimum, pathologizing conspiracy theorists in general is an instance of *inappropriate* pathologizing, since believing in conspiracy theories is not necessarily, or even typically, pathological—even *if* there are particular instances that are (about which I have taken no position). In sum, their argument goes wrong at every turn. No wonder they value "data" and disparage reason—contrasting data, data collection, experimental designs, and empirical research with "armchair" reasoning and various derogatory versions of the same (Dieguez et al. 2016, pp. 22, 25, and 32). But their incantations of holy words such as "data" and "experimental designs" will not turn invalid arguments into valid ones.

After the manuscript for this book was substantially completed, I encountered a newer article by most of the *Le Monde* authors, namely, "Why 'Healthy Conspiracy Theories' Are (Oxy)morons: Statistical, Epistemological, and Psychological Reasons in Favor of the (Ir)Rational View" (Wagner-Egger et al. 2019). It responds to critics, myself included. In that article, these authors have taken a different tack. Abandoning the approach that I had called out as a misrepresentation of their own view, they have reversed course again, now explicitly doubling down on the pathologizing of conspiracy theorists. In a new response to them (Hagen 2020a), I explain why they are still wrong (see also Dentith 2019c). In the process, I also address the question of whether academics ought to debunk conspiracy theories. In short, I argue that academics should not *set out to debunk* conspiracy theories, but that there is nothing wrong with endeavoring to *fairly evaluate* such theories. However, I suggest that, if the writings of these authors provide a basis for judgment, it is unlikely that academics will be particularly helpful in this regard, at least until they better absorb the unconventional wisdom now strongly supported by the philosophical literature on the topic.

10.6. Conclusion

The *Le Monde* authors had, in their *Le Monde* statement, worried that without further scientific study government programs aimed at undermining conspiracy theories might backfire. Indeed, they might have; such programs are fundamentally ill conceived. What is clear is that the *Le Monde* authors' defense of their scientific project in support of this objective has itself backfired. It undermines confidence in their ability to conduct fair and reasonable studies of conspiracy theorists, or of any subject for that matter. And thus their response calls into question the wisdom of their original proposal, even if its objective had been defensible, which even they seem unwilling now to maintain.

This chapter should be troubling, not because it exposes the bad reasoning of a handful of social scientists, but because it raises a more general worry: How pervasive is this kind of thing? Is the research on conspiracy theorists, in particular, significantly biased, perhaps to the point of being seriously misleading? Is the scientific research on other topics likewise biased? While it is well beyond the scope of this chapter to offer answers to these questions, I would like to highlight one particular issue that will be impacted by the answer: the plausibility of conspiracy theories. The less fair and reliable supposed "epistemic authorities" turn out to be, the weaker the argument against conspiracy theories based on their opposition to the views of such (supposed) epistemic authorities. From this I derive two conclusions: (1) So long as these questions have no clear and comforting answer, one should be careful about dismissing conspiracy theories based on the pronouncements of scientists or other (supposed) epistemic authorities. At least, due diligence involving critical review of their findings, rather than uncritical acceptance of them, is in order. (2) Scientists and other would-be epistemic authorities ought to be exceptionally careful to hold their biases in check, lest they unwittingly undermine their own disciplines. Everyone benefits when science is genuinely trustworthy. The costs of unreliable science are high. One cost is that false conspiracy theories cannot easily be dismissed by appealing to the judgment of those who *ought* to be epistemic authorities.[15]

Conclusion

What general conclusions can we draw from this critical analysis of certain academic approaches to conspiracy theories? First, there has been a lively and informative debate that has taken place in the philosophical literature on conspiracy theories, in which progress has been made. A number of philosophers have repudiated the "conventional wisdom" about conspiracy theories—that such theories "should be neither believed nor investigated" (Pigden 2007, p. 219). They reject the view that there is something inherently epistemically problematic about conspiracy theories and that belief in such theories represents some sort of pathology. The position that has emerged dominant is particularism: each conspiracy theory ought to be judged on its own particular merits and problems. Perhaps we can also conclude that a certain amount of epistemic modesty is in order when judging individual conspiracy theories. It may be that agnosticism is often a rational response to conspiracy theories, at least prior to investigation (as suggested in Keeley 2007, p. 137) and perhaps even after (as suggested in Basham 2001, pp. 274–275).

In Part I of this book, I've summarized and augmented many of the arguments in the philosophical literature on conspiracy theories. Supporting the emerging consensus in that literature, I've argued that reflexive disparagement of conspiracy theories is inappropriate. And I've emphasized that this holds true even when conspiracy theories are conceived of as views that conflict with official accounts and that attribute appalling conduct to prominent people in open societies.

A central issue is the reliability of ostensible epistemic authorities. Throughout this book I have given reasons to worry that such "authorities" may not always be as reliable as we would like. I've done this in part by showing, over and over again, examples of published scholarship that is so rife with errors of reasoning, and errors so obvious, that it becomes hard to maintain that there isn't a strong bias in operation. I would add here that if such a bias is manifest in academic work, then official commissions, in which the participants often have much more at stake, cannot reasonably be presumed to have mustered greater objectivity, rigor, and intellectual honesty. (In the appendix, I add a few relevant case studies, addressing the reliability of the National Institute of Standards and Technology and the 9/11 Commission, as well as government officials more generally and the mainstream media, all regarding the events of 9/11.)

I've also shown that many memes disparaging conspiracy theories are either false, not well supported, or misleading. Conspiracy theories generally do not, for example, imply "an almost nihilistic degree of skepticism" (see section 3.1.3), and it is not true that conspiracy theorists have not played a role in revealing real conspiracies—they played a critical role in revealing COINTELPRO, for example (see section 3.2.8). And many cases are ambiguous.

Focusing on the work of Cass Sunstein and Adrian Vermeule, Part II has revealed that, when it comes to discussing conspiracy theories, even highly regarded scholars may be prone to egregious errors of reasoning, to say nothing of poor moral judgments. Providing a highly implausible model for the spread of conspiracy theories, Sunstein and Vermeule propose deceit and conspiracy as a "cure" for conspiracy theories, with no apparent appreciation of the irony of their proposal and despite a clear conflict with the values they claim to espouse. They also beg a central question by declaring certain theories to be demonstrably false without any serious attempt at showing that such demonstrations exist. Instead, they appeal to highly ambiguous and misleadingly described evidence such as "video frames showing Flight 77 approaching the building" (Sunstein and Vermeule 2008, p. 20; cf. Sunstein 2014a, p. 27). And they provide a whitewashed view of the relevant history, thereby skewing prior probability considerations. Adding insult to epistemic injury, they offer misleadingly truncated depictions of conspiracy theorists by ignoring the existence of conspiracy theorists who are knowledgeable and reasonable and possess relevant expertise.

In a revision of Sunstein and Vermeule's article, Sunstein ignores the arguments of critics. Instead, he modifies one of the most offensive aspects of his original proposal, that US officials should lie domestically, and also asserts that the proposed infiltrations are intended only for *foreign* groups, while unconvincingly implying that these were merely *clarifications* of his original intent. As a bonus, he adds a treatment of biased assimilation that is itself an example of biased assimilation.

What could possibly explain all this, if there is not a strong current of bias running through the academy? One must not forget, after all, that Sunstein and Vermeule not only had quite a number of faulty thoughts, they succeeded in getting them prominently published in a venue that refused to even consider rebuttals, specifically, early versions of Chapters 4 and 5 of this book.[1]

Part III, which addresses the social science literature on conspiracy theories, reinforces the notion that some sort of bias may be skewing scholarship on this issue. At least, *something* appears to have led many social scientists down illogical paths. While paying lip service to scientific objectivity as they paint conspiracy theorists as irrational dupes, social scientists with an apparently jaundiced view of conspiracy theorists have engaged in fallacious reasoning so obvious, and yet so widely repeated, that one marvels at the hypocrisy. For example, they equate the attribution of some degree of plausibility, or worthiness of consideration (etc.), with *belief*. They thereby unfairly imply that conspiracy theorists are irrational on account of the following supposed epistemic crime: simultaneously entertaining incompatible conspiracy theories, tending to give them a greater degree of credence than do more conventionalist thinkers. Yet there is nothing irrational about that (see section 7.1)—but there *is* something irrational about finding it to be irrational. Appealing to flawed studies such as the one just described, which encourages the conflation of belief with the attribution of a marginally greater degree of credence (namely Wood et al. 2012), they champion the idea that conspiracy theories form part of a "monological belief system," a concept which they systematically misapply, providing evidence that conspiracy theorists exhibit the ubiquitous and perfectly innocuous aspects of it while insinuating that there is something remarkable and problematic about that (see section 7.3). Admittedly, some social scientists have independently come to recognize at least some of the problems with the "monological belief system" critique of conspiracy theories (Sutton and Douglas 2014).

Many social scientists also conflate "conspiracy theories" with "the paranoid style," which posits a vast network of evil supermen. But most conspiracy theories, especially the interesting ones, do not posit this. These social scientists also fail to acknowledge the degree to which official stories can approximate the paranoid style, and the degree to which history shows that *the kinds of* people and groups often accused of appalling behavior by conspiracy theories—presidents, intelligence and law enforcement agencies or elements within them, and even doctors and psychiatrists—really have sometimes done horrible things. At the same time, these social scientists have difficulty recognizing that many conspiracy theories either do not attribute unusually malevolent intent to the purported conspirators or need not do so to maintain their explanation. Nor do conspiracy theorists typically posit a vast network of preternaturally effective conspirators who exercise perfect control over the flow of history, as the "paranoid style" and the "conspiracy theory of society" suggest.

Finally, the efforts of the *Le Monde* authors (discussed in the previous chapter), and those of Lewandowsky, Lloyd, and Brophy (discussed in Chapter 3), to pathologize conspiracy theorists and "conspiracist ideation" respectively, only serve to undermine confidence in the ability of a number of academics working in this area to reason sensibly about the relevant issues, and further suggest a certain prejudice is in operation in the academy.

On the other hand, as I suggested above, there has been good work done on this subject in philosophy (and some in the social sciences as well).[2] I have commended the work of Charles Pigden, David Coady, M Dentith, Lee Basham, Juha Räikkä, and, with some reservation,[3] Brian Keeley. It is lamentable that most of the social scientists who study conspiracy theories and conspiracy theorists, though they often cite the philosophical literature, do not seem yet to have properly absorbed its implications.

Have I been a little harsh, particularly in this summary? Perhaps. But it should be remembered that, if my critiques have merit, those herein criticized have used their relatively privileged position as (ostensible) epistemic authorities to unfairly pathologize an already marginalized class of people. I think that justifies a stern response.

Appendix

9/11 and Epistemic Authorities

One of the central issues in evaluating the plausibility of conspiracy theories at the general level is the degree to which mainstream media reports and governmental inquiries can be assumed to be reliable. Focusing on 9/11 as a case study, I explore some examples that suggest that these sources may not be as reliable as they should be, and I suggest that it may be unwise to accept their conclusions uncritically. This supports the notion that conspiratorial counternarratives ought to be given fair consideration, and not assumed to be false based on their contrast with official dogma.

It is not my purpose here to support any particular account of 9/11. Rather, using 9/11 as an example, I offer a number of considerations that highlight the difficulties involved in assessing what really happened in controversial events. Namely, the mainstream media will tend to support the official story, sometimes in invalid or manipulative ways. Government agencies tasked with looking into the matter are less than perfect, and there are limited mechanisms for independent verification of their accounts. And, perhaps most significantly, government officials are prone, for whatever reason, to giving false and misleading accounts.

First, I consider the honesty and reliability of the mainstream media. After a rather extreme example involving a rigged vehicle-collision "experiment," I provide two examples of media manipulation regarding 9/11: the first involving a rather misleading and seemingly contrived "experiment," and the other involving a voice-over that appears to negate the point the witness was intending to make. Next, I discuss the government's investigation into the destruction of the Twin Towers and Building 7, focusing on the technical concerns of critics of official accounts and highlighting several confidence-undermining statements by lead investigators. I then address apparently false statements by President Bush and his chief of staff, Andrew Card, regarding their personal experiences at the time of the attacks on the World Trade Center. Finally, I address epistemic problems surrounding the 9/11 Commission, reflecting chiefly on the work of John Farmer, who was senior counsel to the 9/11 Commission, and journalist Philip Shenon, author of *The Commission* (2008). In all of these cases, my concern is the reliability of supposed epistemic authorities, not the underlying truth of what happened. However, if ostensible epistemic authorities are as unreliable as these examples suggest, determining the truth of what happened is not something that should simply be left to such authorities.

A.1. The Mainstream Media

Before providing examples of the mainstream media's questionable treatment of 9/11, let's take a brief trip down memory lane. The question here will simply be: How far might a TV newsmagazine go in creating a misleading presentation?

A.1.1. Dateline's Explosive Exposé

In 1992, the television show *Dateline NBC* "hired the Institute for Safety Analysis to conduct crash demonstrations" in order to "see for ourselves what might happen in a side impact crash" involving GM trucks of a certain vintage. The results were dramatic; indeed, they were explosive. In one of the demonstrations the car impacting the side of the truck was engulfed in flames. But it turns out that *Dateline* had allowed its contractor to take steps to ensure such a result. These steps, needless to say, were concealed from the audience. Forced by a GM lawsuit to publicly confess, Dateline coanchors Jane Pauley and Stone Phillips apologized on air. The confession/apology tells the story:

1. NBC's contractor did put incendiary devices under the trucks to ensure that there would be a fire, if gasoline were released from the truck's gas tank. NBC personnel knew this before we aired the program, but the public was not informed, because consultants at the scene told us the devices did not start the fire. We agree with GM that we should have told our viewers about these

devices. We acknowledge the placing of the incendiary devices under the truck was a bad idea from start to finish.

2. . . . The "Dateline" reporter [said,] "At impact, a small hole was punctured in the tank." That was based on a report from a consultant at the scene. GM has now X-rayed the tank and found no hole. We have performed no such tests but do not dispute their X-rays.
3. We said the speeds at impact were "around 30 mph" and "about 40 mph." GM says the speeds were at least 39 mph and 47 mph, and says we substantially misrepresented the energy involved in each of these side impacts. We do not dispute GM's calculations.
4. Lastly, in reporting on the crash that resulted in a fire, we said the crash "forced gasoline to spew from the fuel cap." GM says since the gas cap—which now cannot be found—was the wrong cap for the GM filler tube, and because the gas tank was overfilled, the cap came off when the impact occurred. The gasoline came out of the open end of the filler tube. We do not dispute this. Therefore, this unscientific demonstration was not representative of an actual side impact collision. (Associated Press 1993)

What does this tell us? It shows that a seemingly reputable TV newsmagazine backed by a major network would collude with a seemingly reputable contractor to produce what may now be justly labeled "fake news."

A.1.2. *National Geographic* on the Power of Thermite

Now let's turn to a *National Geographic* documentary, "9/11 Science and Conspiracy" (Erickson 2009), which attempts to debunk several positions held by at least some members of the 9/11 Truth Movement, with the help of "experts" from the Energetic Materials Research and Testing Center (EMRTC). Since many members of the 9/11 Truth Movement believe that some form of thermite was involved in the destruction of the World Trade Center skyscrapers, EMRTC performed an "experiment." The narrator frames it this way: "Was thermite the main cause of the building collapse? Once again, there is a way to find out. EMRTC calculated [*sic*] a controlled experiment to see if thermite was a viable option in the collapse of the towers." *National Geographic* clearly signals that the experiment had the potential to show that the use of thermite is not a viable explanatory cause of the destruction of the WTC buildings. To do this they used what they acknowledged was the weakest kind of thermite, standard powdered thermite, to address the question "Can thermite of any kind burn through steel beams?"

Exhibiting a remarkable lack of imagination, the hired experts poured large quantities of thermite around a steel beam, with no attempt to concentrate a pressurized stream of thermite on the beam. After the thermite was ignited and burned away, the beam remained essentially undamaged. The experts seemed to imply that they had thereby shown that cutting a steel beam cannot be done with thermite "of any kind." And the "truthers" interviewed on the show (David Ray Griffin, Richard Gage, and Dillon Avery), being unmoved by this demonstration, were depicted as irrationally committed to their position. (In a less interesting experiment, EMRTC shot an aluminum tube through a box and compared the result with blowing the box up. The results, which were hardly surprising, were supposed to have some relevance to whether or not a Boeing 757 hit the Pentagon. Again, the "truthers" were not so subtly ridiculed for not being impressed.)

This documentary prompted engineer Jonathan Cole to do his own experiments. Instead of using ordinary thermite, Cole used a form of thermite called thermate, which includes sulfur and some other chemicals. "It is the added sulfur," Cole explains, "that helps form a eutectic that lowers the melting point of steel" (2013, p. 242). Using a simple makeshift device for directing the flow of the thermate as it reacts, Cole demonstrated that, in fact, relatively small amounts of thermate can cut through steel beams.

Now, let's disregard the question of whether Cole's experiment shows what it seems to show—that some kind of thermite, in some kind of configuration, *can* cut steel beams—and treat this merely as a hypothetical possibility. Does that alone give us reason to regard the *National Geographic* experiment as unfair and deceptive? Yes. The experiment was presented *as if* it seriously undermined the notion that thermite may have played a significant role in bringing down the WTC buildings. But clearly it does not.

National Geographic frames the experiment with the ambiguous question, "Can thermite of any kind burn through steel beams?" Do they mean, "Can thermite of any kind (no matter how strong) burn through steel beams?" No. They didn't test the strongest kind. So we must take the question to mean, "Can thermite of any kind (no matter how weak) burn through steel beams?" But a negative answer to this question tells us little if anything about the relevant issue: whether a role for thermite (of some kind) could have been possible.

It seems that *National Geographic* structured the experiment for the greatest effect—like *Dateline NBC*—rather than as a genuinely informative experiment. They seem to have set things up so that the viewer would conclude that even enormous quantities of thermite could not have been instrumental in bringing the buildings down, thereby undermining the thermite hypothesis. This doesn't mean *National Geographic* was intentionally deceptive. They may have been operating under such a strong bias that awareness of even the most obvious problems with their "experiment" did not rise to the level of consciousness. (Presumably, not everyone involved was in the same state of mind. Some participants may have had *some degree of* deceptive intent, or a willingness to go along with what they recognized as not fully rational, while others may have been completely well intentioned, but either blinded by bias, incompetent, or not paying close attention.)

What is the likely effect of this kind of "experiment" regarding the belief in conspiracy theories? Intensified polarization. Specifically, those who already think that the thermite hypothesis is silly may or may not be fooled by the bogus reasoning underlying the "experiment." Their views will either remain unchanged or will tend to be strengthened for the wrong reasons. Those who already believe the thermite hypothesis, being more motivated to be critical, will remain unconvinced (as they should be, even if they are in fact wrong). Indeed, their belief will likely be strengthened by the very fact that the attempt to refute their position was such a sham. Responding to conspiracy theories with what will rightly be called "hit pieces" is neither a noble nor effective way to dissuade people from belief in conspiracy theories. That is not the high road.

We should not conclude from this that it is a bad idea to address the technical issues involved in conspiracy theories. Rather, it is bad to address them *badly*.

A.1.3. ABC on Rodriguez's Story

Let's consider another example. In this one the problem is subtler and requires background knowledge for the viewer to fully recognize it.

William Rodriguez was a custodian in the North Tower. He had a master key that was used to unlock doors for firemen. He went up through the building, leading the firemen, as he was familiar with the building, and he helped rescue many people. He was a hero and was celebrated and recruited, for a time, by the Republican Party. But that isn't the most interesting part of his story. According to his own harrowing account, Rodriguez was the last person to make it out of the North Tower alive. He dove under a fire truck to avoid falling debris and was rendered unconscious as the area was covered in thick dust. He was then dragged out from under the fire truck before the ground under part of it gave way—he would have been crushed. But that's not the most interesting part of his story either, at least not for our purposes.

Rodriguez insists that there was an explosion in the basement, below him, several seconds prior to the first airplane hitting his building. He reports having testified to this in a closed session of the 9/11 Commission. His claim is interesting because it supports the notion that explosives were involved in bringing the building down. That's the background. Our focus here is on how the media dealt with his story, not the accuracy of his story itself. ABC News ran a report that featured Rodriguez, portraying him favorably, saying, "instead of running from danger, he ran straight to it," and characterizing him as "a changed man . . . an activist fighting for disaster relief for the Hispanic community." It is all very heartwarming and presumably true, as far as it goes. And yet there is something misleading about the report. When it came to Rodriguez's account of what happened, ABC hijacked his story right at the critical point:

> VOICE-OVER: On the morning of 9/11 he [Rodriguez] was in the basement.
>
> RODRIGUEZ: We hear "Boom!"—an explosion so loud that it pushed us upwards in the air, and we started screaming. We didn't know what it was.
>
> VOICE-OVER: It was the first hijacked plane that hit the tower.[1]

Rodriguez had long maintained that there was a loud explosion in the basement *before the plane hit the tower*—he believes he heard the plane impact several seconds after the explosion in the basement. Rodriguez's story seems to have evolved somewhat from his first account on the day of 9/11 itself, which described a "big rumble." And it is always possible that any witness might be lying, exaggerating, or misremembering (or some combination). But the accuracy of Rodriguez's account is not my concern here. My concern is the reliability of such entities as ABC News. Did they edit their report to alter Rodriguez's story in order to make it better conform to the official story? Rodriguez maintains that they did.[2] And it certainly appears from what remained in the report—that the explosion pushed people up in the air—that Rodriguez was trying to relate his account of an explosion coming from below, which he regards as unrelated to and indeed preceding the airplane strike, a point which Rodriguez consistently took pains to emphasize. He *consistently and adamantly denies* that the first "Boom!" he heard "was the first hijacked plane that hit the tower." It is hard to believe the reporters failed to understand this. It seems ABC News decided to override Rodriguez's story with its own (the official story), without indicating to the audience that Rodriguez was in fact *opposing* the official version of events *on that very point.*

This may be "innocent" in the sense that the people at ABC News presumably believed that the "Boom!" that Rodriguez heard must have actually been from the plane impact, even if Rodriguez himself maintained that it preceded it. And so they included what they presumably regarded as a fact in place of what they regarded as a mistaken understanding on the part of Rodriguez. They were, after all, just trying to produce a "feel good" puff piece, and they didn't want anything to distract from their purpose. This is all perfectly understandable, and almost excusable. But it is important to recognize that we may be surrounded by subtle manipulations of this kind that reinforce official accounts of controversial events. By itself, this particular case is insignificant. But it nevertheless exemplifies how the media can manage public perceptions, and it reminds us that they sometimes do exercise this power, even if not *intending* to deceive. Regarding conspiracy theories, it may be reasonable to expect that this kind of manipulation (which may mostly be innocent in intent) is relatively pervasive.[3]

A.2. NIST

The National Institute of Standards and Technology (NIST) is the agency that conducted the most extensive official analysis of the destruction of the three World Trade Center skyscrapers that collapsed on September 11, 2001. How reliable should we take their investigation into the collapse of these buildings to be? We cannot rule out, a priori, the possibility that some degree of finessing of the facts has led to seriously erroneous conclusions.

A.2.1. On the Credibility of Shyam Sunder and John Gross

NIST is an agency under the direction of the executive branch. So for the duration of the investigation in question, it was an arm of the Bush administration, which had been credibly accused of politicizing science—by fifteen thousand scientists.[4] Shyam Sunder was NIST's lead investigator for the investigation of the World Trade Center disaster, and John Gross was a project leader for the structural fire response and collapse analysis involving the Twin Towers and Building 7. As we will see, they have both made false or misleading statements that, at least to some degree, undermine their credibility and thus the credibility of the NIST reports.

Regarding the collapse of the Twin Towers, Shyam Sunder states: "Clearly, buildings are not designed to withstand airplane impacts."[5] But that's not true. *Some* buildings are designed to withstand airplane impacts, and the buildings under discussion are among them. More specifically, the Twin Towers were designed to withstand the impact of a Boeing 707.[6] We can debate the differences between a 707 and a 767, and whether the speeds of the planes that hit the towers were higher than what was contemplated by the designers, and whether or not the designers would have neglected to consider the jet fuel that would be on the plane, and so on. But it is just plain false to suggest that the buildings were not designed to withstand airplane impacts. And as Sunder was a lead investigator into the cause of the WTC collapses, it is hard to believe that he did not know this. Perhaps he just misspoke. But it is a strange thing to mistakenly assert knowing both its relevance and falsity. Thus, we have a reason already to find that Sunder's credibility is at least somewhat questionable.

John Gross has made similarly confidence-undermining statements. For example, Gross denied that he was aware of anyone who had claimed to have seen a pool of molten steel at the base of the towers. A questioner, who later identifies himself as "Mark," engaged him saying, "I'm curious about the pool of molten steel that was found in the bottom of the towers." Gross responded flippantly, "I am too. Please tell me about it. Have you seen it?" After Mark responds, trying to explain the various other kinds of corroborating evidence, such as video of the South Tower shortly before its collapse, Gross makes a specific denial: "Let's go back to your premise that there was a pool of molten steel. I know of absolutely nobody, no eyewitness who says so, nobody who has produced it."[7] The size of the supposed pool is not at all clear, and the word "pool" may not trace back to any verbatim witness report. But there are many well-known eyewitness accounts of molten metal, often identified as steel,[8] as well as corroborating evidence of various kinds. For example, it has been widely reported that Peter Tulley, the president of Tulley Construction, has claimed that he observed "literally molten steel" at the WTC site (Griffin 2007b, p. 181). It is possible that Gross did not know of this particular eyewitness claim, or that he would dispute the contention that Tulley used the word "pool," as he may not have. However, a line of defense that splits hairs in this manner would seem disingenuous, because Gross clearly *implies* that he does not think there is any evidence for molten steel at all. And it seems he does so for a very good reason. Continuing his response, Gross says, "Steel melts at around two thousand six hundred degrees Fahrenheit. I think it is probably pretty difficult to get that kind of temperature in a fire."[9] Now *that* is something that both NIST and its critics can agree on. Such temperatures seem to be inexplicable on the official account. This is precisely why the multiple reports of molten steel—seemingly corroborated by video evidence,[10] iron-rich microspheres in the dust,[11] and so on (see Jones et al. 2008; Ryan et al. 2009; and Harrit et al. 2009)—are regarded as troubling to many people. When faced with a serious question Gross's response was to dismissively deny.

A.2.2. The Free Fall of Building 7

There is one significant fact, however, that NIST found it could no longer deny. So the officials adopted a different strategy: admit it, and then say that it doesn't matter. Let's begin with a Q&A session regarding the "Draft for Public Comment" version of NIST's final report on Building 7. David Chandler, a physics teacher, asks the following question:

> Any number of competent measurements using a variety of methods indicate the northwest corner of WTC 7 fell with an acceleration within a few percent of the acceleration of gravity. Yet your report contradicts this, claiming 40% slower than freefall based on a single data point. How can such a publicly visible, easily measurable, quantity be set aside?

After asking for the question to be repeated, Sunder provides the following curiously irrelevant response (removing the copious hesitations):[12] "Well, first of all, gravity is the loading function that applies to the structure—applies to every body, every—all bodies on this particular—on this planet, not just in ground zero." After eroding confidence with that baffling non sequitur, Sunder becomes more lucid:

> The analysis shows there is a difference in time between a free-fall time. A free-fall time would be an object that has *no structural components below it*. And if you look at the analysis of the video, it shows that the time it takes for the seventeen—[I mean] for the roof line of the video to collapse down the seventeen floors that you can actually see in the video below which you can't see anything in the video—is about 3.9 seconds. What the analysis shows, the structural analysis

> shows—the collapse analysis shows—that same time that it took for the structural model to come down from the roofline all the way for those seventeen floors to disappear is 5.4 seconds. It's about 1.5 seconds, or roughly 40 percent more time, [than for] free fall to happen. And that is not at all unusual because *there was structural resistance that was provided in this particular case.* And you had a sequence of structural failures that had to take place, and everything was not instantaneous.[13]

Sunder is saying that the building would take 3.9 seconds to fall seventeen floors at free-fall acceleration, but that the actual collapse time of that distance was 5.4 seconds, which is 40 percent longer, and their computer model also predicts 5.4 seconds. But this is not really responsive to the question, which admittedly could have been clearer. What Chandler is trying to say is: When one measures *from the corner of the building* (as Chandler did) the acceleration appears to be very close to free fall (at least for a certain period).[14] And Sunder's response is, well, we measure the *overall* collapse time (from a point near the center of the roofline) to be 40 percent slower than freefall. Chandler does not have an opportunity to press the issue further (even his original question is asked through an intermediary), so he makes his case much more precisely on the internet, in a YouTube video.[15]

Chandler seems to be suggesting that NIST used a single (perhaps artificially early) starting point (together with an unambiguous end point), combined with the assumption of constant acceleration throughout[16]—contrary to easily observable fact. Chandler shows that a more detailed analysis reveals that for 2.25 seconds (to use the figure that NIST ultimately adopts) the collapse is *very* close to free fall—indeed, NIST later admitted that during this period the building fell *at* freefall acceleration, 9.81 m/s^2. During these 2.25 seconds the building fell over one hundred feet, or about eight stories, so this is no insignificant duration. In the final version of their report, NIST concedes that there was "a freefall descent over approximately eight stories at gravitational acceleration for approximately 2.25 [seconds]" (NIST 2008a, p. 48). However, free fall does not seem to be consistent with Sunder's own comments at the Q&A session, quoted above. Namely, Sunder had suggested that free fall can only happen when there are "no structural components below," and yet, in Sunder's own words, "there was structural resistance that was provided in this particular case." Later, on a BBC documentary about Building 7, Sunder modifies his position: "It turns out that when you have connections that essentially don't have strength for the loads that they are being subjected to and you have this massive failure of a column, it does not take time. The structure has lost all integrity at that point in time."[17] It seems rather late in the game for Sunder to have come to this realization.

Did NIST's computer model, which predicts a 5.4-second overall collapse, also predict that for a significant portion of that time the collapse would be indistinguishable from free fall? According to critics, "NIST's computer model shows no such period of free fall, nor did NIST attempt to explain how WTC7 could have had 'no structural components below it' for eight stories" (Jones et al. 2016, p. 24). NIST does make a relevant though less precise claim: "[T]he structural analysis model . . . showed the exterior columns buckling and losing their capacity to support the loads from the structure above."[18] But is that sufficient to explain free fall? This is an issue that, one would think, should be subject to peer review. But NIST has not made its model available for independent scrutiny, and many requests for details have been denied. Although NIST has admitted free fall and has addressed its plausibility to some degree, NIST continues to stress that the *overall* collapse time is 40 percent slower than free fall, as though *that* was the most salient fact.

A.2.3. Implications

Does the above entail that Building 7, at least, was a controlled demolition? No. There may be good explanations for the salient "errant" data. If the internal collapse pulled on the external columns causing them to buckle, assuming this can happen rather suddenly on all sides, perhaps they would offer negligible support. It is clear, after all, that the north face did bend considerably just a second before global collapse. Despite appearances, and despite Sunder's Q&A remarks about structural resistance, perhaps there is just nothing amiss here, and the conclusion that Sunder rather belatedly arrives at may be correct.

As for the reports of molten steel, perhaps the eyewitnesses were just mistaken. But what of the corroborating evidence from the dust? There are two main issues there: iron-rich microspheres, thought to be indicative of a thermite reaction, and red/gray chips, thought to be unreacted nanothermite. Mick West, who is described on the dust jacket of his recent book as a "science writer and professional debunker," cites R. J. Lee, who wrote a report documenting the presence of iron microspheres in the dust. Lee explains:

> What about the iron microspheres? The iron has a thin layer of rust flakes that can easily be removed by sticky tape. The iron is heated red hot or hotter and subjected to hurricane force blast furnace-like wind. The iron flakes are liberated as small particles and some iron is vaporized. Like drops of water, the iron flakes form molten spheres that solidify and the fume also condenses into spheres, the most efficient geometrical form. . . . The formation of iron and other type spheres at temperatures obtainable by the combustion of petroleum or coal based fuels is not a new or unique process. (West 2018, p. 147)[19]

While that may not settle the matter for everyone, it is nevertheless the *kind* of response that is appropriate.

As for the red/gray chips, West suggests they are just chips of paint. This was earlier suggested by James Millette and has been contested. (Millette was hired to investigate the claims about these chips by Chris Mohr, who opposes the controlled demolition hypothesis.[20] Millette never published his report in any peer-reviewed venue.) West also points out that the dust was "missing an expected product of thermite combustion. . . . When we burn thermite we find many white aluminum oxide spheres and a large number of hybrid 'eyeball' spheres with a dark iron sphere embedded in a larger white oxide sphere. These were absent in the descriptions of the dust" (p. 147). This is "errant data" relative to the controlled demolition hypothesis. Further, West did a little experiment. He used paint chips from a red painted wheelbarrow that came off when he struck the wheelbarrow with a hammer and looked "identical" to the chips found in the WTC dust. He explains, "When I heated them in a flame they formed microscopic shiny spheres of some iron-rich material (it was magnetic)" (p. 148).

Perhaps there are good responses to these considerations. Jones and others associated with the original red/gray chip paper, for example, reject the notion that the red/gray chips are just paint. Rather than assuming they are impervious to reason, one should treat their arguments seriously, respectfully, and with an open mind. Of course, nobody is *required* to spend any effort on this. But if one thinks that conspiracy theories are a *problem*, and thinks something should be done about them, the honest inquiry approach is much better than deceptive infiltration (such as advocated by Cass Sunstein), bogus "experiments" (such as those performed by National Geographic), or any kind of subtle manipulation.

A.2.4. Two Levels of Critique Regarding NIST's Analysis of Building 7

One can question NIST's results on a general level or a specific level. That is, one may question NIST's *general* conclusion, that the collapse of Building 7 was *due primarily to fire, not controlled demolition*. Or, one may question the *specific sequence of failures* NIST identified. The first line of critique, if successful, has troubling conspiratorial implications, while the second may not. But the second issue is still important. Why else would NIST bother trying to come up with a specific cause of the collapse in the first place? After all, a steel-framed high-rise had never collapsed due to fire before. So the precise reason it did collapse may have significant public safety implications. So, *if* there are good reasons to think their specific explanation is dubious, the whole matter may deserve a second look even if one cannot imagine that controlled demolition could be part of the real answer.

Critics maintain that there are a number of technical reasons to think that NIST's specific explanation is dubious. NIST's explanation for the collapse of Building 7 involves a sequence of events tied to the specific structure of Building 7. The sequence depends critically on a single failure, that of column 79. According to NIST's theory, the key failure occurred when girder A2001 "was pushed off its seat at column 79 by beams framing into it from the east due to thermal expansion of the beams caused by the raging fires" (Griffin and Woodworth 2018, p. 47). All of the rest of NIST's specific sequence of failures depend on that event, or so critics maintain. They argue further that there are a number of reasons to think that that would not have happened, some of which involve flaws in NIST's model. For example, they maintain that NIST's analysis omits the presence of "steel plate stiffeners that provided critical support for girder A2001" (p. 48) and that "The girder flange for column 79 could not bend or fail with the stiffeners present" (p. 48). Architectural drawings for Building 7, released in 2011 as a result of a Freedom of Information Act request, apparently show the stiffeners to be present. Critics also identify other purported errors and omissions that bolster their case against NIST's specifically identified critical failure.

Both general and specific critiques of the collapse of Building 7 have recently been investigated by Leroy Hulsey at the University of Alaska, funded by Architects and Engineers for 9/11 Truth. Hulsey's findings align with the goals of the organization funding him,[21] just as James Millette's did on the other side of the issue, and one could say essentially the same of NIST's findings as well. The new study is unlikely to be the end of the story, nor should it be.

A.3. Government Officials: Bush in the Classroom

President George W. Bush has stated, on at least two recorded occasions, that he saw a plane crash into the World Trade Center on the morning of 9/11 while he was waiting to enter a classroom for a reading lesson. "The TV was obviously on," he explains on one occasion. He then reports thinking, "There's one terrible pilot."[22] On a separate occasion, Bush explains, "[W]hen we walked into the classroom, I had seen this plane fly into the first building. There was a TV set on. And you know, I thought it was pilot error."[23]

However, it was not possible for him to have seen this on TV at that time, since the footage of the first plane hitting the North Tower did not surface until much later. And he cannot have been mistakenly thinking about having seen the second plane hit, because that struck while he was reading *The Pet Goat* with the schoolchildren. Bush was famously informed about this second strike, and that America was "under attack," by his chief of staff, Andrew Card, who reportedly whispered into his ear, "A second plane hit the second tower. America is under attack" (Kean and Hamilton 2004, p. 38).

The president, therefore, could not have seen either impact on TV until after he already knew that "America is under attack." So there is no way he could have thought, "What a terrible pilot," or that it was "pilot error," upon watching one of the towers being struck. Perhaps Bush had become extremely confused about what happened that morning, though this is a bit hard to believe. (For what it is worth, Lillian Glass uses this case in her book, *The Body Language of Liars: From Little White Lies to Pathological Deception*, to exemplify two signals of deception: providing too much extraneous information, by mentioning the TV, and using the word "obviously" in such a context [Glass 2014, pp. 170–171].)[24] In any case, if we cannot rely on the president of the United States to speak accurately about his experience of such a memorable event, it seems that skepticism regarding the pronouncements of high officials, even regarding their first-hand experience, may be reasonable.

The problem is exacerbated by the false retelling of the events of that morning by Andrew Card on the first anniversary of 9/11. In an article, written by Card himself, published in the *San Francisco Chronicle*, Card describes the president's actions after Card told him that "America is under attack":

> [The president] looked up—*it was only a matter of seconds, but it seemed like minutes*—and I thought that he was outstanding in his ability not to scare either the American people that were paying attention to the cameras or, more importantly, the students that were in the classroom. And *he just excused himself very politely to the teacher and to the students, and he left*. (Emphasis added)

Similarly, on September 9, 2002, Card claimed: "I pulled away from the president, and not that many seconds later, the president excused himself from the classroom, and we gathered in the holding room and talked about the situation" (Finnegan 2007, p. 4).

However, as anyone can now see by watching the video of the incident,[25] Card's description is far from accurate. Bush sat there for five agonizing minutes, until the reading session was done. He then reportedly stayed a couple minutes more to chat. He appeared to be gratuitously "stretching out the moment" and "lollygagging as if he didn't want the session to end," according to conservative journalist Bill Sammon, now vice president of Fox News (Sammon 2002, p. 89). The "dawdler in chief," Sammon reports, was "taking his good old time" (p. 90).

As hard as it may be to imagine that Bush was merely mistaken about his account of seeing the first impact on TV, it is harder still to imagine that Card's depiction of Bush's handling of the situation is a case of innocent misremembering. However, even if Card can fairly be described as lying, that doesn't make him part of a sinister conspiracy. We can understand him wanting to put the president in a positive light, even if we wish we could rely on officials to be more honest. Still, such cases demonstrate that some officials will stretch the truth even for relatively minor gains. And so evaluation of official accounts ought to include the recognition that details stemming from the reported experience of officials may have been manipulatively framed or even fabricated.

Further, such accounts may even be supported by false reports given by other witnesses and participants. For example, the day after Card's revision of history came out in the *San Francisco Chronicle*, NYPost.com reportedly published an account given by the teacher in the classroom, Sandra Daniels, which seemed to corroborate Card's depiction, but further exaggerated it, adding inaccurate detail. According to Daniels, "The president bolted right out of here and told me: 'Take over'" (Balfour 2002).[26] That, however, is not at all what the video shows.

This case is informative because we can compare a recording of what took place to what was reported. Unfortunately, most incidents of interest were not recorded, and so we often have only the sometimes-contradictory reports of participants and witnesses to go on. The case of "Bush in the classroom" vividly demonstrates just how far these reports can deviate from reality, even when there are multiple witnesses.

Tangentially, it is worth noting that by the time Card whispers into Bush's ear, the Secret Service, along with most of the country, realized that hijacked airplanes were targeting select buildings. The president's presence at Booker Elementary was widely publicized in advance, and no one knew how many more strikes there would be, or whether there were any other kinds of attacks coming. It was clear at this time, however, that there was a coordinated attack underway. The Secret Service surely assumed that the president might be a target. And yet, strangely, the president remained at the school for another half hour. The *9/11 Commission Report* acknowledges that the Secret Service was "anxious to move the president to a safer location." So why didn't the Secret Service do so? The report addresses this simply by stating that they "did not think it imperative for him to run out the door" (*9/11 Commission Report*, p. 39), as if that was the only alternative.

I mention this not as some sort of proof of complicity, but rather as an example of an apparent anomaly that seems to deserve a fuller explanation but was instead glossed over with a false dichotomy—a choice between "running out the door" and staying at the school for another half hour. The various attempts to make it seem that Bush did not linger in the classroom aid in the glossing over of this issue, by drawing attention away from the fact that the Secret Service just let him sit there while planes were being crashed into buildings. When such issues are given short shrift, it is hard to determine whether or not there truly is something significant underneath.

In this case, there are a number of possibilities. It could turn out that the Secret Service acted properly—leaving the school before a route was cleared may have been judged riskier than staying there.[27] Or it could be that the decisions taken are genuinely criticizable, but not nefarious. Alternatively, if the conspiracy theorists are on to something, further investigation of this issue may lead to more troubling conclusions. Or it may lead to something surprising. For example, perhaps the Secret Service could be reasonably sure that they were safe, at least from airliner attack, because, as Richard Clarke reveals, the "Secret Service had a system that allowed them to see what the FAA's radar was seeing" (Clarke 2004, p. 7). Perhaps the real scandal should be the fact that NORAD apparently did *not* have such a system.[28]

Here is the point: if officials and other witnesses falsely report various details (seemingly with some intention), and apparent anomalies are glossed over, how is one supposed to have justifiable confidence regarding what actually happened, and why it happened?

A.4. The 9/11 Commission

Both senior counsel to the 9/11 Commission, John Farmer, and investigative reporter Philip Shenon have written books that support the conclusions of the 9/11 Commission, but that nevertheless provide analyses that suggests that the commission was confronted by a great deal of inaccurate testimony and dishonesty emanating from various government agencies. While Farmer and Shenon tend to suggest that this should be interpreted as either carelessness or perhaps a "benign" cover-up—a cover-up of relatively innocent failures and incompetence—it is not clear that more troubling interpretations can be easily dismissed.

Conspiracy theories about 9/11 are often classified as either MIHOP (Made It Happen On Purpose) or LIHOP (Let It Happen On Purpose). Although the phrase "let it happen" sounds passive, LIHOP theories need not be entirely passive. They can, and often do, suggest that complicit individuals actively *went out of their way* to make sure the response to the attacks was ineffective and/or to ensure that the hijackers were not apprehended before the attack could take place.

In addition to MIHOP and LIHOP theories, there is also the possibility that, though 9/11 may have been a surprise attack by al-Qaeda, there are nevertheless aspects of the official account that are not correct—perhaps largely covering up incompetence and honest screwups,[29] but not *just* that. For example, there may also be truths about ill-advised programs that ought to be subject to greater oversight.[30] These various possibilities are not mutually exclusive. Cover-up of incompetence, screwups, and/or ill-advised programs does not contradict LIHOP or MIHOP theories—each of which come in various degrees of ambitiousness. And evidence for LIHOP usually does not weigh against MIHOP, at least not strongly. The left hand may have been LIHOPing without knowing that the right hand was actually MIHOPing, perhaps harnessing efforts to cover up screwups and incompetence in order to, in effect, cover up something worse.[31] The empirical issues are not easy to sort out, and it is not my purpose to try to do so here. I aim, more modestly, to suggest that there may be some middle ground between ambitious conspiracy theories and "benign" cover up of honest error, imprudence, or negligence. Exploring this middle ground may lead to the discovery of something worse; and investigating the possibility of something worse may lead to discovery of some middle-ground truth. Either way, both types of explorations ought not be discouraged and dismissed as mere "conspiracy theory."

A.4.1. John Farmer's "Ground Truth"

Is *The 9/11 Commission Report* an accurate account of what happened on 9/11? John Farmer argues that the report is indeed "accurate, and true"—at least as it applies to "the nation's *response* to the 9/11 attacks" (Farmer 2009, p. 2, emphasis added). On the one hand, perhaps we should take Farmer's endorsement of the report with a pinch of salt, since Farmer is not exactly a neutral party—he was senior counsel to the 9/11 Commission and participated in the writing of the report. On the other hand, an insider's perspective is often worth considering.

What is interesting about Farmer's account is that, although he suggests that the commission succeeded in getting to the "ground truth" of 9/11, on his view many other government agencies did rather poorly at discovering and reporting the truth. For example, Farmer argues that reports released by the "Inspector General of the Departments of Defense and Transportation . . . ignore critical evidence in concluding that government officials did not knowingly mislead the public about the events of 9/11" (Farmer 2009, p. 2). In explicit contrast with the inspector general reports, Farmer concludes that "At some level of the government, at some point in time . . . there was a decision not to tell the truth about what happened" (p. 4).

Farmer seems to suggest that the commission's production of an accurate report is a particularly noteworthy accomplishment in light of the degree to which officials had made false claims about what had happened. Farmer writes:

> [T]he 9/11 commission staff discovered that the official version of what had occurred that morning—that is, what government and military officials had told Congress, the commission, the media, and the public about who knew what when—was almost entirely and inexplicably untrue. (Farmer 2009, p. 2)[32]

Just as the with the Warren Commission report (see section 9.3, above), *The 9/11 Commission Report* is regarded by mainstream sources as the "truth," even though it is clear that both commissions were significantly obstructed. While it is certainly possible that these commissions nevertheless produced accurate reports, this notion does suggest something a little odd: that these commissions are *exceptional*—that they did an honest and thorough job in reconstructing what had occurred, *though almost nobody else did*.

Further, the idea that the 9/11 Commission got to the truth of what happened is more plausible for some aspects of what occurred than for others. For example, according to Farmer, the commission was able to discern the "ground truth" regarding the communication between the FAA and the military, despite all of the false testimony presented to them, *because there were tapes and logs that clearly established that truth* (Farmer 2009, p. 3). But what about all the areas for which there are no tapes and logs?

Regarding the statements of detainees, although there were transcripts, they were of low quality (as discussed more below). For one thing, interrogators were not focused on what had already happened. In addition, as the cochairs of the 9/11 Commission, Thomas Kean and Lee Hamilton, admit, "We also had no way of evaluating the credibility of detainee information. How could we tell if someone like Khalid Sheikh Mohammed, a sworn enemy of the United States, was telling us the truth?" (Kean and Hamilton 2006, p. 119).

In any case, Farmer's most significant claim is that 9/11 was not only a result of a "failure of imagination," but also a "collapse of competence" (p. 8). He writes:

> Because the government's version of what occurred on 9/11 overstated the efficiency and effectiveness of our national defense response, it obscured the day's essential reality, and its causes: a radical disconnect between those who were putatively in charge of conducting our nation's defense and those who were on the ground, making operational decisions. Because the government didn't tell the truth, in other words, we lost the "ground truths" of 9/11; (1) there was no connection or collaboration between Washington and the ground-level commanders, as a result of which (2) our national decision-makers in Washington—or, in the case of President Bush, in the air—were irrelevant to how we were defended during that critical time period. (p. 5)

I don't know if this is truly the most important lesson of 9/11, or even if Farmer is right about it. But it is a position worth considering and discussing. And it is not clear why critiques that cut deeper than Farmer's should be left out of that discussion, though Farmer seems to imply that they should. Regarding "conspiracy theories that persist regarding 9/11," Farmer's explicit purpose, in part, is to "lay at least some of these theories to rest by identifying and establishing the deception that actually did occur" (p. 5). But it is hard to see how the deception that Farmer documents undermines the idea that other kinds of deception were also in play. Regardless, conspiracy theorists can at least agree with Farmer on one thing: "The perpetuation of the untrue official version remains a betrayal of every citizen who demanded a truthful answer to the simple question: What happened?" (p. 5).

A.4.2. Philip Shenon on Cover-Up and Conspiracy Theories

In his book *The Commission: The Uncensored History of the 9/11 Investigation*, Philip Shenon, who accepts the general outline of the official account of 9/11, provides many details that support Farmer's characterization of the pattern of inaccurate testimony that the commission had to contend with—often citing Farmer himself. For example, Shenon writes:

> [I]n his testimony, [General Richard] Myers offered the first of what would be several contradictory—and flatly inaccurate—statements from the Pentagon about the military response on September 11. (Shenon 2008, p. 119)

> [Farmer] was startled to realize that NORAD still had its facts wrong in the spring of 2003—and either it did not care or was misstating the record intentionally. (Shenon 2008, p. 120)

Farmer believed the testimony from General Arnold and other NORAD generals should have been referred to the Justice Department for possible prosecution. It would later be determined that almost every one of those assertions by Arnold in May 2003 was flat wrong, most startlingly his claim that the military had close-tracked United 93 and was prepared to intercept it (p. 121).

Further, from Shenon's account, it is not clear that the commission did all that it could have to get to the "ground truth." As Shenon notes, the commission requested but did not receive, access to members of al-Qaeda held in American custody, such as Khalid Sheikh Mohammed. According to Kean and Hamilton's own account, while they pressed as hard as they felt they reasonably could, they ultimately chose not to pursue every avenue to the end. They were advised that their legal authority to demand access was dubious,[33] and they judged that making a public issue out of it was unlikely to be successful (Kean and Hamilton 2006, pp. 122–123). So they contented themselves with CIA and Pentagon interrogation reports that were "incomplete and poorly written" (Shenon 2008, p. 391) and presumed to have been coerced. (Indeed, much of the report is based on what was reportedly said by people we now know were subjected to waterboarding.) Harvard historian Ernest May, who was hired as a consultant to the commission, reports that "what impressed me overall was the poor quality of the summaries as historical evidence" (Shenon 2008, p. 391). Dieter Snell, who served as senior council to the 9/11 Commission and also as team leader regarding the 9/11 plot, agreed with others on his team that the partial reports were inadequate (Shenon 2008, p. 183).

Mike Jacobson, who was working under Snell, "was convinced that FBI officials had tried to hide much of the evidence in its files" that pertain to two of the hijackers, Nawaf al-Hazmi and Khalid al-Mihdhar (Shenon 2008, p. 184)—more will be said about these two hijackers below.[34] Jacobson and at least some of his team members, reports Shenon, "could see few innocent explanations for why so many Saudis and other Arab men living in Southern California had come forward to try to help the two hijackers—to help them find a home, to set up bank accounts, to travel" (p. 184). Regarding such matters, the FBI proved to be uncooperative. It had previously been so uncooperative with the earlier congressional investigation that it "ended with a recommendation that the government seriously consider breaking up the FBI" (Shenon 2008, p. 184). And yet the FBI's level of cooperation with the commission, Shenon suggests, was no better.

Like Farmer, Shenon shows that the level of trustworthiness of government agencies and officials is quite low. Although generally defending the commission's findings, Shenon goes further than Farmer in providing reasons for questioning whether the commission really got to the "ground truth" of all significant aspects of the events. Yet Shenon is nevertheless dismissive of common "conspiracy theories." He writes:

> [B]y the time the 9/11 commission opened its doors in 2003, many of the most outrageous, if well circulated of the [conspiracy] theories—that the attacks were an inside job by the Bush administration, that the Twin Towers were brought down by preplaced explosives, that

> the Pentagon was hit by a missile and not a plane—had been well debunked. (Shenon 2008, p. 118)

Shenon then proceeds to summarize that debunking:

> The evidence was incontrovertible that al-Qaeda was behind the September 11 attacks; [1] Osama bin Laden had been videotaped bragging to his colleagues about his role in the preparations. [2] There was clear-cut documentation to show that bin Laden had dispatched 19 young Arab men to the United States to carry out the hijackings—he had chosen them personally for the mission—and that [3] those same men were aboard the four planes. [4] There was a well-documented money trail for the plot. [5] Independent scientists and engineers had plausible explanations for the physical collapse of the twin towers and other buildings nearby. (Shenon 2008, p. 118; numeration added)

Shenon provides no specific citations for any of this. Rather, the content for the chapter as a whole is merely attributed to interviews with John Farmer, and with two commission members, Slade Gorton and Richard Ben-Veniste (see Shenon 2008, p. 429). Every part of Shenon's "debunking" is contestable. But perhaps the most significantly misleading aspect of Shenon's characterization is the suggestion that any of these details are even relevant to the claim that 9/11 was an "inside job." After all, to assert that, say, a bank robbery was an "inside job" is not to claim that the robbers were employees. It is to claim, vaguely, that the robbers got some form of insider help. That is, that there was some degree of insider complicity. The same applies to 9/11. The possibility that bin Laden and/or individuals associated with him were manipulated or protected in some way, and to some degree, cannot be dismissed by appealing to the kind of evidence that Shenon points to, even if it were substantial. That is to say, the involvement of bin Laden and his al-Qaeda network is compatible with insider complicity. It is even compatible with very ambitious MIHOP conspiracy theories, such as those positing controlled demolition and/or remote piloting of the aircrafts, as well as less ambitious LIHOP theories.

There are specific reasons to take at least some of such possibilities seriously. These have been elaborated at length by author such as Peter Dale Scott (2013) and Kevin Fenton (2011). Some of it has to do with the possible involvement of Saudi and/or Pakistani intelligence agencies, which may or may not have had some connection to the CIA, or elements within it. Shenon himself acknowledges:

> Even though the CIA had strong reason to believe in 2000 that Hazmi and Mihdhar were both at large somewhere within American borders, the agency [i.e., the CIA] had waited until just days before 9/11 to ask that they be added to the government's terrorist watch lists. (Shenon 2008, p. 140)

And Richard Clarke, former national coordinator for counterterrorism, muses:

> I can understand them [the CIA] possibly saying, we need to develop sources inside al-Qaeda; when we do that, we can't tell anybody about it. And I can understand them perhaps seeing these two guys [al-Hazmi and al-Mihdhar] show up in the United States and thinking, ah-ha, this is our chance to flip them, this is our chance to get guys inside al-Qaeda, and to do that, we can't tell anybody outside CIA until we got them, until they're really giving us information. (McGovern and Leopold 2011)

If something like that had happened, the whole business, including any related activity of foreign intelligence agencies, would presumably have to be covered up. So the denials of former CIA chief George Tenet and his associates, which quickly followed Clarke's musings, would be expected whether there was any truth to these ideas or not. These seem like important issues, and they do not seem to have been resolved.

So was 9/11 simply an intelligence failure, or is there more to the story? Were hijackers under surveillance? Was there an effort to "flip" them? There are many levels at which the official account of 9/11 can be reasonably questioned. Even if the problems with mainstream understandings of what happened on 9/11 are likely to be of the relatively "benign" variety articulated by Farmer and Shenon, that would not justify stigmatizing more critical inquiries into such events.

A.5. Conclusion

We've seen that the mainstream media sometimes manipulate audiences to favor an official story. The response of NIST's lead investigators to the concerns of critics have not always been accurate or convincing, and their reports are questionable at multiple levels. The president's account of his own experience of the morning of 9/11, and claims made of his chief of staff, even though the latter was supported by a third party, are provably untrue. And, the 9/11 Commission was given false information, and low-quality information, from government agencies. All this suggests that official and mainstream accounts ought not simply be accepted uncritically.

Whatever the truth about 9/11, it seems doubtful that either the *9/11 Commission Report* or the mainstream media have truly gotten to the bottom of all the important aspects of it. And that is a tragedy. If the truth is important, epistemic authorities ought to face critical questions and skeptical challenges to test how well they can support their claims. To a large degree, the role of conspiracy theorists is to press these questions and raise these challenges.

Notes

Introduction

1. See Dentith 2018c for more on the application of "the argument to the best explanation" to conspiracy theories.

2. This description is heavily influenced by C. Stephen Layman's *Letters to Doubting Thomas: A Case for the Existence of God* (2007), the introduction to which provides a clear explanation of the argument to the best explanation, with examples. Throughout the book, Layman uses the interplay between explanatory power and prior probability (emphasizing simplicity) as he considers various phenomena requiring explanation.

3. In contrast to Hegel's idea that "Reason rules the world," the slogan "Shit happens" appears as an epigraph at the beginning of Brian Keeley's "Of Conspiracy Theories" (1999). An article entitled "Shit Happens," by Pete Mandik (2007), is discussed in Chapter 2, section 2.6.

4. Jaron Harambam cites Michael Barkun, Daniel Pipes, Jovan Byford, Brian Keeley, and David Aaronovitch as making this criticism, and points out that conspiracy theories are criticized for being "at once too complex and too simple to be true" (Harambam 2020, p. 13).

Chapter 1

1. See Pigden 2007 for a philosophical analysis involving this case.

2. For a longer list of categorized cases, see deHaven-Smith 2013a, p. 210 (Table 5.1). This is a table of what deHaven-Smith describes as "Crimes against American democracy committed or allegedly committed by elements of the U.S. government." It includes "crime or suspicious event" (with time frame and modus operandi), motive or policy implication, "suspected or confirmed perpetrators," and deHaven-Smith's assessment of the "degree of confirmation of government role."

3. Epstein received what some regard as extraordinarily light treatment given the allegations against him. According to a *Sun Sentinel* report, "Feds Deceived Us about Billionaire Sex Offender's 'Sweetheart Deal,' Teen Victims Say": "Experts said it was an unusually lenient punishment and his 'highly unusual treatment' included serving his sentence in a local jail, not prison. He also was allowed to leave the jail six days a week on a work-release program to work at his office in West Palm Beach" (McMahon 2016).

4. In an article titled "Prince Andrew's Link to Sex Offender Jeffrey Epstein Taints Royalty in US," *The Guardian* reports: "The Duke of York's friendship with billionaire convicted paedophile Jeffrey Epstein has put him on American front pages" (Harris 2011). See also *Newsweek*'s "Jeffrey Epstein: The Sex Offender Who Mixes with Princes and Premiers" (Ostler 2015).

5. To name just one contribution, Pigden's 1995 article, "Popper Revisited, or What is Wrong with Conspiracy Theories?," was the first scholarly challenge to the long-standing prejudice against conspiracy theories in the academy. In this article, Pigden responds to Karl Popper's influential critique of "the conspiracy theory of society." I address this in section 2.3.

6. Dentith cites Pigden (in press), which is still at this time in press. That article is an updated version of Pigden 2007, in which Pigden suggests that a "conspiracy theory is simply a theory that posits a conspiracy—a secret plan on the part of some group to influence events by partly secret means" (p. 222). The draft of the newer version more explicitly indicates that this is Pigden's preferred interpretation.

7. Here Stokes cites a 2016 article republished as Dentith 2018c.

8. In fairness to Basham and Dentith, it should be noted that the scholars with whom they are in dialogue occasionally conflate conspiracy theories with conspiratorial thinking themselves. For example, the very social scientists I've quoted here registering this complaint (Dieguez et al.) have themselves equivocated between "conspiracism," "conspiracy theories," and "conspiratorial thinking" (see Basham and Dentith 2016, pp. 17–18). See also Chapter 9 for an extended critique of these social scientists.

9. Strictly speaking, Pipes only uses this phrase to describe "petty" conspiracy theories. But there is every reason to believe he means it to apply to his other category of "world" conspiracy theories as well.

10. For details, see Pigden (in press).

11. Perhaps the most revealing statement concerning Aaronovitch's view of conspiracy theories is the following: "Conspiracy theories are theories that, among other things, offend my understanding of how things happen by positing as a norm how they do not happen" (Aaronovitch 2010, p. 8). In his worldview, conspiratorial activity is just not "how things happen."

12. I can accept Brotherton's characterization of conspiracy theories as "sensationalistic," if understood as meaning that they attribute rather appalling conduct to people or agencies that are "otherwise considered to be the good (e.g. the President), the benign (e.g. the Secretary of Commerce), and the boring (e.g. the International Monetary Fund)" (Uscinski and Parent 2014, p. 41). I argue, in effect, that being "sensationalistic" in this sense does not render conspiracy theories implausible (see Chapter 8).

13. Similarly, writing with Chris French, Brotherton writes, "A conspiracy

theory can be defined as an unverified and relatively implausible allegation of conspiracy" (Brotherton and French 2014, p. 238; citing Brotherton 2013). In addition, using "conspiracist belief" to refer to belief in conspiracy theories, Brotherton, French, and Alan Pickering approvingly cite David Aaronovitch 2009, p. 5 (which apparently corresponds to 2010, p. 6), writing, "A conspiracist belief can be described as 'the unnecessary assumption of conspiracy when other explanations are more probable'" (Brotherton et al., 2013, p. 279). (They follow that sentence with several sentences referring to "conspiracy theories," suggesting that they take "conspiracy theories" as synonymous with "conspiracist belief." They reference *JFK*, 9/11, and the 7/7 London bombings conspiracy theories.)

14. Many critics of conspiracy theories do not want to debate conspiracy theorists, often arguing that doing so would give them undue credibility. Yet disparaging conspiracy theorists from a safe perch while not being willing to stand up to them directly does not seem particularly epistemically virtuous.

15. Keeley ultimately agrees. He writes, "Others (for example, Clarke 2002, 138) have noted that 'nefarious' seems too strong a requirement. My response is to follow Charles Pigden (2006, 157), when he argues that an important proviso on the definition of conspiracy theories is that the conspiracies they posit must at least be morally suspect, although not necessarily morally wrong" (Keeley 2007, p. 141).

16. We must understand "historical event" in a narrow and commonsensical way. For example, the plot to kill JFK was a plot to *kill JFK*, not to bring about all the historical details surrounding that assassination. Otherwise we would end up pretty nearly with what Karl Popper called "the conspiracy theory of society," according to which conspirators are controlling everything. We don't want a definition to result in this because almost nobody believes this kind of a theory. See Chapter 2, section 2.3, of this book for a discussion of Popper, and the same section for a related discussion about the "preternaturally effective" aspect of "the paranoid style."

17. Michael Barkun also takes this position: "Insofar as conspiracy theories always present themselves as alternative narratives, there must be an official version to offer a contrast" (2013, p. 172). We should be careful not to overstate this. Critics of this view will be quick to offer counterexamples and raise ambiguities—as indeed they should. Nevertheless, if imperfect, I still find the generalization useful.

18. Basham, for instance, argues that there are important counterexamples. For instance, he suggests, some conspiracy theories do not contrast with official accounts because there is no relevant official account. Basham gives an example of a theory involving a number of deaths that have been assumed to be unrelated. According to the theory, all were murdered as part of a conspiracy to prevent the revelation of "the real and embarrassing motives for a government or corporate policy or project" (Basham 2011, p. 53). In cases like this, it seems to me, there is something approximating an official account—the view that the deaths are unrelated. In addition, Basham suggests that "Coady's restriction also forces on us a distinction between conspiratorial theories (like the official story of 9/11) and conspiracy theories, which hardly seems helpful" (2011, p. 54 n. 7). On the contrary, this *is* helpful. When people—including ordinary people—refer to 9/11 conspiracy theories, they generally do not mean to include the official account. The exception is when someone is being deliberately provocative. But that just confirms that it is

not *ordinary* usage. In such cases the phrase "the official conspiracy theory" is often used. That can be a useful phrase, but "the official conspiracy theory" is not a conspiracy theory in the ordinary sense.

19. In his masterpiece *The Structure of Scientific Revolutions*, Thomas Kuhn cites Max Planck's remark: "[A] new scientific truth does not triumph by convincing its opponents and making them see the light, but rather because its opponents eventually die" (Kuhn 1962, p. 151). Or, as it is commonly put, "Science progresses one funeral at a time."

20. Basham makes the following relevant observation: "We need not *distrust* public institutions of information to recognize that our epistemic justification for trust in them is, in many situations where critical interests are at stake, very limited" (2011, p. 80).

21. We need not evoke conspiracy here. Better to think of Upton Sinclair's famous remark: "It is difficult to get a man to understand something when his salary depends upon his not understanding it."

22. "Evil" is too strong to accurately characterize theories that typically count as "conspiracy theories." Elsewhere Pigden uses the phrase "morally suspect" rather than "evil": "What they really mean by conspiracy theory is *a theory which posits a secret and morally suspect plan on the part of Western governments or government agencies to influence events by partly covert means*" (Pigden 2006, p. 164).

23. Here is an extended example of a well-known conspiracy theorist, James Corbett, explaining his position on the idea that there is a set of conspirators controlling everything. It seems to me fairly representative: "[T]here is certainly immense amounts of power and wealth that are concentrated in the hands of a few ruling oligarchs. But I don't want that to become the simplistic, you know, 'there are X number of families, they control the world, they dictate every single thing that happens in the world, and everything is going to their plan,' because I think that is the comic book version of all of this. . . . [It is an] extremely simplistic explanation. I think in reality it's a much messier thing. In fact, I don't think it is 'the deep state,' for example. I think there are deep states, or at least warring factions within the deep state. And there really are differences of opinion amongst different oligarchs that are seeking to influence and direct society this way and that. . . . I don't think there are people who literally control every single thing that happens in the world so much as there are people who are trying to steer the change that is just the natural part of existence. . . . And obviously, with enough wealth and enough geopolitical power, that can be done more or less easily. But still, when we take it just that there are controlling families that control everything, again, that takes the agency out of anything we do . . . and if that was my ideology I would, I guess, just lay down and die" (Corbett Report Podcast, Interview 1370, "James Corbett on the New World Order," July 3, 2018, at 00:34:55).

24. See Chapter 2, section 2.11 for an unconvincing comparison between Elvis and 9/11 conspiracy theories.

25. See section 9.3 for an example involving a "benign cover-up" regarding the JFK assassination that was largely nonconspiratorial.

26. Regarding the September 11 attacks, Basham muses: "Would a mass-media investigation revealing that the highest elements of the federal government are

involved in the murder of thousands of American civilians be the story of the century, or corporate suicide? Probably both" (2011, p. 69).

27. See "Judge Calls Hastert 'Serial Child Molester,' Gives Him 15 Months in Prison" (Meisner et al. 2016).

28. For a brief overview, see "The Rape Allegation against Bill Clinton, Explained" (Matthews 2017).

29. Coady argues that when we use the phrases "conspiracy theory" or "conspiracy theorist" or the like, "we are implying, even if we don't mean to, that there is something wrong with believing, wanting to investigate, or giving any credence at all, to the possibility that powerful people (and especially governments or government agencies of Western countries) are engaged in secretive or deceptive behaviour" (2019, p. 173).

Chapter 2

1. Here I am primarily thinking of Cass Sunstein and Adrian Vermeule, both law professors, who coauthored an article on conspiracy theories that was published in the *Journal of Political Philosophy*. This is discussed briefly below and is the subject of Chapters 4–6. I am also thinking of Jeffrey Bale, who is included below because of his attempt at making a "clear analytical distinction," which is a philosophical task. In Chapter 3, I address an article by psychologist Stephan Lewandowsky, who, writing with two philosophers, published an article on "conspiracist ideation" in a philosophy journal.

2. COMPACT (Comparative Analysis of Conspiracy Theories) is associated with a funding agency called European Cooperation in Science and Technology (COST). For a description see https://www.cost.eu/actions/CA15101/#tabs|Name:overview

3. While Buenting and Taylor provided both the label and first elaboration of the distinction, David Coady seems to anticipate it in the first paragraph of his "An Introduction to the Philosophical Debate about Conspiracy Theories" (2006b, p. 1).

4. See introduction, section I.2, above, for a discussion of prior probability.

5. Juha Räikkä (2009a, 2018) seems also to align with the particularists, with some qualification. Although he takes a particularist stance regarding less expansive conspiracy theories, he remarks, "An implication that hundreds of people and dozens of institutions are involved in (genuine) conspiracy is strange enough to warrant rejection" (2009a, p. 195). This would rule out very few interesting conspiracy theories, especially because of the word "genuine." As I discuss in Chapters 8 and 9, much of what *seems* to be cover-up activity, some of which probably *is* cover-up activity, is nevertheless not genuinely conspiratorial.

6. Regarding the appropriateness of this question, see Palamara 2015, which provides an extensive collection of early news reports and statements by doctors at Parkland Memorial Hospital, where President Kennedy was treated after being shot. The statements of these "epistemic authorities," taken as a whole, support the view that Kennedy was shot from the front. The late Dr. Charles Crenshaw was one of the attending physicians, and he explicitly maintained that Kennedy was shot from the front. In his book *JFK Has Been Shot*, Crenshaw writes, "[W]hen

Drs. Perry, Baxter, Jenkins, Carrico and others wrote their statements on Friday afternoon, they had no knowledge of 'single-bullet theory,' Lee Harvey Oswald, the grassy knoll, the School Book Depository or other evidentiary factors to affect opinions. They also stand as the only recorded medical opinions about Kennedy's wounds before the body was forcibly and illegally (by gunpoint) taken from Dallas by the Secret Service" (Crenshaw 2013). After Kennedy's body was taken from Parkland Hospital, the story gets increasingly complex and contested, and the relevant issues cannot be resolved here.

7. Bale writes, "If certain parties were to say, for example, that a secret Masonic lodge in Italy had infiltrated all of the state's security agencies and was involved in promoting or at least exploiting acts of neo-fascist terrorism in order to help condition the political system and strengthen its own influence in the corridors of government, most readers would probably assume that they were joking or accuse them of having taken leave of their senses. Twenty-five years ago this author might have had the very same reaction. Nevertheless, although the above statement greatly oversimplifies a far more complex pattern of interaction between the public and private spheres, not to mention between visible political institutions ('the overground' or 'the Establishment') and covert political groups ('the underground'), such a lodge did in fact exist" (2007, p. 57).

8. As Daniele Ganser explains, "The US-funded anti-communist parallel government P2 and the US-funded anti-Communist parallel army Gladio cooperated closely during Italy's First Republic" (2005, p. 75). The original purpose, in the wake of World War II, of the network of secret "stay-behind" armies, now loosely referred to as "Gladio," was to have organized insurgent forces ready in case of a Soviet invasion of Western Europe. However, when the invasion didn't happen, the networks engaged in other less legitimate anticommunist activities. As Ganser explains in the introduction of his book, *NATO's Secret Armies*: "As the Cold War ended, following juridical investigations into mysterious acts of terrorism in Italy, Italian Prime Minister Giulio Andreotti was forced to confirm in August 1990 that a secret army existed in Italy and other countries across Western Europe that were part of the North Atlantic Treaty Organization (NATO). Coordinated by the unorthodox warfare section of NATO, the secret army had been set up by the US secret service Central Intelligence Agency (CIA) and the British Secret Intelligence Service (MI6 or SIS) after the end of the Second World War to fight Communism in Western Europe. The clandestine network . . . is now understood to have been code-named 'Gladio' (the sword) in Italy. . . . The secret armies, as secondary sources now available suggest, were involved in a whole series of terrorist operations and human rights violations that they wrongly blamed on the Communists in order to discredit the left at the polls. The operations always aimed at spreading maximum fear among the population and ranged from bomb massacres in trains and market squares (Italy), the use of systematic torture of opponents of the regime (Turkey), the support for right-wing coup d'états (Greece and Turkey), to the smashing of opposition groups (Portugal and Spain)" (2005, pp. 1–2).

9. Bale explains, "[I]f someone were to claim that an Afrikaner secret society founded in the early decades of this [meaning the twentieth] century had played a key role in promoting the system of apartheid in South Africa, and in the process

helped to ensure the preservation of ultraconservative Afrikaner cultural values and Afrikaner political dominance until the early 1990s, some readers would undoubtedly believe that that person was exaggerating. Yet this organization also existed" (2007, p. 57).

10. While I do not wish to defend the books listed by Bale, the fact that he can declare a list of books to be "nonsense" in a peer-reviewed journal, with no details and no argument, apparently just because those books espouse conspiracy theories about 9/11, suggests both that Bale was right about the bias in the academy and that he himself is implicated in it.

11. In 1979, Louie Steven Witt came forward and identified himself as Umbrella Man. His explanation was that the opening of his umbrella, on a sunny day, at nearly the exact time and place at which the shots that killed the president began ringing out, was an ill-timed "bad joke." Witt stated, "If the 'Guinness Book of World Records' had a category for people doing the wrong thing at the wrong time in the wrong place, I would be No. 1 in that position with not even a close runner-up" (Jonsson 2013). To some, this settles the matter; others find problems with Witt's account. I have not pursued the issue further.

12. This article, and a follow-up article (Keeley 2003), along with the relevant section of Popper, and Pigden's response, as well as pre-2006 articles by Basham, Coady, and Clarke, were all republished in *Conspiracy Theories: The Philosophical Debate* (Coady 2006a), which included additional contributions by most of these authors as well.

13. See, for example, Brotherton and French 2014.

14. See Chapter 9, section 9.7, for a discussion of the Final Solution as a conspiracy.

15. See Coady 2003 for a critique of Keeley's analysis of UCTs.

16. If we take the phrase "intended the historical events in question to occur" too strictly—they intended the historical events *precisely as they occurred*—we would end up with an unreasonable theory that almost nobody would believe. This would involve the same problems as the "conspiracy theory of society" and the "paranoid style." For a related discussion, see section 8.4 in Chapter 8.

17. I also address the idea that the conspirators "keep their intentions and actions secret," without making this particular distinction, in section 4.2. Suffice it to say here that conspiracies do not generally require indefinite secrecy, and, in any case, there are examples of conspiracies that have remained secret for long periods. Perhaps there are some that continue to be secret.

18. For more on the dynamics of cover-ups, see section 9.3.

19. See "The AE911Truth Petition," http://www.ae911truth.org/signatures/#/AE

20. See also the lists provided in Griffin 2011, pp. 32–42.

21. Similarly, Pigden argues, "*[U]nless we adopt a question-begging definition of conspiracy theories* (which *defines* conspiracy theories as somehow suspect or unbelievable, making argument unnecessary), there is nothing *inherently* vicious about believing or being disposed to believe conspiracy theories" (2017, p. 121).

22. This much even Sunstein and Vermeule acknowledge (2009, p. 204), though they assume that this information must be limited and skewed.

23. Relatedly, Coady remarks, "Those who say there is a widespread tendency to commit the fundamental attribution error themselves seem to be committing that very error" (2003, p. 208).

24. In response, Basham remarks, "[Clarke] offers no usable standard of when a research program has become truly 'degenerate' [for that] is a limitation of Lakatos' general approach" (2006, p. 136). And he points out that "the real problem with appealing to it in the context of conspiracy theory is that while nature does not, presumably, fake the data essential to our physical theories, in the case of our social theories, *people do*" (p. 136).

25. As Pigden writes, putting his words in the mouth of Coriolanus: "So when I say that some men do conspire / I read no reason into this our world / Than that we know exists: no purposes / But those of men we *know* have purposes" (2006, p. 144).

26. Also, interrupting a talk by M Dentith, Clarke declared, "I retracted that," in reference to a slide associating him with the arguments made in Clarke 2002.

27. Note also that the subtitle of Griffin's *The New Pearl Harbor* (2004) is *Disturbing Questions about the Bush Administration and 9/11*.

28. It is "unambiguous," that is, to anyone with some background knowledge, such as familiarity with the introduction to Griffin's *The New Pearl Harbor*. The reference, more specifically, is to a comment in *Rebuilding America's Defenses*. The authors and endorsers of that document, which includes prominent neoconservatives, advocate a "military transformation" in order to "preserve American military preeminence" (Donnelly 2000, p. 50). But they worry: "[T]he process of transformation, even if it brings revolutionary change, is likely to be a long one, absent some catastrophic and catalyzing event—like a new Pearl Harbor" (p. 51).

29. Not having read this book before, this quotation took me approximately one minute to find, so I doubt it is anomalous. In any case, Griffin's latest book (with Elizabeth Woodworth) says that the "9/11 Truth Movement" has identified evidence that goes "beyond showing problems in the official account, also suggest[ing] that the attacks were organized by people within the US government, specifically the Bush-Cheney administration and its Pentagon" (Griffin and Woodworth 2018, p. vii—this is the first page of the introduction).

30. If conspiracy theorists do not make a specific allegation against particular suspects, they are (inappropriately) written off as too vague. But when they do make specific allegations, they are faulted for that too. For example, Stokes, quoting Juha Räikkä, asserts "when it comes to political conspiracy theorizing, publishing suspicions can be morally problematic, because harsh claims are made in too early a stage of the investigations" (Stokes 2018b, p. 196, quoting Räikkä 2009b, p. 466). See the following section for an analysis of Stokes's position.

31. One ought to consider reasonable liability concerns, which exist even if one is both right and justified. When E. Howard Hunt was accused of a role in the JFK assassination, he sued. Hunt won initially, but the decision was overturned on appeal, and then Hunt lost on retrial. (See "Revisiting Watergate: Key Players: E. Howard Hunt," *Washington Post*, http://www.washingtonpost.com/wp-srv/onpolitics/watergate/howardhunt.html.) Incidentally, much later Hunt made a deathbed confession of sorts, claiming to have been merely a "bench warmer." (See "Confes-

sion of Howard Hunt," Mary Ferrell Foundation, https://www.maryferrell.org/pages/Confession_of_Howard_Hunt.html.)

32. When a theory postulates a frame-up of sorts, this presents a special difficulty for evaluating whether the theory is progressive or degenerating.

33. A fuller quotation reads: "[I]n and of itself this activity [criticizing the received view] cannot be sufficient to overthrow a received view in favour of a conspiracy theory. The most it can do is cause people to suspend judgement. Before we can reasonably expect a conspiracy theory to replace a received view, that conspiracy theory must be judged to provide a superior explanation of the relevant phenomena than the received view. But this cannot happen unless the conspiracy theory in question is fleshed out to the extent that it is in a position to figure in specific explanations of the events in question" (Clarke 2007, p. 177). These are the concluding sentences of Clarke's article. Note the exaggeration in the way this is framed. He has treated a degree of emphasis as if it was an absolute "all or nothing." It is not the case that the theories that Clarke has criticized provide *absolutely no* positive claims. But so long as there are *some* positive claims, the first sentence, and the second, which derives from it, are non sequiturs. The last sentence is not transparently clear, but if interpreted according to the requirements mentioned in the paper, it is false. The third sentence I grant as both true and relevant, but not particularly helpful by itself.

34. Relevant references can be found in note 12 to Chapter 4, below.

35. Perhaps Julia agrees with Noam Chomsky about one thing: it would not make much sense to rig the Twin Towers for demolition and then just trust a group of idiots with marginal flight skills to successfully hijack at least two planes and manage to hit the Twin Towers with them. So if the buildings were demolished with preplaced explosives and/or incendiaries, the planes were controlled remotely (or computer controlled). Controlled demolition of the Twin Towers implies remote control of the planes. Otherwise, too much unpredictable "shit" could happen.

36. The anthology, *Taking Conspiracy Theories Seriously* (Dentith 2018b), also includes the responses cited in this section by Basham, Dentith, and Pigden.

37. I address Hofstadter's "paranoid style" in detail in Chapter 8.

38. In 1964 debate took place between Mark Lane, on one side, and Joseph Ball, Herman Selvin, and A. L. Wirin, on the other. Wirin, a prominent lawyer, criticizes Lane, an early and prominent critic of the Warren Commission, for making comments similar to those made by some "rightists." (The full debate can be heard on Black Op Radio, show #527, replayed in 2018 on show #902. Wirin's comments are at 2:35:00.)

39. The content of "CIA Document 1035-960: Concerning Criticism of the Warren Report" can be read online at, for example, https://archive.fo/5cq47. See Olmsted 2009, p. 137, for more on the "Communist propagandists" allegation.

40. The most severe accusation is that conspiracy theorists are responsible for 330,000 AIDS deaths in South Africa. To this, Basham responds, "An extensive review of the literature shows these deaths were not caused by a warrantless popular belief in a Western anti-African conspiracy, thundering through the administration and then the cities and villages" (2018b, p. 48). Rather, it was "Motivated by scientifically interesting, if mistaken early doubts about HIV causation" (p. 49).

41. After the manuscript for this book was substantially completed, another set of attempts to show that there is something epistemically problematic about conspiracy theories has been published: "What's Epistemically Wrong with Conspiracy Theorising?" (Harris 2019). I show that these attempts don't work in Hagen 2020b, "Is Conspiracy Theorizing Really Epistemically Problematic?"

Chapter 3

1. LL&B offer an unsparing critique of Donald Trump in an overreaching attempt to paint him as irrational. In contrast, when using an example that is clearly based on a situation involving Hillary Clinton, they (1) simply refer to her as "a politician" instead of naming her, and (2) inaccurately characterize the situation. They describe "the FBI exonerating a politician from allegations of having misused a personal email server" (p. 222). In this way they minimize any negative impression regarding Clinton. But it is misleading, for the FBI never *exonerated* Clinton. They aren't even in that business.

2. Henceforth, in this chapter, when referring to this article, I will simply give the page number, rather than repeating "Lewandowsky et al. 2018."

3. In a February 5, 2013, blogpost entitled "The Involvement of Conspiracist Ideation in Science Denial," Lewandowsky equates the two: "conspiratorial thinking, also known as conspiracist ideation." http://www.shapingtomorrowsworld.org /lewandowskyrecfury.html

4. Elsewhere Lewandowsky remarks that there is "evidence to suggest, of course, that this [i.e., misleading literature denying climate change] is orchestrated and financed and all that." See "Climate Psychology & Solutions—Stephan Lewandowsky." YouTube. https://www.youtube.com/watch?v=scy1tAjc4Xc at 00:09:25.

5. Although Lewandowsky admits he has "no direct evidence" of coordination regarding his own case, he cites precedent in order to suggest the reasonableness of this suspicion: "In the case of tobacco research there is clear evidence that complaints about academics to their institutions are not random but highly organized" (Lewandowsky 2019, p. 153). Fair enough.

6. It certainly seems that Lewandowsky has indeed been the victim of an abusive campaign. And he is absolutely right to expose and condemn abusive tactics. I also agree with some of his related claims: He not only has a right, but may well have an obligation, to make assertions that some people will find offensive—for offensiveness does not imply invalidity (see Lewandowsky 2019, p. 170). (Indeed, he may legitimately consider himself a bit of a hero for doing so.) And his "presumed antagonism toward 'climate denial'" (p. 169), which he does not deny (indeed the choice of that phrase, 'climate denial,' is patently antagonistic), does not disqualify him from weighing in on the debate. However, when he complains about charges of "bias" and "activism" (p. 169), he is on weaker ground. His prejudicial word choices, and indications of partisanship, as well as his collaboration with activists such as John Cook, who runs a transparently activist climate science blog, suggest at least some degree of bias and activism. And that's okay. We all have biases, and sincere advocacy is laudable. We are all entitled to free speech as activists, and even to express views that are biased. But it is also fair for this apparent bias and activism to be pointed out and taken into consideration.

7. See the long quotation from Räikkä 2009a in section 2.5.

8. Ironically, the original version of Chapter 7 was published in the same issue of *Argumenta* as Lewandowsky et al. 2018.

9. According to PBS's *Frontline*, General Hayden was at first worried about the legality of the programs he was being encouraged to implement. Hayden reports that President Bush told him, "I understand your concerns, but there are some things we're going to have to do. And I think I have the authority to authorize you to do things that you've outlined" (Kirk 2014).

10. LL&B seem to mean: "There must be one or more individuals *from this agency* operating in complete secrecy." That is, the "individuals" are *from this agency* and operating in secrecy, not operating in secrecy *from this agency*.

11. Any "happy talk" praising their reliability can be taken with a grain of salt.

12. It may be tempting to quibble that Trump's numbers don't add up. But there are possible worlds in which they do add up, such as a world in which all undocumented voters would vote for Clinton. Further, it is not clear that LL&B's numbers add up very well either. They figure: "It is . . . relevant to examine the voting pattern among legitimate Hispanic voters. The data suggest that a third of Hispanics seemed to have voted for Trump. In other words, even if there had been 3 million illegal Hispanic voters (there were not), Trump likely would have received a third of their vote" (p. 227). Is it safe to assume that illegal Hispanic voters would have voted for Trump in the same proportion as their legal counterparts?

13. Likewise, LL&B also remark: "THUNC is fairly labeled as irrational thinking, contrary to the claims of philosophers like Coady (2007), because the reality-denying or counterfactual parts of Trump's belief system conflict with his other cherished beliefs, and are not coherent with them" (p. 227). What makes this version of their claim not only questionable but bizarre is the reference to Coady 2007, which predates both the coining of the term "THUNC" and Trump's conspiracy theories by about eight or nine years. So it is hard to understand how Coady could have claimed that THUNC is not fairly labeled as irrational. The position Coady took in the article in question was this: "[O]n the best definitions of what it is to be a conspiracy theorist, conspiracy theorists do not deserve their reputation for irrationality . . . [T]here may be occasions on which we should settle for an inferior definition which entails that conspiracy theorists are after all irrational" (Coady 2007, p. 193). So, putting the anachronism of LL&B's claim aside, it is not at all clear that Coady would disagree with the claim that "THUNC is fairly labeled as irrational thinking." Indeed, almost certainly he would agree. That is hardly in question. It is the association between THUNC and the typical thinking of conspiracy theorists that is questionable.

14. The assertion that Trump's theories are wrong is just my relatively uninformed opinion on the matter. I haven't actually investigated these particular claims at all, since I am willing to concede this point for the sake of argument anyway. If my opinion about this is mistaken, *so much the better* for my overall thesis.

15. It is not clear why the authors do not include Watergate in their table of results, as they had mentioned it as being uncovered by "conventional reasoning and cognition—rather than epistemically impoverished conspiracist cognition" (p. 219). In any case, here too it is not clear that there was no element of conspiracist cognition involved. And it is not clear that Woodward and Bernstein should not count, in that context, as conspiracy theorists, as indeed they probably should.

16. Sheehan is reporting his firsthand experience and does not supply supporting references.

17. The theory that Sheehan is lying seems to imply a bit of a conspiracy, since Sheehan's wife was involved in the investigation and she gives every indication of vouching for the authenticity of his autobiography as a whole. And, further, to my knowledge, none of his former associates have contradicted his account. Are they all in cahoots with him?

18. I picked Alex Jones because there can be no denying that he counts as a "conspiracy theorist," and I picked "2012" simply because that was the year before the Snowden revelations.

19. According to James Jones, author of *Bad Blood*, "[Preventing] the men from receiving treatment had always been a violation of Alabama's public health statues requiring public reporting and prompt treatment of venereal diseases" (cited in Thomas and Quinn 1991). Note the relevance of this to LL&B's analysis of Trump's theory regarding the wiretapping of Trump Tower. Sometimes people do follow directives rather than the law.

20. Bracketed interpolation in original article.

21. Dr. John Heller exemplifies this callousness when he defends the study saying, "They were subjects, not patients; clinical material, not sick people" (Thomas and Quinn 1991). Heller was director of Venereal Diseases at the Public Health Service from 1943 to 1948.

22. Not wanting to engage in long diversions, I've kept my analyses of these cases brief. However, further exploration of the NSA surveillance issue and the Iran-Contra affair, in particular, do lead to "rabbit holes" that make simple characterizations of these cases especially problematic.

23. For example, Bamford concludes *A Pretext for War* writing, "[T]he Bush administration's massive disinformation campaign, abetted by a lazy and timid press, succeeded spectacularly in driving the public to support its long-planned war. . . . [O]nly once before, in Vietnam, have so few manipulated so many at such a great price" (2004, p. 377). See also the long quotation in section 5.2.1.

24. Betty Medsger, who wrote the original article about the stolen files for the *Washington Post*, explains: "Two members of Congress—Sen. George S. McGovern (D-S.D.) and Rep. Parren J. Mitchell (D-Md.)—and the Washington bureaus of the New York Times and Los Angeles Times had also received the files. All four of those recipients immediately handed them over to the FBI" (2014).

Chapter 4

1. Even when conspiracy theories do not directly produce harmful actions, Sunstein and Vermeule explain, "[Conspiracy] theories can still have pernicious effects from the government's point of view, either by inducing unjustifiably widespread public skepticism about the government's assertions, or by dampening public mobilization and participation in government-led efforts, or both" (Sunstein and Vermeule 2009, p. 220). However, whether widespread skepticism of a particular official story is bad depends, in part, on whether there are legitimate grounds for such skepticism.

2. By "extremist groups" I mean groups who espouse extreme views. Who decides what is an extreme view, and how it is decided, are significant problems. I

am concerned in this chapter only with the issue of whether or not it is justifiable to infiltrate such groups *solely* on the basis that they espouse what are regarded as extreme views. Whether or not *violent* extremist groups should be infiltrated, on account of their tendency toward violence, is a different issue and is not considered here. Presumably, at least sometimes, that can be justified.

3. As Emma Jane and Chris Fleming put it: "[T]he best way to rectify the fear that powerful people and institutions are engaged in conspiratorial and covert schemes designed to manipulate the gaumless masses, [Sunstein and Vermeule] suggest, is to have powerful people and institutions embark on a conspiratorial and covert scheme designed to manipulate the gaumless masses" (2014, p. 5).

4. While informational and reputational cascades are addressed in Chapter 5, two points can be made briefly here: First, such cascades can explain the relative success of official stories at least as well, and arguably better, than they can explain the success of counternarratives. Second, Sunstein and Vermeule's depiction of conspiracy theorists is a crude caricature. For example, it ignores the hundreds of scholars and professionals who publicly question the official account of 9/11. (See, for example, patriotsquestion911.com.) The notion that reputational cascades, in particular, can account for *them* is not very plausible.

5. Upton Sinclair's famous remark is once again relevant: "It is difficult to get a man to understand something when his salary depends upon his not understanding it." Paid shills, by the nature of their role, have a conflict of interest, and thus are highly likely to be biased.

6. Although Sunstein and Vermeule soft-pedal the deceit involved in their proposal, deceit is nonetheless explicit when it comes to concealing one's identity or falsifying the source of "information." Further, *in practice*, deceit is very likely when it comes to the content of that information. This issue is addressed in more detail later in the chapter.

7. If it "*really was* proposed," in what sense is it merely a *rumored* plan? See section 5.2.1 in Chapter 5 for more on the Northwoods document, including an extended quotation.

8. See the end of Chapter 8 for more about the Final Solution as a conspiracy theory.

9. See, for example, "2004 Scientist Statement on Restoring Scientific Integrity to Federal Policy Making," published by the Union of Concerned Scientists, and signed by fifteen thousand US scientists. http://www.ucsusa.org/scientific_integrity/abuses_of_science/scientists-sign-on-statement.html

10. Investigative journalist Jennifer Abel questioned a NIST spokesperson, Michael E. Neuman, about checking for evidence of explosives. Abel inquires, "What about that letter where NIST said it didn't look for evidence of explosives?" Neuman responds, "Right, because there was no evidence of that." Abel asks the obvious follow-up, "But how can you know there's no evidence if you don't look for it first?" Then Neuman delivers the punch line: "If you're looking for something that isn't there, you're wasting your time . . . and the taxpayers' money" (see Griffin 2010, p. 73).

11. See section A.2.1 in the appendix.

12. For several credible eyewitness accounts, see Griffin 2010, pp. 36–37. In addition, a "flow of an orange-glowing liquid" can be seen on any video of the

South Tower, near the impact zone, shortly before the collapse. This is also documented in Jones et al. 2008, p. 38. One line of corroboration of extremely high temperatures comes via a report by the RJ Lee Group, "WTC Dust Signature Report: Composition and Morphology" (December 2003), p. 18. It documents "spherical iron particles" in the WTC dust, in quantities sufficient to include them as part of the "WTC dust signature." The report treats the existence of these spheres as unproblematic, even "expected." See the section A.2.3 in the appendix, for RJ Lee's explanation.

13. Much of the evidence I am referring to is found in the works cited in the previous note, as well as the books mentioned by name in note 23 of this chapter, below. A common non sequitur response to this issue is to suggest that it is not necessary to assume that the columns actually melted in order to explain the collapse. But the issue here is not at all about explaining the collapse. It is about the multifaceted independent evidence that there was in fact a large amount of melted steel (or iron). The question is: How can the official story account for that? And yet a common counternarrative does account for it: molten iron is the by-product of a thermite reaction. So, on the theory that thermite of some kind was used to help bring down the towers (and Building 7), molten iron is to be expected. Indeed, when comparing videos of thermite reactions with video of the orange liquid substance flowing from the South Tower shortly before its collapse, the similarity is striking. This is direct prima facie evidence of the use of thermite. But it is the *combination* of this kind of prima facie evidence with scientific studies documenting significant quantities of iron-rich spheres in the dust, and so on, that makes casual or flippant denials inappropriate. What is appropriate is an explanation that accounts for the various phenomena; see the appendix for a sample of what this should look like.

14. The draft report states, "[T]he actual time for the upper 18 stories to collapse, based on video evidence, was approximately 40 percent longer than the computed free fall time and was consistent with physical principles." See NIST 2008b, p. 41.

15. See: "WTC7: NIST Finally Admits Freefall (part I) by David Chandler." YouTube. https://www.youtube.com/watch?v=iGMvnwjUizY&type=youtube&title=WTC7

16. NIST 2008a, p. 45. NIST continues to stress that the overall collapse time was 40 percent longer than free fall, as if a plausible overall (or average) acceleration removes the need to explain 2.25 seconds of absolute free fall.

17. As another example of evasion, consider the dialogue documented in Griffin 2010, p. 69.

18. While Baer has wavered, and seems now to support the official account, on the Thom Hartmann Show (June 9, 2006), Baer answered a question about whether he thought there was an "aspect of inside job" to the 9/11 events saying, "There is that possibility, the evidence points at it," and then asking a string of suggestive questions of his own. Later, in a commentary entitled "The CIA's Gift to Conspiracy Theorists," time.com (December 7, 2007), he writes, "I myself have felt the pull of the conspiracy theorists—who believe that 9/11 was an inside job, somehow pulled off by the U.S. government." He stresses that he does not believe theories about controlled demolition or missiles, but he concludes, "More than anything what we need right now is complete and total transparency on 9/11." http://content.time.com/time/politics/article/0,8599,1692518,00.html

19. During his twenty-seven years in the CIA, McGovern was a senior intelligence analyst and took part in White House intelligence briefings during the Reagan and Bush Sr. administrations. He has called the *9/11 Commission Report* "a joke," charging, "[T]here's a cover-up." While he is cautious about making explicit speculations about what happened on 9/11, one does not need to be a mind reader to intuit the seriousness with which he approaches the topic. On the inside cover of the book *9/11 and American Empire: Intellectuals Speak Out* (Griffin and Scott 2006), McGovern writes, "This book . . . confronts the American people—indeed the people of the world as a whole—with an issue second to none in importance and urgency. I give this book, which in no way can be dismissed as the ravings of 'paranoid conspiracy theorists,' my highest possible recommendation."

20. Videos of Jowenko's reaction to the collapse of Building 7 can easily be found on YouTube.

21. See "The AE911Truth Petition": http://www.ae911truth.org/signatures/#/AE

22. See Sunstein and Vermeule 2009, p. 223. Strictly speaking, Sunstein and Vermeule don't even consider open and fair inquiries involving proponents of various perspectives. They choose to focus on the likely response to "government rebuttal." Given the history, even just recent history, of government deceit (such as that involving the existence of WMDs in Iraq), we should all be skeptical of official pronouncements until they are tested in the crucible of independent scrutiny.

23. This could include a careful study of, and response to, the work of Peter Dale Scott (2007), David Ray Griffin, Nafeez Ahmed (2005), and Michael Ruppert (2004), as well as the relevant articles published in technical journals (for example, Ryan et al. 2009; Jones et al. 2008; and Harrit et al. 2009). Alternatively, one could reasonably focus on two more recent summaries of the evidence: *The 9/11 Toronto Report: International Hearings on the Events of September 11, 2001* (Gourley 2013), and *9/11 Unmasked: An International Review Panel Investigation* (Griffin and Woodworth 2018).

24. Sunstein and Vermeule write, "Throughout, we assume a well-motivated government that aims to eliminate conspiracy theories, or draw their poison, if and only if social welfare is improved by doing so. . . . This is a standard assumption in policy analysis" (2009, p. 219). Whether or not this assumption is in some way "standard," applying it in a case *like this* demonstrates an astonishing lack of good sense. The question is whether the proposed policy would be a good idea in real life, not in some idealized world. In an ideal world we could just give all-encompassing dictatorial powers to a trustworthy and benevolent ruler. But that tells us nothing about what we should *actually* do.

25. The words "often" and "usually" are significantly different. Someone who often lies may nonetheless tell the truth a great majority of the time—lying only a few times a week, for example. But if he *usually* lies, then we know that, at least, he lies more than half the time. Sunstein and Vermeule can't really establish that secrets are quickly revealed *most* of the time. But if all they mean is that *sometimes* (some kinds of) secrets get exposed, the inference that government officials are too incompetent to ever succeed in holding secrets is quite tenuous.

26. Let us not forget that the *New York Times* sat on the wiretapping story for a full year before deciding that they would go ahead and run it.

27. Morally dubious black ops range from the Phoenix Program, which involved

torture and assassination on a large scale, to contributing to the overthrow of democratic governments, such as Iran in 1953. A more recent example is the rendition program of the George W. Bush administration. As David Coady notes, "[I]t is no longer a secret that the American government regularly kidnaps terrorism suspects and sends them abroad to be tortured (Thompson and Paglen, 2006). But although this is not a secret *now*, it is an operation that was originally planned and conducted in secret, since it contravenes national and international laws" (Coady 2012, p. 120).

28. The jury found: "YES—Loyd Jowers participated in a conspiracy to do harm to Martin Luther King" and "YES—Others including governmental agencies were parties to this conspiracy as alleged by the defendant." The judge apportioned 30 percent of the liability to Jowers and 70 percent to "all other co-conspirators." See Pepper 2003, p. 147.

29. The justice department did its own investigation and concluded that the jury's verdict "is incompatible with the weight of all relevant information" (US Department of Justice 2015). So is this an example of a conspiracy being exposed, or much ado about nothing? I don't know. But I do think it is an example of the "self-sealing" quality of official stories. Even if a jury says that there was a conspiracy, the government can just pronounce the jury wrong—and the media won't inform us of any of it.

30. Sunstein and Vermeule write, "The first-line response to conspiracy theories is to maintain an open society, in which those who might be tempted to subscribe to such theories are unlikely to distrust all knowledge-creating institutions" (2009, p. 218).

31. Police, presumably, lie all the time when dealing with suspected criminals. I have no comment on the appropriateness of that. But here we are not talking about criminals. Further, the issue is how to deal with so-called crippled epistemologies. Responding to a supposed epistemic problem with lies is, to say the least, epistemically suspect.

32. While the original statement, by James Burtchaell, was about murder rather than dishonesty, the logic is the same. See Doerflinger 2010, p. 319.

33. Is this really an accurate depiction of "1960s-style infiltration," or is it a whitewash of history? It seems to be the latter.

34. For the extended quotation, and more on deceit and utilitarian concerns, see section 6.2 in Chapter 6.

35. I admit that I am reading between the lines here, but it is hard not to draw this inference.

36. See also Buncombe 2006, which provides some samples of the planted stories in question, accompanied by some fact-checking. It also quotes John Pike, director of a defense think tank called GlobalSecurity.org: "Anybody who knows about propaganda knows the first rule of propaganda is that it should not look like propaganda. . . . It's embarrassing enough that [the US military] got caught . . . but then for their product to be so cheesy . . . It's just embarrassing."

37. *Daodejing* 38. Translation adapted from Lau 1963, p. 99, and modified to fit the current context.

38. Although Karl Popper famously attacked "the conspiracy theory of society" (see Popper 2006), he would nevertheless likely oppose infiltration, for reasons

similar to those that I have given. In *The Open Society and Its Enemies*, he remarks, "Only political power, when it is used to suppress free criticism, or when it fails to protect it, can impair the functioning of [the various social institutions which have been designed to further scientific objectivity and criticism], on which all progress, scientific, technological, and political, ultimately depends" (1962, p. 218). For a response to Popper's view of conspiracy theories, see Pigden 1995 (which is discussed in section 2.3, above).

Chapter 5

1. One wonders what Sunstein and Vermeule would have said about allegations of sexual abuse at Abu Ghraib if graphic stories of naked human pyramids, and worse, had surfaced and swirled, before the actual pictures came out. The accusation that American soldiers were engaged in *that kind of* heinous abuse may well have counted as an extreme idea. Not all ideas branded "extreme" are false, though we don't always have pictures to clearly establish the truth.

2. Sunstein and Vermeule acknowledge a similar conspiracy theory. They write, "[R]eal-world governments can themselves be purveyors of conspiracy theories, as when the Bush administration suggested that Saddam Hussein had conspired with Al Qaeda to support the 9/11 attacks" (2009, p. 219). But this was not generally dismissed as a mere "conspiracy theory."

3. Notable publications include three anthologies—*Conspiracy Theories: The Philosophical Debate* (Coady 2006a), *Taking Conspiracy Theory Seriously* (Dentith 2018b), and *Conspiracy Theories and the People Who Believe Them* (Uscinski 2019)—and also two special issues of philosophy journals—*Episteme* 4.2 (2007), and *Argumenta* 3.2 (issue 6) (2018).

4. For more examples of historical conspiracies, see the long quotations by Olmsted and by deHaven-Smith (both in section 8.7), and by Sunstein and Vermeule themselves (in section 4.1), as well as the quotations by Charles Homans (in note 3 to Chapter 8).

5. According to journalist Charles Lewis, "[I]n the two years after the terrorist attacks of September 11, 2001, President George W. Bush and seven of his administration's top officials made at least 935 false statements about the national security threat posed by Iraq. The carefully orchestrated campaign of untruths about Iraq's alleged threat to US national security from its WMDs or links to al Qaeda (also specious) galvanized public opinion and led the nation to war under decidedly false pretenses" (2014, p. xiii).

6. Many of Sunstein and Vermeule's claims about conspiracy theorists are true of conspiracy skeptics and agnostics as well. For example: "[S]ome people who [reject] ~~accept~~ conspiracy theories are mentally ill and subject to delusions" (2009, p. 211). That is true too, of course. While Sunstein and Vermeule's point here is that it is not plausible that "all or even most" conspiracy theorists are mentally ill, they make this point in such a way as to suggest there is some significant correlation.

7. An example that has been cited by conspiracy theorists is footage from Fox News showing (ostensibly) a random bystander wearing a Harley Davidson shirt who was interviewed shortly after the towers had collapsed. The "bystander" says, "I witnessed both towers collapse, one first and then the second, mostly due to structural failure because the fire was just too intense." See "9/11: Clues You Might

Have Missed," https://www.youtube.com/watch?v=jVtdrhnYc0E. The allegation is that presenting this on the news was supposed to begin or reinforce a cascade of opinion that the towers collapsed due to fire (in addition to the damage from the plane), not from something more elaborate. Once this opinion became entrenched, most people continued to believe it, despite the discovery of significant quantities of red/gray chips in the dust that appear to be bits of unreacted nanothermite (see Harrit 2009). Or so it could be argued.

8. For a very brief set of examples, see Griffin 2011, pp. 68–70. For an exhaustive treatment of the scientific evidence relevant to the collapse of Building 7, see Griffin 2010. For more recent summaries of a variety of lines of evidence, see Gourley 2013 and Griffin and Woodworth 2018.

9. An article entitled "Another Scholar under Fire for 9/11 Views" provides further details and a similar case: "State legislators chimed in, demanding Woodward's dismissal and threatening to consider the issue when they next review the university's budget. In some respects, the political reactions mirror those in Wisconsin, where lawmakers lined up to urge the University of Wisconsin at Madison to fire Kevin Barrett, who shared Woodward's view" (*Inside Higher Ed*, August 29, 2006).

10. Other cases similar to Woodward's could be offered. For example, shortly after 9/11, noted philosopher John McMurtry, according to his own account, suggested that there was "evidence of pretext and falsehood" regarding the preparation for war with Afghanistan and its relation to 9/11. This is how he describes what followed: "The Wall Street Journal called me Osama McMurtry. The New York Times blog-hated me. Fox News ranted. The Globe and Mail wrote editorials of fierce if inane indignation from John Ibbitson (twice) and Margaret Wente. The eminent neo-Darwinian Michael Ruse wrote a letter of condemnation from a Florida university joining the flood of e-mails to the department chair and the university president demanding my head" (McMurtry 2015). See also the documentary, *9/11 and the Academic Community*, which covers both the Woodward and McMurtry cases, and includes an interview with McMurtry.

11. See Basham 2018a for a detailed analysis of "toxic truths."

12. See Hagen 2011b for my review of Griffin's book *Cognitive Infiltration*.

13. I have no objection to this proposal (number 3), but Sunstein and Vermeule do not emphasize it.

14. In his 2014 revision, Sunstein writes, "Perhaps only a handful of kooks believe that US government officials had any kind of role in the events of 9/11" (p. 24). He writes this despite having already cited a poll finding that "49 percent of New York City residents believed that officials of the US government 'knew in advance that attacks were planned on or around September 11, 2001, and that they consciously failed to act'" (p. 1).

15. According to a report by Senators Carl Levin and John McCain, as summarized in the *New York Times*: "[T]op Bush administration officials, including Donald H. Rumsfeld, the former defense secretary, bore major responsibility for the abuses committed by American troops in interrogations at Abu Ghraib in Iraq; Guantánamo Bay, Cuba; and other military detention centers. . . . The abuse of prisoners at Abu Ghraib was 'not simply the result of a few soldiers acting on their own' but grew out of interrogation policies approved by Mr. Rumsfeld and other top

officials, who 'conveyed the message that physical pressures and degradation were appropriate treatment for detainees'" (Shane and Mazzetti 2008).

16. Although, I claim, their proposal *treats* conspiracy theorists as irrational, strictly speaking, Sunstein and Vermeule do not *characterize* conspiracy theorists as (fully) irrational. They explain that the theories that they are concerned with are "unjustified (not in the sense of being irrationally held by those individuals who hold them, but from the standpoint of the information available in the society as a whole)" (2009, p. 207). In other words, while making a contemptuous proposal they pay lip service to the (limited) rationality of conspiracy theorists. This is similar to expressing support for "maintain[ing] an open society" (2009, p. 218) while, as I have argued in the previous chapter, "their recommendations involve moving *in the direction of* a more closed one" (section 4.2 above).

17. As mentioned at the beginning of the previous chapter, Sunstein and Vermeule write, "[W]e suggest a distinctive tactic for breaking up the hard core of extremists who supply conspiracy theories: cognitive infiltration of extremist groups, whereby government agents or their allies (acting either virtually or in real space, and either openly or anonymously) will undermine the crippled epistemology of believers by planting doubts about the theories and stylized facts that circulate within such groups, thereby introducing beneficial cognitive diversity" (2009, p. 219).

18. As David Coady points out, "[T]he defining insight of liberal political thought [is] that you can't just assume that governments are well intentioned" (Coady 2018b, p. 185).

19. Northwoods Document pp. 10–11. See Ruppert 2004, "Appendix A: Joint Chiefs of Staff 'Northwoods' Document" to see photocopies of the entire document, which can also be found online.

20. See Chapter 4 for my defense of the claim that their proposal is deceit-countenancing, antidemocratic, and epistemically suspect.

21. Charles Pigden also describes a particularly appalling COINTELPRO practice: "Suppose the husband of a civil rights worker received an anonymous letter suggesting that his wife had been having an affair. The obvious explanation would be that the letter was genuine (if malicious) and had been written by a mutual acquaintance. As for the idea that the letter was a forgery planted by the FBI to undermine his marriage—well, that would be just too fantastic for words. Would the US government, or even the great but sinister J. Edgar Hoover, descend to such petty malice? The husband, like a true disciple of Occam, would opt for the simpler hypothesis and institute divorce proceedings. But in some cases, the FBI mounted just such a conspiracy" (Pigden 1995, pp. 24–25). This quotation is from an article that should have been cited by Sunstein and Vermeule. Mentioning Pigden's critique of Popper's account of conspiracy theories, Sunstein and Vermeule cite an article not found in the book they indicate. However, that book does include Pigden's article "Popper Revisited, or What Is Wrong with Conspiracy Theories," in which the above quotation appears.

22. See, for example, Griffin 2010; Scott 2007 and 2013; Griffin and Scott 2006; Ahmed 2005; MacQueen 2013; and deHaven-Smith 2013b.

23. Relevant publications include Jones et al. 2008; Ryan et al. 2009; and Harrit et al. 2009.

24. See list of petition signers at http://www.ae911truth.org/signatures/#/AE

25. McGovern was a high-level intelligence analyst for the CIA and is the founder of Veteran Intelligence Professionals for Sanity. See note 19 to Chapter 4 above, for his endorsement of the book, *9/11 and American Empire*.

26. See patriotsquestion911.com. See also Griffin 2011, pp. 29–42, for a list of several dozen accomplished professionals who question the official account of the events of September 11.

27. In particular, "reputational cascades" are implausible explanations for the scholars and professionals who take counternarratives about September 11 seriously. Further, both informational and reputational cascades can explain the success of dubious official stories at least as well as they can explain the success of conspiracy theories, as explained above.

28. Of course, even if theories that deny Flight 77 hit the Pentagon were demonstrably false, it would not follow that all theories positing insider complicity must be false. But there may be a perceived "guilt by association."

29. Charles Pigden—whom Sunstein and Vermeule cite twice (2009, pp. 206 n. 15 and 208 n. 24)—has rightly characterized this bias against conspiracy theories as "dangerous and idiotic." He writes, "The idea that there *is* something suspect about conspiracy theories is one of the most dangerous and idiotic superstitions to disgrace our political culture. . . . It is an *idiotic* superstition since a modicum of critical reflection reveals that it is false. And it is *dangerous* superstition since it invests the lies, evasions and self-deceptions of torturers and warmongers with a spurious air of methodological sophistication" (Pigden 2006, pp. 139, 165).

30. See deHaven-Smith 2010 for an analysis of SCADs.

Chapter 6

1. Luke Rudkowski of We Are Change, identifying himself, tongue in cheek, as Bill DeBerg (Bilderberg). See "Obama Information Czar Cass Sunstein Confronted on Cognitive Infiltration of Groups." https://www.youtube.com/watch?v=4OIiOztc52g

2. Sibel Edmonds, author of *Classified Woman* (2012), was reportedly described by the ACLU as "the most gagged person in the history of the United States of America" (Friedman 2007). See also ACLU (undated).

3. The full title of the blog post is "Got Fascism? Obama Advisor Promotes 'Cognitive Infiltration.'" It appeared on *The Rag Blog*, January 11, 2010. Incidentally, this was the blog that sparked the controversy on the internet about Sunstein's proposal (see Marantz 2017). Estrin was prompted to write this blog having just read a draft of "Is Infiltration of 'Extremist Groups' Justified?," Chapter 4 of this book.

4. Charles Pigden and David Coady have recently added critiques: "Are Conspiracy Theorists Epistemically Vicious?" (Pigden 2017) and "Cass Sunstein and Adrian Vermeule on Conspiracy Theories" (Coady 2018a).

5. On the same page, Sunstein writes, "Under this approach, government agents and their allies might enter *foreign* chat rooms, online social networks, or even real-space groups and attempt to undermine percolating conspiracy theories" (2014a, p. 29, emphasis added; cf. Sunstein and Vermeule 2009, pp. 224–225). The word "foreign" has been added in the revised version.

6. Sunstein also mentions in the preface that "The focus [of the chapter on conspiracy theories] is on threats—especially terrorist threats—that arise when people in other nations believe false conspiracy theories about the United States" (2014a, p. X). However, this remark immediately follows the remark that the chapter "has been significantly revised and updated for this book." This makes it unclear whether the "focus" is supposed to have been there all along (though unstated), or whether this is a new focus introduced in the revision.

7. Similarly, when Sunstein mentions that banning and taxing conspiracy theories would likely be unconstitutional (2014a, pp. 22–23), he is clearly referring to the *American* Constitution.

8. These were included in Sunstein and Vermeule's 2008 online draft, but not in the shorter 2009 article published in the *Journal of Political Philosophy*. They were reintroduced in Sunstein's 2014 book chapter.

9. Here I have borrowed a move from David Ray Griffin's book *Cognitive Infiltration*, namely, checking the articles cited by Sunstein to see whether they undermine rather than support his claims. See, for example, Griffin 2011, pp. 4–7.

Chapter 7

1. Unlike the others mentioned here, Moore actually notices the problem with the study. He notes, rather too forgivingly, that while the respondents *may* hold contradictory beliefs, "[I]t is also possible that respondents are not assenting to the literal truth of a proposition, but rather are saying they are 'prepared to entertain the possibility' of X, or might 'not rule out the hypothesis' that X, or simply 'doubt the official account' of X, all of which would weaken the force of the supposed contradiction" (Moore 2019, pp. 112–113).

2. This is not an isolated slip, or a one-time remark. Sunstein has asserted this repeatedly. For example, in the *New York Times* he writes, "In fact, people who embrace one conspiracy theory are also inclined to embrace another conspiracy theory that cannot simultaneously be true. In one study, people who said they believed that Osama bin Laden is alive and well were more likely to believe that he was dead before U.S. forces invaded his compound" (Sunstein 2015). Likewise, in a *Vox* interview that can be viewed on YouTube under the title "Why Conspiracy Theories Are Rational to Believe," Sunstein again makes the same claim. He says, "[A] good predictor of whether people will believe in conspiracy theories is whether they believe in other conspiracy theories." That much is true. Sunstein continues, "And that effect is so intense that people will believe in logically incompatible conspiracy theories. So, if you think princess Diana was murdered, you are more likely to think she is still alive, and these can't simultaneously be true." https://www.youtube.com/watch?v=xnYT5Fp6w_M (at 1:15—1:45). The advantage of actually watching such an interview, rather than merely reading the words, is that Sunstein's successful conveyance of ridicule, which can be read clearly on his interlocutor's smirking face, cannot plausibly be denied. The irony is that this ridicule is based on a misinterpretation of the study, and a failure to make a critical and rather obvious distinction.

3. Sunstein and his coauthor Adrian Vermeule take a more nuanced position in an article on this subject (2009), arguing that conspiracy theorists are not irrational per se, but merely are hampered by a "crippled epistemology." That thesis,

and Sunstein and Vermeule's paper more generally, has been shown to be highly problematic for a variety of reasons. See Chapters 4 and 5, as well as Pigden 2017; Coady 2018a; and Griffin 2011.

4. Wynne Parry, writing for *LiveScience*, got the message. Quoting the study's conclusion, Parry writes, "The central idea—that authorities are engaged in massive deceptions intended to further their malevolent goals—supports any individual theory, to the point that theorists can endorse contradictory ones, according to the team. 'Believing that Osama bin Laden is still alive is apparently no obstacle to believing that he has been dead for years,' they write in a study published online Wednesday (Jan. 25 [2012]) in the journal *Social Psychological and Personality Science*" (Parry 2012, referring to Wood et al. 2012).

5. Other combinations could have been taken as equally contradictory, but for some reason were not indicated as such in Table 1 in Wood et al. 2012 (p. 769). But it does not matter. If they had been, the same critique would apply.

6. Elsewhere two of the authors of the study write, "If indeed Princess Diana was assassinated, the royal family and Al-Fayed's business partners could not both have been independently responsible for her death" (Sutton and Douglas 2014, p. 258). This is a little misleading. The study in question does not require that the various hypothetical culprits had to act *independently*. (It also doesn't include "the royal family did it" as an option, though one could imagine that MI6 did it on behalf of the royal family.)

It is curious what these authors regard as incompatible. They write, "Consider, for example, the left-leaning theory that oil companies plot to discredit evidence that use of their product is warming the planet, and the right-leaning theory that governments and scientists deliberately exaggerate the risk of anthropogenic climate change. With such different and often mutually incompatible content, it is by no means obvious why these theories tend to come together in the minds of some people" (Sutton and Douglas 2014, p. 256). These examples, though contrasting, are clearly not mutually incompatible.

7. Their analysis reveals that "the correlation in endorsement of the two contradictory theories is explainable entirely by their connection with belief in a deceptive cover-up by authority" (Wood et al. 2012, p. 771).

8. Basham and I independently noticed this flaw in Wood's paper, as did Alfred Moore (2019, pp. 112–113, quoted in note 1 to this chapter). That should not be surprising; the problem is rather obvious. What is surprising, at least to me, is that so many others failed to notice.

9. The authors come close when they noted, "[H]igher-order beliefs may be so strongly held that any conspiracy theory that stands in opposition to the official narrative will gain *some degree of endorsement* from someone who holds a conspiracist worldview, even if it directly contradicts other conspiracy theories that they also find *credible*" (Wood et al. 2012, p. 768, emphasis added).

10. Here is another proposed explanation for what "holds beliefs in various conspiracy theories together." Sutton and Douglas explain, "[C]onspiracy theories imply that powerful elites are willing and able to conspire. This central belief in the existence of conspiracies may be the essential glue that holds beliefs in various conspiracy theories together" (Sutton and Douglas 2014, p. 264). Does any serious person actually question "the existence of conspiracies" or that "powerful elites

are willing and able to conspire"? I should think that the only real questions are questions of extent: *How* prevalent are conspiracies? *How* willing and *how* able to conspire are powerful elites?

11. Like "Dead and Alive," this study does not clearly measure belief per se. Ostensibly, it measures the degree of truth or falsity. But this seems to be an inappropriate measure in this context. How can one assess the degree of accuracy (which is what degree of truth or falsity means, if taken literally) in a case about which one knows nothing? Thus, the subjects have little choice but to interpret and answer in terms of the degree of plausibility or likelihood, rather than degree of truth or falsity.

12. I'm following Wood and Douglas 2013 in adopting the word "conventionalist" for those who stick with conventional, orthodox interpretations, rejecting alternative "conspiracy theories."

13. The authors never explicitly say that their result reflects poorly on conspiracy theorists. If challenged, they may even deny that they intended such an inference. Uscinski and Parent (2014), for example, have strenuously insisted that they do not mean to be judgmental. But it is one thing to claim to mean no offense, and another to actually conduct even-handed research—"By their fruits ye shall know them." Despite the problems of "Dead and Alive" outlined above, Michael Wood's more recent work has been more even-handed (see Wood et al. 2013 and Wood 2016). More will be said below about the mixed but improving record of Sutton and Douglas, Wood's coauthors for "Dead and Alive." Goertzel's intent, on the other hand, is clear. This is also discussed below.

14. Brotherton favorably cites the studies that I have criticized in this article, and he himself conducts similar psychological studies, sometimes in collaboration with Christopher French. He is generally dismissive regarding controversial conspiracy theories, though he does claim no intention to be insulting. After all, he is just trying to find the psychological peculiarities that drive people to beliefs that he dismissively rejects. Why should anyone be offended by that?

15. More recently, Wood and Douglas explain it this way: "Conspiracy skeptics . . . (incorrectly) apply their skepticism to things like MKULTRA experiments (the CIA's Cold War mind control and brainwashing program) or various instances of government corruption, while conspiracy theorists will either be more aware of them or just think they sound more likely to be true" (2019, p. 248).

16. See Basham 2011 pp. 64–68, for a useful analogy (involving a "good family" and a "bad [Mafia] family") illustrating how considerations of prior probability influence, and *ought* to influence, the evaluation of conspiracy theories.

17. To say that it is mountainous is not to say that it is of good quality (though I happen to think much of it is good). The claim that it is mountainous can be established by pointing to a large pile of books; whether on the whole it is of good or poor quality can *only* be established by reading and critically evaluating those books.

18. Similarly, Swami et al. write, "This provides additional support for Goertzel's (1994) argument that a person who believes in one conspiracy theory is more likely to believe in others, including entirely fictitious ones perceived as real" (2011, p. 459).

19. What follows is the complete context, which also includes claims similar to

those quoted from Swami et al. 2011: "Perhaps one of the most important conclusions to emerge from the handful of studies to focus explicitly on the individual antecedents of belief in conspiracy theories was Goertzel's (1994) assertion that conspiracy beliefs form part of a 'monological belief system.' This allows conspiracy theorists to easily assimilate explanations for new phenomena that would otherwise be difficult to understand or would threaten their existing beliefs. Recent work supports this, showing that those who more strongly endorsed 9/11 conspiracy theories were also more likely to believe in other, seemingly unrelated conspiracy theories" (Swami and Coles 2010, p. 562).

20. Here Swami cites Hardin 2002 and Sunstein and Vermeule 2009.

21. Swami cites Douglas and Sutton 2008, though its relevance does not seem particularly strong.

22. This seems to be an unstated corollary of their study. The study suggests that a given conspiracy theory is more likely to be accepted by those who already accept other conspiracy theories than by conventionalists. So conventionalists are less likely to accept the conspiracy theory, and thus more likely to stick with the conventional account. And thus, it would seem, a prior predominance of conventionalist ideation would be predictive of acceptance of future conventional accounts.

23. Elsewhere, Swami and Coles characterize the theories in question as "*seemingly* unrelated." They write, "[T]hose who more strongly endorsed 9/11 conspiracy theories were also more likely to believe in other, seemingly unrelated conspiracy theories" (2010, p. 562, citing then-forthcoming Swami et al. 2011). This qualification, of course, makes a big difference.

24. Clearly these beliefs must be about the *degree* of deception, and *degree* of effectiveness, since no serious person could deny that authorities *sometimes* deceive, and that intimidation and bribery have *some degree* of effectiveness.

25. Sutton and Douglas continue, "This is an important finding even if it does not show that conspiracy theorists have a monological mind-set" (2014, pp. 266–267). Why exactly is this thought to be an important finding? After all, people who already believe in existing conspiracy theories *should* tend to be receptive to new conspiracy theories, precisely because one belief *should* serve as evidence for another. So it is hard to understand why that finding is thought to be significant. Perhaps other psychologists should study why some psychologists think that ordinary reasoning is particularly interesting when it happens to be employed by conspiracy theorists.

26. I have had encounters with other (white) academics who are perfectly aware of the Tuskegee experiment (and similar cases), but *until I remind them of it* (or of similar cases) they seem to reason as though they did not know.

27. The pejorative ring seems to have been intentional. Goertzel reportedly declared the pejorative use of the term "conspiracy theory" to be "one of our accomplishments" (Walker 2015). Also, consider the following characterization of conspiracy theorists as monological thinkers: "Monological conspiracy thinkers do not search for factual evidence to test their theories. Instead, they offer the same hackneyed explanation for every problem—it's the conspiracy of the Jews, the capitalists, the patriarchy, the communists, the medical establishment, or whatever. In these cases, the proof which is offered is not evidence about the specific incident or

issue, but the general pattern; for example, the X conspiracy has been responsible for all our other problems, so it is obvious that X must be responsible for this one as well" (Goertzel 1994, p. 741). That is quite a caricature. And further, isn't Goertzel hypocritically employing the same strategy that he criticizes? He seems to be condemning all particular conspiracy theories based on a questionable generalization.

28. Citing Goertzel 1994, Swami et al. 2010, 2011, and Clarke 2002, Wood et al. describe the worldview of conspiracy theorists as "a unitary, closed-off worldview in which beliefs come together in a mutually supportive network known as a *monological belief system*" (2012, p. 767).

29. I am not aware of a good reason to think that Joe Conventionalist knows any more or less than Joe Conspiracist regarding the empirical facts surrounding the conspiracy theories about which they both have firm, if ungrounded, convictions. But it is rather clear that *serious* conspiracy theorists (like most of those who give talks at JFK assassination conferences, for example) know a lot more about the relevant empirical evidence than do the academics who study conspiracy theorists, who largely try to remain above the fray.

30. Regarding the push to address the pertinent empirical facts regarding 9/11, see David Ray Griffin's "9/11: Let's Get Empirical" (2007a), which can be found at http://www.brianrwright.com/Griffin.pdf, or watch his corresponding lecture, which can be found in various places online, including YouTube.

31. Referring to the 9/11 Truth Movement's apparent attachment to case-specific facts as "rhetorical" seems to be itself rhetorical.

32. The suggestion that conspiracy theorists operate from a position of "limited information" was made intellectually respectable by the assertions of Cass Sunstein and Adrian Vermeule (2009), who did not actually support it with any evidence. Sutton and Douglas seem to subtly endorse this view when, having just alluded to Sunstein and Vermeule, they write, "[I]t may be necessary to address not only the limited information available to the disaffected but also the sociostructural bases of the disaffection itself" (2014, p. 264). To what end these interventions are necessary is not explicitly stated, but since the context of the statement links disaffection with inclination toward conspiracy theories, it seems that there is an assumption that conspiracy theorists have limited information, and that that is partially responsible for their presumed errors. Although Sutton and Douglas's implied "cure" is less problematic than Sunstein and Vermeule's proposed "cognitive infiltration," they nevertheless seem to have adopted the view, explicit proclamations to the contrary notwithstanding, that conspiracy theories represent a *problem* needing to be "addressed."

33. While the claim that conspiracy theories tend to be (*to some degree*) "resistant to falsification" seems correct, this may not be as telling an epistemic defect as many seem to assume. Indeed, as philosopher Keith Harris correctly notes, "If there is an epistemic problem with conspiracy theories, and belief in them, it is not that such theories are unfalsifiable" (2019, p. 246; see also Hagen 2020b, p. 3). Addressing this issue, philosopher Brian Keeley has made important observations: "[U]nfalsifiability is only a reasonable criterion in cases where we do not have reason to believe that there are powerful agents seeking to steer our investigation away from the truth of the matter" (2019, p. 429). However, "[W]e have a reason to believe that there exist forces with both motive and capacity to carry out effec-

tive disinformation campaigns" (p. 429). Regarding the exposure of Watergate and Iran-Contra, Keeley points out, "Strictly hewing to the dogma of falsifiability in these cases would have led to a rejection of conspiracy theories at too early a point in the investigations, and may have left the conspiracies undiscovered" (p. 429).

34. There is probably a way of interpreting "conspiracy theories," and of setting the bar for what counts as "lacking" evidential support, according to which this claim would be true. But this would probably necessitate either caricaturing conspiracy theories or employing an unusual and unfair sense of "lacking."

35. Ironically, it is Sunstein himself who seems to be peddling "information" that is wrong. He claims that video frames released by the Department of Defense show Flight 77 approaching the Pentagon (see Sunstein 2014a, p. 27). But the frames in question do not clearly show what is approaching the Pentagon. So, to be clear, although the footage might have captured Flight 77, or part of it, approaching the Pentagon, it is wrong to say these frames *show* that happening. Since Sunstein does not provide much evidence about the events of 9/11, it is particularly significant that what he does provide is wrong and misleading. It misleads by giving the impression that this was clear and incontrovertible evidence of Flight 77 approaching the building, when it is not.

36. As discussed in Chapter 1, Rob Brotherton defines a conspiracy as, among other characteristics, "an unverified claim of conspiracy which is not the most plausible account of an event or situation . . . based on weak kinds of evidence" (2013, p. 9). Citing this, Brotherton and Chris French write, "A conspiracy theory can be defined as an unverified and relatively implausible allegation of conspiracy" (Brotherton and French 2014, p. 238).

37. With the exception of the above-mentioned appeal to flagrantly mischaracterized photographs (see section 5.2.2 and note 35 of Chapter 7), Sunstein gives almost no argument at all against 9/11 conspiracy theories, which are nevertheless used as a "running example" of a "demonstrably false" conspiracy theory. As another example, this time regarding surveys that document increasing belief in conspiracy theories regarding the assassination of JFK, Goertzel remarks, "This increase in the belief in conspiracy has taken place despite the fact that the accumulation of evidence has increasingly supported the lone-assassin theory" (1994, p. 731). Goertzel cites one source (namely, Moore 1990). Well, I guess that settles it then!

38. There have been attempts to find shortcuts, such as Clarke 2002 and Keeley 1999. But these have been effectively refuted and/or acknowledged to be unsuccessful. See Chapter 2 and Coady 2006a.

39. Expressing frustration with the cluelessness and gullibility of most Germans while the Nazis perpetrated outrages, a leaflet from the White Rose Society reads, "The German people are again sleeping on in obtuse, stupid sleep, giving these fascist criminals the temerity and opportunity to continue to rage" (cited in Hoffmann 2014). The title of the article from which this is quoted, "The Good Germans: Inside the Resistance to the Nazis," turns the ordinary use of the phrase "good Germans" on its head. Regardless, I mean it in the sense expressed in the quotation—referring to the people who went along with the program, even to the point of believing the propaganda.

Chapter 8

1. This view is essentially what Karl Popper called "the conspiracy theory of society." See Popper (reprinted in Coady 2006a). For an effective response to Popper, see Pigden 1995 (also reprinted in Coady 2006a). For a brief discussion of both see section 2.3 of this book.

2. Hofstadter continues: "All political behavior requires strategy, many strategic acts depend for their effect upon a period of secrecy, and anything that is secret may be described, often with but little exaggeration, as conspiratorial" (1965, p. 29).

3. Michael Barkun writes, "[P]aranoid has an unmistakably pejorative connotation. Indeed, it seems clear that Hofstadter utilized it precisely because of its judgmental quality. Its overtones are such that its use, even in careful hands, runs the risk of merely labeling people whose ideas we disapprove of" (2013, p. 9). And yet, as Charles Homans, writing for the *New York Times*, points out, "Dispatching conspiracism as a pathology is tricky. We now know that by the time Hofstadter's essay was published, the Joint Chiefs of Staff had proposed faking Cuban terrorist attacks, with real casualties, on United States soil; the C.I.A. had experimented on unsuspecting Americans with LSD; the F.B.I. had infiltrated and manipulated domestic activist groups." Providing an even more interesting twist, Homans adds, "Some of Hofstadter's own research on the far right was funded by an organization that was, apparently unknown to him, a C.I.A. front" (Homans 2016).

4. Very similarly, Viren Swami and Adrian Furnham write, "Traditionally, many scholars have relied on Hofstadter's ([1965], pp. 14, 29) definition, first provided in his seminal work *The Paranoid Style in American Politics and Other Essays*, of a conspiracy theory as any belief in the existence of a 'vast, insidious, preternaturally effective international conspiratorial network designed to perpetrate acts of the most fiendish character' and that aims to 'undermine and destroy a way of life'" (2014, p. 219).

5. Rather than objecting to theories that happen to involve charges of pernicious conspiracy, Hofstadter was concerned with a *worldview* in which powerful forces were out to get us. For example, Hofstadter writes, "There can be no doubt that [William Morgan] was abducted by a small group of Masons [in 1826], and it was widely and quite understandably believed that he had been murdered, though no certainly identifiable body was ever found. . . . Not all of these [other similar charges] and fears need be dismissed as entirely without foundation. What must be emphasized here, however, is the *apocalyptic and absolutist framework* in which this hostility to Masonry was usually expressed" (1965, pp. 15 and 17, emphasis added).

6. Strictly speaking, this is only true on a very broad definition of "conspiracy theory." If, to count as a "conspiracy theory," a theory has to be opposed to some official story (as I prefer to understand it, following David Coady), then one need not be a conspiracy theorist to exhibit the paranoid style. Official theories involving purported conspiracies can also exhibit the paranoid style, or at least some of its characteristics. Arguably, the official story of 9/11 sometimes takes on these qualities. The "evildoers" in a vast network of terrorist cells around the world are so effective, as demonstrated by nineteen Islamic 007s with box-cutters, that it will

require a war without end, and the sacrifice of precious rights, to keep them in check. For more on this theme, see section 8.7 below.

7. Dentith makes essentially the same point: "[B]y focusing their attention on the class of suspect theories, [scholars] often draw broad conclusions about the rationality of belief in conspiracy theories *generally*. Yet it is not clear this subset of suspect conspiracy theories is representative of the wider kind" (2019a, p. 102). See Dentith 2018a for further analysis.

8. Determining the stronger proponents of a conspiracy theory is not a terribly difficult thing to do, nor is it an unusual expectation. Scholars do this sort of thing all the time on other topics. If one needs a hint, consider that David Ray Griffin is often referred to as "the Dean of the 9/11 Truth Movement." One might also notice that Peter Dale Scott publishes books, such as *The Road to 9/11* (2007), at the University of California Press.

9. Ed Tatro, interviewed in Terrence Raymond's *Evidence of Revision*. Quotation found at about 1:07:00 on cited YouTube version (Raymond 2012).

10. As one online conspiracy site notes: "There only needed to be a select few people at the top of the chain of command who knew exactly what was being carried out, where and how to create confusion, and why." (See "Too Grand a Conspiracy?," *9/11 Hard Facts*, http://911hardfacts.weebly.com/xx-too-grand-a-conspiracy.html.) It is my impression that prominent 9/11 conspiracy theorists such as David Ray Griffin, Steven Jones, Graeme McQueen, and Peter Dale Scott (to name just a few) have a similar understanding, though that does not mean that they don't suspect more extensive complicity *in the cover-up*.

11. See Swami et al. 2011, p. 444, for an example. Relevant quotations from Swami et al. 2011 can be found in section 8.7.

12. Dentith and Orr have provided a similar analysis, arguing that "monolithic conspiracies [in which the participants all fall under a single governing structure] can involve large numbers of conspirators, and yet be kept secret because such conspiratorial activity is governed not just by the conspirators 'at the top,' but by virtue of the fact the conspirators have control over the entire conspiratorial 'pyramid'; they employ, co-opt, or dupe people to do their bidding, knowing full well that they have near total control over the conspiracy" (2018, p. 439). Dentith and Orr cite the Volkswagen emissions testing conspiracy and cover-up as an example.

13. Philip Shenon finds that the FBI (Shenon 2013, pp. 25, 26, 540–542, 548), the CIA (pp. 221, 533, 534, 544, 546, 548), the Secret Service (p. 374), the Dallas police (p. 342), and the Warren Commission itself (pp. 146, 549) all engaged in cover-up activities. Some details are covered in the next chapter; see section 9.3.

14. Shenon may himself be regarded as a "conspiracy theorist," but only in a very weak sense. In pointing out cover-up activities, Shenon acknowledges the existence of minor conspiracies—which are now so well documented that they are rarely denied, and are in some cases admitted, and thus, in effect, at least arguably, cease to contrast with the official story as it is currently promulgated. Regarding the assassination itself, Shenon acknowledges the *possibility* of some sort of *minimal* conspiracy involving Cubans who may have encouraged Oswald. More significantly, however, his work serves an essentially *anti*-conspiracy-theory function, by offering a "fallback" position in the face of undeniable evidence of cover-up activities.

15. It should be acknowledged that these doctors were not exactly "silent" about what they observed. The claim is, roughly, that they did not go out of their way to expose the discrepancies between what they observed and the official story that was taking form, and some even modified their views over time to less clearly contradict the official position. The most exhaustive treatment of the statements by witnesses to the body of President Kennedy is Palamara 2015.

16. Cass Sunstein and Adrian Vermeule's proposed "cognitive infiltration," discussed in Part II of this book, provides an example. Infiltrators, though believing in what they are doing, will likely have a hard time being perfectly fair and intellectually honest as they attempt, with presumably good intentions, to undermine the group that they have infiltrated.

17. While size does matter, it is important to consider organizational structure along with it. As Dentith and Orr have argued: "[I]t is not size *per se* but, rather, the diversity of the set of conspirators; presumably (but definitely not always) the larger a conspiracy grows, the more diverse the set of conspirators becomes, leading to the conspiracy becoming more uncontrollable (and thus—once again presumably but not always—revealed). Size matters, but it is not the only thing" (2018, p. 440 n. 9).

18. I here consider the ideas that conspiracy theorists assert or believe related to a purported conspiracy, and what those assertions and beliefs imply, to constitute that individual's particular conspiracy theory.

19. Similarly, in *State Secret: Wiretapping in Mexico City, Double Agents, and the Framing of Lee Oswald*, Bill Simpich writes, "Oswald was lucky to survive that day. But he did. His survival illustrates how the whole assassination operation was in danger of falling apart. That was not part of the plan" (Simpich 2014, chap. 6). Such examples are ubiquitous. They are the rule, not the exception. Here is another representative example regarding the JFK assassination: referring to Lee Harvey Oswald, (conspiracy theorist) Jacob Hornberger tells (conspiracy theorist) James Corbett, "As good as you can make a case when a guy is being framed, it's hardly ever perfect. So, they can put the rifle there. They can have the little boxes set up where the assassination supposedly takes place. They can hide the rifle. They can do all sorts of things—communism as his background—to frame the guy. But it's never going to be perfect. Things are going to go wrong. And it is clear that when you start looking at the Mexico City side of this thing—everything went wrong with the Mexico City side" (Hornberger 2017).

20. While Fenton does not regard himself as a member of the 9/11 Truth Movement per se, the facts presented in his book *Disconnecting the Dots* (2011) are nevertheless strongly suggestive of insider complicity in the events of 9/11.

21. "Kevin Fenton 9–11 Interview 'Disconnecting the Dots'" (Kevin Fenton interview with Erik Larson, part 1), https://www.youtube.com/watch?v=cErMYgfBJ60, quotation begins at 36:03.

22. Analogously, regarding the Oklahoma City bombing, (conspiracy theorist) Holland Van den Nieuwenhof remarks: "The fact is, this was not a perfect operation. It was a screw-up. . . . That Ryder truck was supposed to be parked in the parking garage. But McVeigh rented a truck that was too big to fit into the parking garage. And it was seen—people saw it that morning trying to pull in at different angles and driving around and trying to pull in. . . . But had that truck exploded

inside the parking garage it would have been much more difficult to demonstrate conclusively, as we have, it was not the sole cause of destruction" (The Opperman Report podcast, "Aberration in the Heartland of the Real by Wendy S. Painting," January 25, 2019, at 1:07:00).

23. Sunstein tags on the comment: "[This is] despite abundant evidence that in open societies, government action does not usually remain secret for very long" (2014, p. 5). For my response to this idea, see section 4.2, above.

24. If Uscinski and Parent do not mean to assert that the supposed offenders are "*reasonably* presumed" to be good, but merely "considered," perhaps unreasonably, to be good, then it is all the more clear that no normative conclusion follows.

25. If we had made the conclusion stronger by saying that we "ought to reject" conspiracy theories, then the argument would be invalid. And so, the conclusion would not follow even if the premises were true. For it assumes, wrongly, that one ought to reject views that appear unlikely, to some unspecified degree, *given only one dimension of the case*.

26. See section 8.7 in this chapter.

27. It is not clear whether this characterization was meant to apply to the various examples Uscinski and Parent give over a couple of pages or only to the one example about the corruption and monopolization attributed to Monsanto, which is accused of having "captured the FDA," and doing likewise elsewhere. In any case, ruthless capitalism is not quite the same as genuine evil. Some of their other examples are even more tame, such as "The D's and R's are finally working together . . . to keep Gary Johnson out of the debates" (Uscinski and Parent 2014, p. 149). Though it may be underhanded, this does not seem strictly evil. And it is not clear why this would be thought to be implausible. Perhaps the closest they come to a comment suggestive of genuine evil in this section is the following: "[Obama] doesn't care about our health. We are a nuisance. An overgrown population that needs to be destroyed. Agenda 21. New world order" (Uscinski and Parent 2014, p. 148). Even this example, with a seemingly hysterical comment about the population needing to be destroyed, still does not unambiguously suggest that the conspirators are thought to be genuinely evil, rather than merely ruthlessly selfish for desiring to save the world for themselves at the expense of the rest of us. The commenter does come across as unhinged, but not quite because he or she posits an evil motive. Sometimes people do sacrifice others for their own benefit.

28. The position that "the President was killed because he should have been" is articulated at length, though not endorsed, by Daniel Sheehan in his course on the JFK assassination at UC Santa Cruz. See Sheehan 2013b (quotation at 36:45).

29. Another suggestive passage is a long quotation from Dulles's wife, Clover (see Talbot 2015, p. 127; see also pp. 121–122.) Also, Clover told Mary Bancroft, Dulles's wartime mistress, "that she had once heard the Dulles brothers referred to as sharks. 'And I do think they are,' said the wife to the mistress." After that they referred to Allen as "The Shark" (Talbot 2015, p. 130).

30. I've left the original citations so they can be easily checked with reference to the unmodified version of the quotation. Brotherton is asserting that "most conspiracy theories" have the various qualities described in the (unmodified) quotation. Seemingly to back up this claim, Brotherton provides these various citations (as indicated in the quotation). However, this evidence is weak. For example,

Lewandowsky et al. 2013, a retracted article that analyzes "the response of the climate blogosphere" to another paper that addresses the association between "conspiracist ideation" and the rejection of scientific findings, is cited as supporting the idea that most of "the motives of the alleged conspirators often verge on pure maniacal evil." The paper does extensively discuss charges of "malfeasance," but it is a stretch to characterize this as verging on "pure maniacal evil," even if some examples can be found that might be thought to arguably qualify. The stronger the malevolent intent that is asserted to apply, the weaker the claim that this level of malevolent intent is characteristic of conspiracy theories generally. In any case, as I have stressed, what is "often" the case, or even what is *characteristic* of such theories, is not epistemically important. We ought to be interested in evaluating the *best* versions.

31. Arguably one could include Presidents Richard Nixon, Lyndon Johnson, and Harry Truman as having committed war crimes as well (Goodman 2014). At the very least, we now know that Nixon treasonously conspired to "monkey wrench" peace negotiations regarding the Vietnam War (Farrell 2016).

32. More context reveals that McNamara, at least in retrospect, saw himself as "part of a mechanism" that recommended "evil" actions in the utilitarian pursuit of "good":

> Mr. Morris [director of the film *The Fog of War*] unearthed spine-curdling government reports showing the raw calculus undertaken to speed America's victory. "In order to do good," Mr. McNamara says, articulating the film's ninth lesson, "recognize that at times you will have to engage in evil." In a single bombing raid, he recalls, "We burned to death 100,000 Japanese civilians in Tokyo—men, women and children." Some 900,000 Japanese civilians were killed overall. Was he aware this would happen? "Well, I was part of a mechanism that in a sense recommended it," Mr. McNamara tells Mr. Morris. "LeMay said, 'If we'd lost the war, we'd all have been prosecuted as war criminals.' And I think he's right." (Blight and Lang 2005, p. 211)

33. Olmsted notes, "Illegal mail-opening, extensive infiltration of peaceful protest groups, covert actions against democratically elected governments: all were documented by this official U.S. government investigation" (2019, p. 293).

34. See Wikipedia's "List of American Federal Politicians Convicted of Crimes." https://en.wikipedia.org/wiki/List_of_American_federal_politicians_convicted_of_crimes

35. The charge comes from Juanita Broaddrick, a nursing home administrator and volunteer for Clinton's first gubernatorial campaign, who was regarded by the staffers from the House Judiciary Committee as "the most credible witness that either one of them have ever talked to" (Kurtz 1999). See the case study in section 1.4.2 in Chapter 1.

36. Clinton emphatically misled the American people with his now-infamous claim, "I did not have sexual relations with that woman." And, having made a similar statement under oath, he was held to be in contempt of court, was fined $90,686, and later had his license to practice law suspended. While Clinton maintained that his statement was strictly true, based on his interpretation of the defini-

tion of "sexual relations" that was provided to him, he also denied ever having been alone with Lewinski in the location that he later admitted the oral sex occurred. So *that*, at least, appears to have been a lie even by the strictest definition. See Jackson 1999, "Clinton Fined $90,686 for Lying in Paula Jones Case."

37. The dialogue from which these quotations were taken is revealing:

> William Sullivan [head of the Domestic Intelligence Division during the investigation of Dr. King] was asked by the Committee whether tactics, such as placing female "plants," were common practices of the FBI. Sullivan testified that they were: "common practice among intelligence services all over the world. This is not an isolated phenomenon. . . . This is a common practice, rough, tough, dirty business. Whether we should be in it or not, that is for you folks to decide. We are in it. . . . No holds were barred. We have used that technique against Soviet agents. They have used it against us. . . . [The same methods used overseas were] brought home against any organization against which we were targeted. We did not differentiate. This is a rough, tough business." (US Senate 1976b, pp. 134–135)

Senator Mondale then asks, "Would it be safe to say that the techniques we learned in fighting . . . true espionage in World War II came to be used against some of our own American citizens?" Sullivan responds, "That would be a correct deduction."

38. A personal experience is illustrative. A highly regarded senior colleague and friend (who will remain nameless here) once tried to make a point about integrity, citing President Lyndon Johnson as exemplifying that virtue. When I objected, scoffingly, he responded by saying that we don't want to have so strict an understanding of moral terms that *even our presidents do not qualify*. At this point, I was backed up by another colleague who pointed out, in effect, that the cart had been put before the horse: *first* we should define what the word means, and only *then* should we appraise the conduct, judging it according to that fixed standard. In this way, an attempt at presidential apologetics was successfully thwarted in this instance. Unfortunately, however, such apologetics, though usually a bit more subtle, is all too common.

39. As Charles Pigden has commented, "[P]articularism is quite compatible with the view that many conspiracy theories can be dismissed in short order as *devoid* of intellectual merit" (2018, p. 202).

40. This chapter is derived from an article published in *Social Epistemology* in 2017, available online: https://www.tandfonline.com/doi/full/10.1080/02691728.2017.1352625

Chapter 9

1. For my own critique of Russell's position, see Hagen 2016, pp. 175–176.

2. Russell has been criticized for supposedly longing for a world in which "Diet, injections, and injunctions will combine, from a very early age, to produce the sort of character and the sort of beliefs that the authorities consider desirable, and any serious criticism of the powers that be will become psychologically impossible" (Russell 1953, p. 50). But Russell seems only to be foreseeing such a circum-

stance developing, and warning of that danger. I don't know of any evidence that he *wanted* this to come about.

3. So far, forty years on, things have not turn out as badly as predicted, and so in hindsight the authors appear somewhat overly pessimistic, if not hysterical. Indeed, they may be thought to exhibit one aspect of the paranoid style that I have not discussed in this article: "apocalypticism" (see Hofstadter 1965, p. 30).

4. Walker also mentions the Willowbrook hepatitis studies: "For a decade and a half [from 1956 to 1971 at the Willowbrook State School], scientists used students at a New York school for the developmentally disabled as guinea pigs, deliberately infecting them with hepatitis in hopes of finding ways to combat the sickness" (Walker 2014). While controversial, this case is arguably less outrageous in its details than the Tuskegee and Guatemala experiments. Another scandal involves studies on AIDS drugs on foster kids. See Solomon 2005. Just how inappropriate such studies were depends on which allegations about them are true. Much is contested.

5. I have simply stumbled over examples like this in the course of writing this book. The terrain is littered with such examples. Those who claim or imply that conspiracy theories uniformly attribute evil intent to a vast network of preternaturally effective conspirators quite literally do not know what they are talking about. This particular quotation can be heard at CorbettReport.com, Featured Interviews #1178, "Jerry Day Examines the Science of Social Control" (13:00).

As another example, (conspiracy theorist) William Pepper discusses the "extraordinary complicated and complex" psychology he attributes to a man he regards as having a role in the MLK assassination, John Downie. In Pepper's view, Downie was probably a racist. And yet, regarding the carnage taking place in Vietnam, Pepper regards Downie as having "compassion and concern" and "show[ing] such humanity." Pepper concludes, "You can't just blatantly condemn all these people as being evil. And I certainly don't." CorbettReport.com, Featured Interviews #1353—"Dr. William Pepper on the Assassination of Martin Luther King" (45:42–48:43).

6. It is worth noting here, as Olmsted elsewhere mentions, that much later "The House Assassinations Committee concluded that there was probably a broad conspiracy behind Kennedy's murder" (2019, p. 293). So the idea that there was a conspiracy of *some kind* should not be regarded as far-fetched or completely at odds with the findings of ostensible epistemic authorities. (By pointing this out, I do not mean to suggest that the House Assassinations Committee should be regarded as a particularly well-constituted set of epistemic authorities.)

7. As a page entitled "Warren Commission" on the Mary Farrell Foundation website explains: "In some of the phone calls that week, LBJ alluded to the possibility of 40 million Americans dying in a nuclear exchange with the Soviet Union. This fear was apparently related to the issue of Oswald's visit to Mexico City and the allegations of Communist conspiracy emanating from that city." https://www.maryferrell.org/pages/Warren_Commission.html. Note that even defenders of the Warren Commission, such as Philip Shenon, acknowledge that Warren was persuaded to head the commission for these reasons (see Shenon 2013, pp. 58–61, for details; see also the note immediately below).

8. Responding to criticism for having agreed to head the Commission, Warren

wrote a letter stating, "In this situation, the President so impressed me with the gravity of the situation at the moment that I felt in good conscience I could not refuse" (Shenon 2013, p. 401).

9. See the quotation by Jim Garrison in the main text below. Although (conspiracy theorist) Bill Simpich emphasizes other relatively innocent motives for cover-up, he also acknowledges Warren's fear of war: "The national security cover-up was driven by Helms and Hoover, determined to protect not only their individual agencies but their personal careers from oblivion. Unlike men like Earl Warren and many political leaders, I don't believe that either Helms or Hoover were motivated by the fear of war" (Simpich 2014, chap. 6).

10. There is *strong* evidence that Oswald was impersonated on phone calls to the Soviet Embassy in Mexico City. There is *some* evidence, such as Silvia Duran's description of "Oswald," that suggests Oswald may have been impersonated in person as well, as he seems to have been elsewhere. Also, there are no pictures of Oswald entering or exiting the Soviet Embassy even though it was under camera surveillance. This, along with other evidence, leads some researchers to doubt that the real Oswald even went to Mexico City. The very credible testimony of Silvia Odio is particularly significant in this regard. It places Oswald (or an impostor) at a location and time that would make it very difficult for Oswald to have gotten to Mexico City in a plausible way.

11. One can now easily listen to Johnson saying this, and the previous quotation, with the surrounding context. See "LBJ-Russell 11-29-63, 2nd call," *History Matters*, http://www.history-matters.com/archive/jfk/lbjlib/phone_calls/Nov_1963/audio/LBJ-Russell_11-29-63_2nd.htm (excerpt 3 has both comments; the bit about crying and "I'll just do whatever you say" comes at the very end).

12. The quotation occurs at the beginning, right after the title shot, of Nigel Turner's documentary *The Men Who Killed Kennedy* (Episode 3: "The Cover Up"). More coverage of Hosty can be seen at 38:00. See also Shenon 2013, pp. 11, 25–26, 210–212.

13. Having been directed to "Get rid of this," Hosty tore up a note to him from Oswald and flushed it down the toilet (see Shenon 2013, pp. 25–26 for details). Incidentally, those who accept Shenon's analysis may conclude that Oswald, acting alone, shot Kennedy. But they cannot easily reach this conclusion *by appeal to the implausibility of a large-scale cover-up*, since Shenon's account also includes a variety of cover-up activity.

14. Kennedy's back-channel communications with Fidel Castro and Nikita Khrushchev, for example, which were likely not *entirely* secret, would have been alarming to staunch cold warriors.

15. More specifically, in Talbot's view, Nelson Rockefeller thought that "JFK's unassertive leadership had encouraged our enemies and demoralized our allies, and had made the world more dangerous" and "men like [James] Angleton and [Curtis] LeMay regarded the president as a degenerate, and very likely a traitor" (Talbot 2015, p. 559).

16. This is not Sheehan's own position. He is presenting how the situation may plausibly have appeared from a certain point of view. See Sheehan 2013b (quotations at 33:35 and 36:45). This was part of Sheehan's course on the JFK assassination at UC Santa Cruz.

17. Strictly speaking, when Prouty uses the term "vast" he is *not* referring to the team itself, but rather to "an intragovernmental undercover infrastructure" to which it has access. For example, "The power of the Team derives from its vast intragovernmental undercover infrastructure and its direct relationship with great private industries, mutual funds and investment houses, universities, and the news media, including foreign and domestic publishing houses" (Prouty 2011, p. 3; cf. pp. xxii, 33, 166, 248, 319, and 369).

18. Prouty explains: "[T]he choice of the word 'Team' is most significant. It is well known that the members of a team, as in baseball or football, are skilled professionals under the direct control of someone higher up. They do not create their own game plan. They work for their coach and their owner. There is always some group that manages them and 'calls the plays.' Team members are like lawyers and agents, they work for someone. They generally do not plan their work. They do what their client tells them to do. For example: this is true of agents in the Central Intelligence Agency. It is an 'Agency' and not a 'Department' and its employees are highly skilled professionals who perform the functions their craft demands of them. Thus, the members of the highest level 'Secret Team' work for their masters despite the fact that their own high office may make it appear to others that they, themselves are not only the Team but the Power Elite" (Prouty 2011, pp. 1–2).

19. Prouty comments: "Whether or not the Secret Team had anything whatsoever to do with the deaths of Rafael Trujillo, Ngo Dinh Diem, Ngo Dinh Nhu, Dag Hammarskjöld, John F. Kennedy, Robert F. Kennedy, Martin Luther King, and others may never be revealed" (2011, p. 3).

20. Unlike the Church Committee, the Pike Committee produced no final report. But the audio of Colby's testimony can be heard at https://www.youtube.com/watch?v=Z6Hm0Sla65E. The quotation starts at 15:40.

21. The work of both Baumeister and Bartlett was brought to my attention by Brian Martin. See Martin 2019, "Technology and Evil," for a summary of Bartlett's work on evil.

22. David Talbot describes an incident that may help us think concretely about whether the Holocaust, or a significant aspect of it, is properly regarded has having been a conspiracy theory:

> The Final Solution was meant to remain secret, with most of the death camps located in remote outposts of the Nazi empire. But as the systematic killing got under way, many people became aware of the mounting barbarity. One day in early 1942, an IG Farben official named Ernst Struss was returning home on a train after inspecting the company's factory that was affiliated with Auschwitz. A German worker also riding on the train began talking loudly about the nightmare at the camp. Great numbers of people were being burned in the compound's crematoria, he said. The smell of incinerated flesh was everywhere. Struss jumped up in a rage. "These are lies! You should not spread such lies." But the worker quietly corrected him: "No, these are not lies." There were thousands of workers like himself at Auschwitz, he said. "And all know it." . . . U.S. minister to Switzerland Leland Harrison, an old colleague of [Allen] Dulles's who was soon to be reunited with him in Bern, took a decidedly skeptical view of the account; in his dispatch to Washington, Harrison dismissed it as

nothing more than "war rumor inspired by fear"—although he did concede that some Jews were dying due to "physical maltreatment . . . malnutrition, and disease." (Talbot 2015, pp. 50–51)

23. Referring to an announcement that stated that "the planting . . . of a dummy explosive device was part of an ongoing *interregional* exercise" (emphasis added), the Ryazan Regional FSB issued a statement saying: "This announcement came as a surprise to us and appeared at a moment when the department of the FSB [the Ryazan FSB, that is] had identified the residences in Ryazan of those involved in planting the explosive device and was preparing to detain them" (Litvinenko and Felshtinsky 2007, p. 72). FSB director Nicolai Patrushev ordered that the suspects not be arrested, but Ryazan FSB agents proceeded to arrest them anyway, "considerably roughing them up in the process" (p. 74). According to Litvinenko and Felshtinsky, they were detained "until the arrival from Moscow of an officer of the central administration with documents which permitted him to take the FSB operatives . . . back to Moscow with him" (2007, pp. 74–75).

24. Clips from this Donahue-style TV program, in which citizens of Ryazan confront a representative of the FSB, can be seen in a documentary *The Assassination of Russia*, which is at present available online at https://www.youtube.com/watch?v=9sx2YmSXDy8

25. For example, the bags of apparent explosive material were later said to have contained nothing but ordinary sugar, even though they had been tested with a gas analyzer that indicated "fumes of a hexogen-type explosive substance," and reportedly did not seem to those on the scene to be sugar (Litvinenko and Felshtinsky 2007, p. 56). Nevertheless, the FSB later insisted it was nothing but sugar, and to prove it they tried to blow it up. "The detonation failed because it was ordinary sugar, the FSB reported triumphantly. 'One wonders what sort of idiot would try to explode three sacks of ordinary sugar at an artillery range,' the newspaper *Versiya* commented ironically. Why, indeed, did the FSB send the sacks to the artillery range if it knew that 'exercises' were being conducted in Ryazan, and the sacks contained sugar bought at the local bazaar by *Vympel* operatives?" (Litvinenko and Felshtinsky 2007, p. 81). In addition, the fact that the car that was used to transport the bags of "sugar" was found and identified as a stolen vehicle does not sit well with the notion that it was part of a legitimate exercise (see 2007 pp. 97–98).

Edward Lucas sums up the problems as follows: "The official version does not properly account for any of the strange details: the people, the 'bomb,' the car, the evacuation or the initial official explanation. The handful of 'culprits' who were eventually caught, tried and sentenced for the bombings had no convincing links to the crimes" (2014, p. 37). See also Dawisha 2014, pp. 212–214, for a summary of the evidence.

Chapter 10

1. See Coady 2019 for another critique of the psychological literature on conspiracy theories. Coady points out: "[Ted] Goertzel [1994] claims that his study found that minorities and young people are particularly prone to believing conspiracy theories. This 'finding' is often cited in the psychology literature. In fact,

Goertzel seems to have simply chosen examples of conspiracy theories which minorities and young people are particularly likely to believe. In at least one case, they are particularly likely to believe them because they are particularly likely to be aware that they are true" (p. 174 n. 12).

2. A translation of the *Le Monde* statement is affixed to the end of Basham and Dentith 2016 (which is open access). References are to this English translation. (Page numbers to *Social Epistemology Review and Reply Collective* articles, such as this one, refer to the pdf versions.)

3. Regarding the original critique of the *Le Monde* statement (namely, Basham and Dentith 2016), it should be noted that while eight scholars, including myself, endorsed the critique, only two of us, Basham and Dentith, actually did the writing. Just to be perfectly clear, while I am proud to be associated with the critique, and refer to it as "our" response, I did not substantially contribute to it, other than offering some comments on a couple drafts. So it seems to me perfectly sensible for it to be published as, and referenced as, "Basham and Dentith 2016," giving credit where it is due.

4. See Basham 2017 and Dentith and Orr 2017 for other responses. Like the present chapter, Dentith and Orr also argue that the *Le Monde* authors have mischaracterized our position.

5. These flaws are discussed in more detail in other articles. For example, a special issue of *Argumenta* (2018), which focuses on the ethics and epistemology of conspiracy theories, includes articles by Dentith ("The Problem of Conspiracism"), Lee Basham ("Joining the Conspiracy"), and myself ("Conspiracy Theories and Monological Belief Systems," which is an earlier version of Chapter 7 of this book). Basham 2017 also briefly addresses some of these problems.

While the critiques of these studies were not yet published at the time of the *Le Monde* authors' response, and thus they can be excused for being unaware of these particular explanations of the problems with the studies in question, they are not excused from failing to notice the flaws themselves, as they were fairly obvious.

6. I thank Brian Keeley for pointing this out to me.

7. Elaborating on this point, I provide an extensive critique of both Wood et al. 2012 and Swami et al. 2011 in Chapter 7. In Chapter 8, I take issue with Swami et al. 2011 (and others) on a related issue, a prejudicial conflation of "conspiracy theories," "conspiracist ideation," and "the paranoid style."

8. Citing Goertzel they write, "[R]egardless of the facts available in the outside world, the *mind* of some people attracts and is attracted by conspiracist cognitions, which come to form a monological belief system involving conspiracies" (Dieguez et al. 2016, p. 22). Ted Goertzel first articulated the notion that conspiracy theorists operate within a "monological belief system." This topic is critically analyzed in Chapter 7.

9. This is, admittedly, a strange case of marginalization, since almost everyone is a conspiracy theorist to some degree. But it can hardly be denied that the epithet "conspiracy theorist" is often used for the purpose of marginalization and discreditation.

10. In my view, this sentence should read, "On this view, conspiracy theories are judged irrational just because they are conspiracy theories." Buenting and Taylor's

version of the sentence seems to imply what I call the "constituent parts" definition of "conspiracy theory," which Buenting and Taylor do not seem to endorse, and neither do I. See section 1.2.1, for a discussion of this.

11. The *Le Monde* authors imply that we are "build[ing] artificial rivalries and point[ing] to imaginary enemies" (Dieguez et al. 2016, p. 23 n4). But the rivalry is real. Consider the title of David Coady's book *Conspiracy Theories: The Philosophical Debate*. The "debate" is essentially between the particularists, represented by Coady and Charles Pigden, and the generalists, represented by, Karl Popper, Steve Clarke, and (seemingly) Brian Keeley, with Lee Basham being hard to nail down at that point. (Basham is now clearly in the particularist camp, of course.) In addition to Keeley (1999) and Clarke (2002), Buenting and Taylor cite Pete Mandik as representing the generalist camp (Buenting and Taylor 2010, p. 568 n. 3). Neil Levy (2007) could also be added as offering a position that is generalist in spirit. And the distinction is not merely applicable to scholars who publish on this issue. Snooty generalist dismissal of conspiracy theories is commonplace, especially by media personalities and government officials, as well as in the academy. (For more detail on the generalists and their views, see Chapter 2.)

12. I am alluding here to a position taken by Patrick Stokes (2016, 2018a, and 2018b). Stokes's discussion of "defeasible generalism" and "reluctant particularism" encourages us to think of generalism and particularism as two ends of a spectrum, inviting the articulation of a middle ground. But I'm not sure that the ground pointed to by Stokes is not already taken. I believe that those of us in the "particularist" camp already agree that there is some degree to which general observations can give us guidance in judging particular cases. That is just to say that prior probability considerations are *relevant*, which I don't think we ever denied. Indeed, I think particularists explicitly insist on this. We simply deny that such considerations *by themselves* could lead to *the blanket dismissal of all conspiracy theories*, which is the generalist move. (See section 2.12 in Chapter 2 for more on Stokes's position.)

13. "Conspiracy theorizing" here means theorizing about something typically regarded as a "conspiracy theory" (e.g., the JFK assassination, or 9/11). Whether or not such a theory is warranted must be judged on the relevant evidence; that is the point of particularism. And it must be acknowledged that such theories are often vigorously, and in some cases cogently, defended. In contrast, "conspiracism" implies being unwarranted, "a form of misinformation." Conflating the two ideas is a way of begging a central question.

14. "Cock-up" is a British expression equivalent to the American expression "screw-up." Conspiracy theories have often been contrasted with "cock-up theories," in which unplanned blundering, rather than secretly coordinated activity, is pointed to as an explanation for some phenomenon.

15. This chapter was originally published by Rowman and Littlefield International in *Taking Conspiracy Theories Seriously*, edited by M R. X. Dentith (2018). It is reprinted here with permission from the publisher.

Conclusion

1. Chapter 4 is very similar to the manuscript rejected by the editor of the *Journal of Political Philosophy*. Chapter 5, however, is considerably longer than the original manuscript, which was short precisely so that it might more likely be given consideration.

2. The work of Ginna Husting and Martin Orr comes to mind.

3. Keeley's seminal 1999 essay "Of Conspiracy Theories," exploring new territory as it does, has some significant problems, in my view, which have caused it to be almost universally misunderstood. It is often cited as though it successfully demonstrates that there is something wrong with conspiracy theories that makes them unwarranted, which it does not. For Keeley *explicitly rejects* that conclusion. And, whereas he does offer reasons to think that a subset of conspiracy theories with five particular characteristics *are* unwarranted, a category he ill-advisedly labels "*unwarranted conspiracy theories* (UCTs)," he does *not* succeed and he essentially admits this (see section 2.5).

Appendix

1. "William Rodriguez on the News on 9-11-06," https://www.youtube.com/watch?v=uQgEAEb4YKQ (00:59 & 01:40).

2. See "Rosie O'Donnell and 9/11 Hero William Rodriguez," https://www.youtube.com/watch?v=8U45WYZqOWU&feature=related

3. I further address the reliability of the mainstream media, when it comes to dealing with conspiracy theories generally, in Hagen 2020b, section 4, entitled "Critique #2b: The View Promoted in the Mainstream Is Unbiased." This section, which is somewhat of a digression from the main argument of that paper, was added to respond to a reviewer who seemed to think the mainstream media can be trusted to get things right.

4. "2004 Scientist Statement on Restoring Scientific Integrity to Federal Policy Making," published by the Union of Concerned Scientists, and signed by fifteen thousand US scientists. http://www.ucsusa.org/scientific_integrity/abuses_of_science/scientists-sign-on-statement.html

5. "Investigation of World Trade Center Building 7," C-Span, August 21, 2008, https://www.c-span.org/video/?280569-1/investigation-world-trade-center-building-7 (at 00:32:45).

6. In an interview that took place on January 25, 2001, Frank A. De Martini, an architect who was the WTC on-site construction manager, discussed the robustness of the Twin Towers in reference to the 1993 bombing of the North Tower. Kevin Flynn and Jim Dwyer explain:

> "The building was designed to have a fully loaded 707 crash into it," De Martini said, "That was the largest plane at the time." . . . "I believe that the building probably could sustain multiple impacts of jetliners because this structure is like the mosquito netting on your screen door, this intense grid," De Martini said. "And the jet plane is just a pencil puncturing that screen netting. It really does nothing to the screen netting." (Flynn and Dwyer 2005, p. 149)

Video of De Martini saying this is included in many documentaries available on the internet. (For an excerpt from one, see: "WTC Construction Manager—Twin Towers Can Take Multiple Plane Impacts! Filmed January 2001," http://www.youtube.com/watch?v=aAVd2txjNEc). DeMartini died on 9/11 in the collapse of the North Tower.

7. See: "9/11: NIST Engineer John Gross Denies WTC Molten Steel (extended)," http://www.youtube.com/watch?v=3SLIzSCt_cg

8. Richard Gage offers a concise summary in a formal letter to NIST (January 3, 2008): "Mark Loizeaux, president of Controlled Demolition, Inc., who was hired for the Building 7 cleanup, said that 'molten steel was found at 7 WTC.' Leslie Robertson, World Trade Center structural engineer, stated that on October 5, '21 days after the attacks, the fires were still burning and molten steel was still running.' Fire department personnel, recorded on video, reported seeing 'molten steel running down the channel rails . . . like you're in a foundry—like lava from a volcano.' Joe O'Toole, a Bronx firefighter, saw a crane lifting a steel beam vertically from deep within a pile. He said, 'it was dripping from the molten steel.' Bart Voorsanger, an architect hired to save 'relics from the rubble,' stated about the multi-ton 'meteorite' that it was a 'fused element of molten steel and concrete'" (Gage 2008).

9. At this point in their dialogue, Mark asserts, "NASA pictures, thermal images show those sorts of temperatures in the basement." Gross responds, "Will you send them to me." Mark says, "Okay." Mark claims that after the talk he approached Gross and asked for his email address for the purpose of sending him the thermal images that Gross had asked for. But Gross refused to provide it. Mark comments, "I think this is important because it reveals the attitude of the NIST investigators, which is one of willful ignorance of what really happened on 9/11." See the abovementioned video: "9/11: NIST Engineer John Gross Denies WTC Molten Steel (extended)."

10. Seemingly corroborating video evidence includes video of the South Tower shortly before its collapse which shows what looks like molten steel or iron flowing out of the building.

11. This has been found not only by Steven Jones and his colleagues, but also by the RJ Lee group, who have no ax to grind on this issue. For more about the RJ Lee report see section A.2.3 and note 12 to Chapter 4, as well as note 19 to this chapter.

12. Here is a transcription with all the hesitations: "Well . . . um . . . the . . . first of all gravity . . . um . . . gravity is the loading function that applies to the structure . . . um . . . at . . . um . . . applies . . . to every body . . . every . . . uh . . . on . . . all bodies on . . . ah . . . on . . . um . . . this particular . . . on this planet not just . . . um . . . uh . . . in ground zero . . . um . . . the . . . uh . . ."

13. See "WTC7: NIST Finally Admits Freefall (Part I)," https://www.youtube.com/watch?time_continue=3&v=Rkp-4sm5Ypc at 3:15; emphasis added. Chandler's question is heard at 1:45.

14. The period that Chandler is considering differs from the period on which NIST's "40 percent slower" claim is based. This is because Chandler starts his measurement when the corner begins to descend, while NIST starts its measurement when a pixel changes color near the middle of the roofline, as the roof begins to bend inward, which is over one second before the corner begins its descent. That accounts for most of the difference from free-fall acceleration. (The mostly inward initial bend of the roofline looks like downward motion from the perspective of an upward-looking camera. NIST's analysis seems to leverage this ambiguity.)

15. A short version is "Freefall Acceleration of WTC7," http://www.youtube.com/watch?v=POUSJm--tgw. This is part of a much longer presentation, "(1 of 9) David Chandler at Visalia Friends Meeting": http://www.youtube.com/watch?v=v8jeodwwaKg

16. The "Draft for Public Comment" version of the report actually refers to

"constant speed," rather than acceleration. But this was presumably just a simple mistake.

17. "WTC 7—BBC The Third Tower—Conspiracy Files," https://www.youtube.com/watch?v=vZbMfTtHkYM at 53:22.

18. "FAQs—NIST WTC 7 Investigation," NIST, https://www.nist.gov/topics/disaster-failure-studies/faqs-nist-wtc-7-investigation

19. I had also inquired with the RJ Lee Group, years ago, but did not get such a helpful reply. Their report states that iron-rich spheres "would be expected" in WTC dust. But their explanation for this expectation seemed inadequate to me. Regarding this inadequacy, I sent the following query to the RJ Lee group:

> I have a question about your "WTC Dust Signature Report" of December 2003, pertaining to the Damage Assessment of the 130 Liberty Street Property. It says that iron-rich spheres "would be expected to be present in WTC Dust" (p. 16). Was this "expectation" developed before or after any dust samples were studied? In other words, having already observed the spheres in the dust, were you just claiming that it seemed plausible that they could be produced by the WTC collapse, while their existence was otherwise inexplicable? Or, did you actually make the prediction that such metallic spheres would be found in advance of any observation of their existence? If so, on what basis?
>
> To my way of thinking, the existence of "high temperatures" and "iron bearing building components," by themselves, may justify the first line of reasoning, but not the second. I wonder if someone could confirm for me what the actual reasoning process was. Your help in clarifying this would be most appreciated.

I received the following reply: "Kurtis, We're not sure how you received this report. We did the work for a third party and are not at liberty to discuss any further details. Tricia Egry, Manager Corporate Marketing."

20. Mohr, incidentally, has a series of videos in which he responds to Richard Gage's case for controlled demolition. By dealing respectfully with the evidence, and demonstrating an understanding of the relevant arguments, these videos provide a model for how conspiracy theories ought to be responded to. See Mohr's YouTube channel, https://www.youtube.com/user/chrismohr911

21. While NIST had maintained that Building 7 collapsed "primarily due to fires" (NIST 2008a, p. xxxv), Hulsey et al. conclude "fire did not cause the collapse of WTC 7" (2020, p. ii). Further contrasting their conclusion with NIST's, Hulsey et al. write, "It is our conclusion that the collapse of WTC 7 was a global failure involving the near-simultaneous failure of all columns in the building and not a progressive collapse involving the sequential failure of columns throughout the building" (p. 110).

22. See "On 9/11, Bush Watched the First Plane Impact," https://www.youtube.com/watch?v=BUXglJU2w6U

23. The president was asked, "What was the first thing that went through your head when you heard that a plane crashed into the first building?" Bush responded:

> Yes. Well, I was sitting in a schoolhouse in Florida . . . and my Chief of Staff—well, first of all, when we walked into the classroom, I had seen this plane fly

into the first building. There was a TV set on. And you know, I thought it was pilot error, and I was amazed that anybody could make such a terrible mistake, and something was wrong with the plane, or—anyway, I'm sitting there, listening to the briefing, and Andy Card came and said, "America is under attack." (Bush 2004, p. 12)

24. Glass's analysis has its own problems. Her debunking of Bush's statement assumes that it refers to the second plane, while Bush seems to be talking about the first. In any case, as I have indicated, it can't actually be about either.

25. "Bush's reaction to 9/11 (Full Classroom Footage)," https://www.youtube.com/watch?v=suB5wNSNBjs

26. Though NYPost.com's web page for September 12, 2002, is still available, this article is not (or is no longer) listed on it, though (at least at the time I am writing this) it still can be found using the Wayback Machine. The content of the article in question, "Tragic Lesson" by Malcom Balfour, can also be found at http://s3.amazonaws.com/911timeline/2002/nypost091202.html

27. I recall encountering this notion on a website dedicated to debunking conspiracy theories. If one wanted to follow this down a rabbit hole, one could ask, "Is it routine for the president to be put in situations from which it is not safe to speedily extract him?"

28. As John Farmer explains, "NORAD utilized only radars located around the perimeter of the United States, and was not capable of seeing much of the nation's interior" (Farmer 2009, pp. 111–112).

29. Perhaps the attempt to cover up facts that may look bad in retrospect is sufficient to explain "former National Security Advisor Sandy Berger hiding documents in his socks, so desperate to prevent the public from seeing certain papers that he sneaked out of the National Archives and hid the documents at a nearby construction site so that he could shred them later" (Farmer 2009, p. 289). But exactly what Berger was covering up is not entirely clear. And the lengths someone of his stature would be willing to go to cover things up should not be forgotten.

30. The possible use of Ali Mohamed as an "informant" is an example the kind of thing that could benefit from greater oversight. See Scott 2007, pp. 151–160.

31. There are elaborate theories about how this kind of dynamic may have occurred in the JFK assassination. According to one relatively popular theory, Mob boss Santo Trafficante was encouraged by Nixon, through Robert Mahue, Sam Giancana, and Johnny Roselli, to arrange an assassination of Castro. In order to ensure that, if anything went wrong, the CIA would be motivated to cover everything up, Trafficante intentionally hired assassins with ties to the CIA. This group then, either with or without CIA help and encouragement, turned on JFK instead of Castro. Both Nixon and the CIA would have had reasons for not wanting any of this discovered.

32. For example, in addition to misleading statements implying the Bush administration was not warned about domestic al-Qaeda cells, the author of *Wake-Up Call: The Political Education of a 9/11 Widow*, Kristen Breitweiser, writes: "[Condoleezza] Rice's assertion that no one knew planes might be used as weapons and flown into buildings was also patently false. She had attended the G-8 summit meeting with the president and other world leaders a mere two months before

9/11, when the possibility of an assassination attempt on President Bush had been mentioned to the group—an assassination that would be carried out by planes. Not to mention the long and well-documented intelligence history of US government awareness and preparedness regarding planes being used as weapons" (2006, p. 164).

33. Given the fact that the commission was provided with false information repeatedly by government agencies, it is noteworthy that the law creating the 9/11 Commission only allowed it to "ascertain, evaluate, and report on the *evidence developed by all relevant government agencies* regarding the facts and circumstances surrounding the attacks" (Kean and Hamilton 2006, p. 123, emphasis added). This is why, according to Kean and Hamilton, they chose not to legally press the CIA regarding access to detainees.

34. Jacobson was the principal author of the classified portion of the congressional report commonly referred to as "The 28 Pages," which was finally declassified (though still partially redacted) in 2016. It is highly suggestive of Saudi involvement in 9/11. Shenon reports that "It was Jacobson's work that had led Senator Bob Graham and others in Congress to accuse the White House of covering up a Saudi role in the attacks" (Shenon 2008, p. 110).

According to Shenon, the twenty-eight pages "were certainly supposed to be available to the commission" (p. 110). But Philip Zelikow, the executive director of the 9/11 Commission, refused to provide such documents, and also severely limited the number of interviews the staff could conduct. Dana Lesemann confronted him on this general issue more than once. Shenon reports one confrontation as follows: "'Philip, this is ridiculous,' she said, almost pounding him in the chest with her outstretched index finger. 'We need the interviews. We need these documents. Why are you trying to limit our investigation?'" (Shenon 2008, p. 111). Lesemann went ahead and obtained the twenty-eight pages in her own (improper) way. She had the relevant security clearances and, according to Shenon, "There was no doubt that she needed [the documents] to do her job" (p. 112). But Zelikow wasted no time in firing her for her transgression.

Works Cited

Aaronovitch, David. 2010. *Voodoo Histories: The Role of the Conspiracy Theory in Shaping Modern History*. New York: Riverhead Books.

ACLU (undated). "Sibel Edmonds: A Patriot Silenced, Unjustly Fired but Fighting Back to Help Keep America Safe." https://www.aclu.org/other/sibel-edmonds-patriot-silenced-unjustly-fired-fighting-back-help-keep-america-safe

Ahmed, Nafeez Mosaddeq. 2005. *The War on Truth: 9/11, Disinformation, and the Anatomy of Terrorism*. Northampton, MA: Olive Branch Press.

Associated Press. 1993. "The Text of NBC's Statement Tuesday on the Settlement of GM's Lawsuit against It, as Read on 'Dateline NBC' by Co-anchors Jane Pauley and Stone Phillips." (February 10). https://www.apnews.com/72180d14c165619981cd6e25c1a597f1

Bale, Jeffery M. 2007. "Political Paranoia v. Political Realism: On Distinguishing Between Bogus Conspiracy Theories and Genuine Conspiratorial Politics." *Patterns of Prejudice* 41.1: 45–60.

Balfour, Malcolm. 2002. "9/11 One Year Later: Tragic Lesson." *New York Post* (September 12).

Bamford, James. 2001. *Body of Secrets: Anatomy of the Ultra-secret National Security Agency from the Cold War through the Dawn of a New Century*. New York: Doubleday.

Bamford, James. 2004. *A Pretext for War: 9/11, Iraq, and the Abuse of America's Intelligence Agencies*. New York: Doubleday.

Barbour, John (director). 1992. *The JFK Assassination: The Jim Garrison Tapes*. (The cited segment can be seen on the "John Barbour's World" YouTube channel at https://www.youtube.com/watch?v=5Ocfr2VdcpU at 46:55.)

Barkun, Michael. 2003. *A Culture of Conspiracy: Apocalyptic Visions in Contemporary America*. Berkeley: University of California Press.

Barkun, Michael. 2013. *A Culture of Conspiracy: Apocalyptic Visions in Contemporary America*. 2nd ed. Berkeley: University of California Press.

Bartlett, Steven James. 2005. *The Pathology of Man: A Study of Human Evil*. Springfield, IL: Charles C. Thomas Publisher.

Basham, Lee. 2001. "Living with the Conspiracy." *Philosophical Forum* 32.3 (Fall): 265–280. (Reprinted in Coady 2006a: 61–75.)

Basham, Lee. 2003. "Malevolent Global Conspiracy." *Journal of Social Philosophy* 34.1 (Spring): 91–103. (Reprinted in Coady 2006a: 93–105.)

Basham, Lee. 2006. "Afterthoughts on Conspiracy Theory: Resilience and Ubiquity." In Coady 2006a: 133–137.

Basham, Lee. 2011. "Conspiracy Theory and Rationality." In *Beyond Rationality: Contemporary Issues*, ed. Carl Jensen and Rom Harré, 49–87. Newcastle upon Tyne: Cambridge Scholars Publishing.

Basham, Lee. 2016. "The Need for Accountable Witnesses: A Reply to Dentith." *Social Epistemology Review and Reply Collective* 5.7: 6–13.

Basham, Lee. 2017. "Pathologizing Open Societies: A Reply to the Le Monde Social Scientists." *Social Epistemology Review and Reply Collective* 6.2: 59–68.

Basham, Lee. 2018a. "Joining the Conspiracy." *Argumenta* 3.2 (issue 6): 271–290.

Basham, Lee. 2018b. "Conspiracy Theory Particularism, Both Epistemic and Moral, versus Generalism." In Dentith 2018b (39–58).

Basham, Lee, and Matthew R. X. Dentith. 2015. "Bad Thinkers? Don't Be So Gullible!" *Three Quarks Daily* (August 17). https://www.3quarksdaily.com/3quarksdaily/2015/08/bad-thinkers-dont-be-so-gullible.html

Basham, Lee, and Matthew R. X. Dentith. 2016. "Social Science's Conspiracy-Theory Panic: Now They Want to Cure Everyone." *Social Epistemology Review and Reply Collective* 5.10: 12–19. (An English version of the *Le Monde* publication, "Let's Fight Conspiracy Theories Effectively," is also contained herein.)

Baumeister, Roy F. 1997. *Evil: Inside Human Cruelty and Violence*. New York: Henry Holt.

Bernstein, Carl. 1977. "The CIA and the Media." *Rolling Stone* (October).

Blight, James G., and Janet M. Lang. 2005. *The Fog of War: Lessons from the Life of Robert S. McNamara*. Lanham, MD: Rowman & Littlefield Publishers. (Based on the Film by Errol Morris.) Transcript for the film: http://www.errolmorris.com/film/fow_transcript.html

Blumenthal, Sidney. 2006. "George Bush's Rough Justice." *The Guardian* (January 11). https://www.theguardian.com/world/2006/jan/12/usa.comment

Boddy, Clive R. 2011. "The Corporate Psychopaths Theory of the Global Financial Crisis." *Journal of Business Ethics* 102: 255–259.

Bost, Preston R. 2019. "The Truth Is around Here Somewhere: Integrating the Research on Conspiracy Beliefs." In Uscinski 2019 (269–282).

Breitweiser, Kristin. 2006. *Wake-Up Call: The Political Education of a 9/11 Widow*. New York: Warner Books.

Broaddrick, Juanita. 2017. *You'd Better Put Some Ice on That: How I Survived Being Raped by Bill Clinton*. CreateSpace.

Brotherton, Robert. 2013. "Towards a Definition of 'Conspiracy Theory.'" *PsyPAG Quarterly* 88 (September): 9–14.

Brotherton, Robert. 2016. "The Logic behind Conspiracy Theories." *LA Times* (January 19). http://www.latimes.com/opinion/op-ed/la-oe-0119-brotherton-conspiracy-theory-20160119-story.html

Brotherton, Robert, and Christopher C. French. 2014. "Belief in Conspiracy Theories and Susceptibility to the Conjunction Fallacy." *Applied Cognitive Psychology* 28: 238–248.

Brotherton, Robert, Christopher C. French, and Alan D. Pickering. 2013. "Measuring Belief in Conspiracy Theories: The Generic Conspiracist Beliefs Scale." *Frontiers in Psychology* 4.279 (May): 1–15.

Bruder, Martin, Peter Haffke, Nick Neave, Nina Nouripanah, and Roland Imhoff. 2013. "Measuring Individual Differences in Generic Beliefs in Conspiracy Theories across Cultures: Conspiracy Mentality Questionnaire." *Frontiers in Psychology* 4.225 (April 30): 1–15.

Buenting, Joel, and Jason Taylor. 2010. "Conspiracy Theories and Fortuitous Data." *Philosophy of the Social Sciences* 40.4: 567–578.

Bugliosi, Vincent. 2008. *The Prosecution of George W. Bush for Murder*. Cambridge, MA: Vanguard Press.

Buncombe, Andrew. 2006. "The US Propaganda Machine: Oh, What a Lovely War." *The Independent* (March 30). https://www.independent.co.uk/news/world/americas/the-us-propaganda-machine-oh-what-a-lovely-war-6105134.html

Burrough, Bryan. 2007. "Or No Conspiracy?" *New York Times* (May 20). https://www.nytimes.com/2007/05/20/books/review/Burrough-t.html?mtrref=www.google.com

Bush, George W. 2001. "Text: President Bush Addresses the Nation." *Washington Post* (September 20). http://www.washingtonpost.com/wp-srv/nation/specials/attacked/transcripts/bushaddress_092001.html

Bush, George W. 2004. *Public Papers of the President of the United States: George W. Bush*, Book I. Washington, DC: US Government Printing Office.

Bush, George W. 2006. "Excerpt from Bush's Remarks." *New York Times* (September 16). http://www.nytimes.com/2006/09/16/us/16text.html?_r=0

Butter, Michael, and Peter Knight. 2019. "The History of Conspiracy Theory Research." In Uscinski 2019 (33–46).

Cassam, Quassim. 2015. "Bad Thinkers." *Aeon* (March 13). https://aeon.co/essays/the-intellectual-character-of-conspiracy-theorists

Cassam, Quassim. 2016. "Vice Epistemology." *The Monist* 99: 159–180.

Cassam, Quassim. 2018. "Epistemic Vices and Conspiracy Theories." *Blog of the APA* (February 28). https://blog.apaonline.org/2018/02/28/epistemic-vices-and-conspiracy-theories/

Church Committee. 1976. "COINTELPRO: The FBI's Covert Action Programs against American Citizens." In *Supplementary Detailed Staff Reports on Intelligence Activities and the Rights of Americans*, Book III, *Final Report of the Select Committee to Study Governmental Operations with Respect to Intelligence Activities United States Senate* (April 23).

Clarke, Richard. 2004. *Against All Enemies: Inside America's War on Terror*. New York: Free Press.

Clarke, Steve. 2002. "Conspiracy Theories and Conspiracy Theorizing." *Philosophy of the Social Sciences* 32: 131–150. (Reprinted in Coady 2006a: 77–92.)

Clarke, Steve. 2006. "Appealing to the Fundamental Attribution Error: Was It All a Big Mistake?" In Coady 2006a: 129–132.

Clarke, Steve. 2007. "Conspiracy Theories and the Internet: Controlled Demolition and Arrested Development." *Episteme* 4.2: 167–180.

Coady, David. 2003. "Conspiracy Theories and Official Stories." *International Journal of Applied Philosophy* 17.2 (Fall): 197–209. (Reprinted in Coady 2006a: 115–127.)

Coady, David, ed. 2006a. *Conspiracy Theories: The Philosophical Debate*. Burlington, VT: Ashgate Publishing.

Coady, David. 2006b. "An Introduction to the Philosophical Debate about Conspiracy Theories." In Coady 2006a: 1–11.

Coady, David. 2007. "Are Conspiracy Theorists Irrational?" *Episteme* 4.2: 193–204.

Coady, David. 2012. *What to Believe Now: Applying Epistemology to Contemporary Issues*. Malden, MA: Wiley-Blackwell.

Coady, David. 2018a. "Cass Sunstein and Adrian Vermeule on Conspiracy Theories." *Argumenta* 3.2 (issue 6): 291–302.

Coady, David. 2018b. "Anti-rumor Campaigns and Conspiracy-Bating as Propaganda." In Dentith 2018b (171–187).

Coady, David. 2019. "Psychology and Conspiracy Theories." In *Routledge Handbook on Applied Epistemology*, ed. David Coady and James Chase, 166–175. New York: Routledge.

Cole, Jonathan. 2013. "The Official Collapse Narrative and the Experimental Method." In Gourley 2013 (233–252).

Crenshaw, Charles A. 2013. *JFK Has Been Shot*. Kindle ed. New York: Kensington.

deHaven-Smith, Lance. 2010. "Beyond Conspiracy Theory: Patterns of High Crime in American Government." *American Behavioral Scientist* 53.6: 795–825.

deHaven-Smith, Lance. 2013a. *Conspiracy Theory in America*. Austin: University of Texas Press.

deHaven-Smith, Lance. 2013b. "Seeing 9/11 from Above: A Comparative Analysis of State Crimes against Democracy." In Gourley 2013 (61–106).

Dentith, Matthew R. X. 2014. *The Philosophy of Conspiracy Theories*. New York: Palgrave Macmillan.

Dentith, Matthew R. X. 2015. "I'm Not a Conspiracy Theorist, but . . ." *Fortean-Times* 324 (February): 36–39.

Dentith, Matthew R. X. 2016. "When Inferring to a Conspiracy Might Be the Best Explanation." *Social Epistemology* 30.5–6: 572–591.

Dentith, Matthew [M] R. X. 2018a. "The Problem of Conspiracism." *Argumenta* 3.2 (issue 6): 327–343.

Dentith, M R. X., ed. 2018b. *Taking Conspiracy Theories Seriously*. New York: Rowman & Littlefield.

Dentith, M R. X. 2018c. "When Inferring to a Conspiracy Might Be the Best Explanation." In Dentith 2018b (3–23).

Dentith, M R. X. 2018d. "What Particularism about Conspiracy Theories Entails." In Dentith 2018b (59–69).

Dentith, M R. X. 2019a. "Conspiracy Theories and Philosophy: Bringing the Epistemology of a Freighted Term into the Social Sciences." In Uscinski 2019 (94–108).

Dentith, M R. X. 2019b. "Conspiracy Theories on the Basis of the Evidence." *Synthese* 196: 2243–2261.

Dentith, M R. X. 2019c. "The Iniquity of the Conspiracy Inquirers." *Social Epistemology Review and Reply Collective* 8.8: 1–11. https://wp.me/p1Bfg0-4hQ

Dentith, M R. X., and Brian L. Keeley. 2019. "The Applied Epistemology of Conspiracy Theories: An Overview." In *Routledge Handbook on Applied Epistemology*, ed. David Coady and James Chase, 284–294. New York: Routledge.

Dentith, Matthew R. X., and Martin Orr. 2017. "Clearing Up Some Conceptual Confusions about Conspiracy Theory Theorising." *Social Epistemology Review and Reply Collective* 6.1: 9–16.

Dentith, Matthew R. X., and Martin Orr. 2018. "Secrecy and Conspiracy." *Episteme* 15.4: 433–450.

Dieguez, Sebastian, Gérald Bronner, Véronque Campion-Vincent, Sylvain Delouvée, Nicolas Gauvrit, Anthony Lantian, and Pascal Wagner-Egger. 2016. "'They' Respond: Comments on Basham *et al.*'s 'Social Science's Conspiracy-Theory Panic: Now They Want to Cure Everyone.'" *Social Epistemology Review and Reply Collective* 5.12: 20–39.

DiEugenio, James. 2012. *Destiny Betrayed*. 2nd ed. New York: Skyhorse Publishing.

DiFonzo, Nicholas. 2019. "Conspiracy Rumor Psychology." In Uscinski 2019 (257–268).

Doerflinger, Richard. 2010. "Assisted Suicide: Pro-choice or Anti-life?" In *The Right Thing to Do: Basic Readings in Moral Philosophy*, 5th ed., ed. James Rachels and Stuart Rachels, 311–320. Boston: McGraw-Hill.

Donnelly, Thomas. 2000. *Rebuilding America's Defenses: Strategy, Forces and Resources*. Washington, DC: The Project for the New American Century.

Douglas, Karen M., and Robbie M. Sutton. 2008. "The Hidden Impact of Conspiracy Theories: Perceived and Actual Influence of Theories Surrounding the Death of Princess Diana." *Journal of Social Psychology* 148: 210–222.

Douglas, Karen. M., and Robbie M. Sutton. 2011. "Does It Take One to Know One? Endorsement of Conspiracy Theories Is Influenced by Personal Willingness to Conspire." *British Journal of Social Psychology* 50: 542–552.

Dunbar, David, and Brad Reagan, eds. 2006. *Debunking 9/11 Myths: Why Conspiracies Can't Stand Up to the Facts*. New York: Hearst Books.

Edmonds, Sibel. 2012. *Classified Woman: The Sibel Edmonds Story, a Memoir*. Alexandria, VA: Sibel Edmonds.

Eggen, Dan. 2006. "Cheney's Remarks Fuel Torture Debate." *Washington Post* (October 27). http://www.washingtonpost.com/wp-dyn/content/article/2006/10/26/AR2006102601521.html

Ehrlich, Paul, Anne Ehrlich, and John P. Holdren. 1977. *Ecoscience: Population, Resources, Environment*. San Francisco: W.H. Freeman.

Erickson, Robert (director). 2009. *9/11 Science and Conspiracy*. National Geographic.

Farmer, John. 2009. *The Ground Truth: The Untold Story of America under Attack on 9/11*. New York: Penguin.

Farrell, John A. 2016. "Nixon's Vietnam Treachery." *New York Times* (December 31). https://www.nytimes.com/2016/12/31/opinion/sunday/nixons-vietnam-treachery.html

Feldman, Susan. 2011. "Counterfact Conspiracy Theories." *International Journal of Applied Philosophy* 25.1: 15–24.

Fenton, Kevin. 2011. *Disconnecting the Dots: How CIA and FBI Officials Helped Enable 9/11 and Evaded Government Investigations*. Walterville, OR: TrineDay.

Finnegan, Lisa. 2007. *No Questions Asked: News Coverage since 9/11*. Westport, CT: Praeger Publishers.

Flynn, Kevin, and Jim Dwyer. 2005. *102 Minutes: The Untold Story of the Fight to Survive inside the Twin Towers*. New York: Times Books.

French, Christopher. 2015. "Why Do Some People Believe in Conspiracy Theories?" *Scientific American* (July 1). http://www.scientificamerican.com/article/why-do-some-people-believe-in-conspiracy-theories/

Friedan, Terry. 1998. "FBI Whistle-Blower Leaves, Gets $1.16 million." *CNN* (February 27).

Friedersdorf, Conor. 2015. "CSI Is a Lie." *The Atlantic* (April 20). http://www.theatlantic.com/politics/archive/2015/04/csi-is-a-lie/390897/

Friedman, Brad. 2007. "FBI Whistleblower Sibel Edmonds Will Now Tell All, Name New Names—to Any Major Television Network That Will Let Her." *HuffPost* (October 30). https://www.huffingtonpost.com/brad-friedman/fbi-whistleblower-sibel-e_b_70383.html

Gage, Richard. 2008. "Undisputed Facts Point to the Controlled Demolition of WTC 7: Response to NIST's Invitation for Written Comments." *Global Research*. http://www.globalresearch.ca/PrintArticle.php?articleId=8472

Ganser, Daniele. 2005. *NATO's Secret Armies: Operation GLADIO and Terrorism in Western Europe*. New York: Frank Cass.

Gellmon, Barton, and Jerry Markon. 2013. "Edward Snowden Says Motive behind Leaks Was to Expose 'Surveillance State.'" *Washington Post* (June 10). https://www.washingtonpost.com/politics/edward-snowden-says-motive-behind-leaks-was-to-expose-surveillance-state/2013/06/09/aa3f0804-d13b-11e2-a73e-826d299ff459_story.html?utm_term=.a860b5a13c8a

Glass, Lillian. 2014. *The Body Language of Liars: From Little White Lies to Pathological Deception*. Pompton Plains, NJ: Career Press.

Goertzel, Ted. 1994. "Belief in Conspiracy Theories." *Political Psychology* 15.4: 731–742.

Goldberg, Robert Alan. 2001. *Enemies Within: The Culture of Conspiracy in Modern America*. New Haven: Yale University Press.

Goldman, Adam, and Peyton Craighill. 2014. "New Poll Finds Majority of Americans Think Torture Was Justified after 9/11 Attacks." *Washington Post* (December 16). https://www.washingtonpost.com/world/national-security/new-poll-finds-majority-of-americans-believe-torture-justified-after-911-attacks/2014/12/16/f6ee1208-847c-11e4-9534-f79a23c40e6c_story.html

Goodman, H. A. 2014. "If Bush Is a War Criminal, Then So Are Truman, LBJ, Nixon and Obama." *Huffington Post* (updated July 30). http://www.huffingtonpost.com/h-a-goodman/if-bush-is-a-war-criminal_b_5416535.html

Gourley, James, ed. 2013. *The 9/11 Toronto Report: International Hearings on the Events of September 11, 2001*. Dallas, TX: International Center for 9/11 Studies.

Greenwald, Glenn. 2010. "Obama Confidant's Spine-Chilling Proposal: Cass Sunstein Wants the Government to 'Cognitively Infiltrate' Anti-government Groups." *Salon.com* (January 15).

Greenwald, Glenn, Ewen MacAskill, and Laura Poitras. 2013. "Edward Snowden: The Whistleblower behind the NSA Surveillance Revelations." *The Guardian* (June 11). https://www.theguardian.com/world/2013/jun/09/edward-snowden-nsa-whistleblower-surveillance

Griffin, David Ray. 2004. *The New Pearl Harbor: Disturbing Questions about the Bush Administration and 9/11*. Northampton, MA: Olive Branch Press.

Griffin, David Ray. 2005. *The 9/11 Commission Report: Omissions and Distortions*. Northampton, MA: Olive Branch Press.

Griffin, David Ray. 2006. *Christian Faith and the Truth behind 9/11*. Louisville: Westminster John Knox Press.

Griffin, David Ray. 2007a. "9/11: Let's Get Empirical." http://www.brianrwright.com/Griffin.pdf

Griffin, David Ray. 2007b. *Debunking 9/11 Debunking: An Answer to Popular Mechanics and Other Defenders of the Official Conspiracy Theory. Northampton*, MA: Olive Branch Press.

Griffin, David Ray. 2010. *The Mysterious Collapse of World Trade Center 7: Why the Final Official Report about 9/11 Is Unscientific and False*. Northampton, MA: Olive Branch Press.

Griffin, David Ray. 2011. *Cognitive Infiltration: An Obama Appointee's Plan to Undermine the 9/11 Conspiracy Theory*. Northampton, MA: Olive Branch Press.

Griffin, David Ray, and Peter Dale Scott, eds. 2006. *9/11 and American Empire: Intellectuals Speak Out*. Northampton, MA: Olive Branch Press.

Griffin, David Ray, and Elizabeth Woodworth. 2018. *9/11 Unmasked: An International Review Panel Investigation*. Northampton, MA: Olive Branch Press.

Hagen, Kurtis. 2010. "Is Infiltration of 'Extremist Groups' Justified?" *International Journal of Applied Philosophy* 24.2 (Fall): 153–168.

Hagen, Kurrtis. 2011a. "Conspiracy Theories and Stylized Facts." *Journal for Peace and Justice Studies* 21.2 (Fall): 3–22.

Hagen, Kurtis. 2011b. "Review of David Ray Griffin's *Cognitive Infiltration*." *Florida Philosophical Review* 11.1. https://fpr.cah.ucf.edu/article/review-of-david-ray-griffins-cognitive-infiltration-an-obama-appointees-plan-to-undermine-the-9-11-conspiracy-theory/

Hagen, Kurtis. 2016. "Project for a New Confucian Century." In *A Future without Borders? Theories and Practices of Cosmopolitan Peacebuilding*, ed. Eddy M. Souffrant, 168–189. Boston: Brill/Rodopi.

Hagen, Kurtis. 2017. "What Are They Really Up To? Activist Social Scientists Backpedal on Conspiracy Theory Agenda." *Social Epistemology Review and Reply Collective* 6.3: 89–95.

Hagen, Kurtis. 2018a. "Conspiracy Theorists and Monological Belief Systems." *Argumenta* 3.2 (issue 6): 303–326.

Hagen, Kurtis. 2018b. "Conspiracy Theories and the Paranoid Style: Do Conspiracy Theories Posit Implausibly Vast and Evil Conspiracies?" *Social Epistemology* 32.1: 24–40.

Hagen, Kurtis. 2018c. "Conspiracy Theorists and Social Scientists." In Dentith 2018b (125–140).

Hagen, Kurtis. 2018d. *Conspiracy Theory: A Philosophical Defense*. Burlington, VT: Fomite.

Hagen, Kurtis. 2020a. "Should Academics Debunk Conspiracy Theories?" *Social Epistemology* 34.5: 423–439.

Hagen, Kurtis. 2020b. "Is Conspiracy Theorizing Really Epistemically Problematic?" *Episteme*, 1–23. https://doi.org/10.1017/epi.2020.19

Harambam, Jaron. 2020. *Contemporary Conspiracy Culture: Truth and Knowledge in an Era of Epistemic Instability*. New York: Routledge.

Hardin, Russell. 2002. "The Crippled Epistemology of Extremism." In *Political Extremism and Rationality*, ed. Albert Breton, Gianluigi Galeotti, Pierre Salmon, and Ronald Wintrobe, 3–22. Cambridge: Cambridge University Press.

Harris, Keith. 2019. "What's Epistemically Wrong with Conspiracy Theorising?" In *Harms and Wrongs in Epistemic Practice* (Royal Institute of Philosophy Supplement 84), ed. Simon Barker, Charlie Crerar, and Trystan S. Goetze, 235–257. Cambridge: Cambridge University Press.

Harris, Paul. 2011. "Prince Andrew's Link to Sex Offender Jeffrey Epstein Taints Royalty in US." *The Guardian* (March 12). https://www.theguardian.com/uk/2011/mar/13/prince-andrew-jeffrey-epstein

Harrit, Niels H., Jeffrey Farrer, Steven E. Jones, Kevin R. Ryan, Frank M. Legge, Daniel Farnsworth, Gregg Roberts, James R. Gourley, and Bradley R. Larson. 2009. "Active Thermitic Material Discovered in Dust from the 9/11 World Trade Center Catastrophe." *Open Chemical Physics Journal* 2: 7–31.

Hertsgaard, Mark. 2016. “How the Pentagon Punished NSA Whistleblowers.” *The Guardian* (May 22). https://www.theguardian.com/us-news/2016/may/22/how-pentagon-punished-nsa-whistleblowers

Hitchens, Christopher. 2002. *The Trial of Henry Kissinger*. New York: Verso.

Hoffmann, Peter. 2014. “The Good Germans: Inside the Resistance to the Nazis.” *Foreign Affairs* (July–August). https://www.foreignaffairs.com/reviews/review-essay/2014-06-18/good-germans

Hofstadter, Richard. 1965. *The Paranoid Style in American Politics*. Chicago: University of Chicago Press.

Homans, Charles. 2016. “The Conspiracy Theorists’ Election.” *New York Times* (September 27). http://www.nytimes.com/2016/10/02/magazine/the-conspiracy-theorists-election.html?_r=0

Hornberger, Jacob. 2017. “Corbett Report Interview 1318—Jacob Hornberger on the Ongoing JFK Cover Up.” *The Corbett Report*. https://www.corbettreport.com/interview-1318-jacob-hornberger-on-the-ongoing-jfk-cover-up/ (Cited quotation at 16:30.)

Hulsey, Leroy J., Zhili Quan, and Feng Xiao. 2020. *A Structural Reevaluation of the Collapse of World Trade Center 7: Final Report*. Fairbanks: Department of Civil and Environmental Engineering, University of Alaska.

Jackson, Robert L. 1999. “Clinton Fined $90,686 for Lying in Paula Jones Case.” *Los Angeles Times* (July 30). https://www.latimes.com/archives/la-xpm-1999-jul-30-mn-61021-story.html

Jane, Emma A., and Chris Fleming. 2014. *Modern Conspiracy: The Importance of Being Paranoid*. New York: Bloomsbury Publishing.

Johnston, David. 1995. “F.B.I. Chemist Says Experts Are Pressured to Skew Tests.” *New York Times* (September 15). https://www.nytimes.com/1995/09/15/us/fbi-chemist-says-experts-are-pressured-to-skew-tests.html

Jones, Nigel. 2008. *Countdown to Valkyrie: The July Plot to Assassinate Hitler*. London: Frontline Books.

Jones, Steven E., Frank M. Legge, Kevin R. Ryan, Anthony F. Szamboti, and James R. Gourley. 2008. “Fourteen Points of Agreement with Official Government Reports on the World Trade Center Destruction.” *Open Civil Engineering Journal* 2: 35–40.

Jones, Steven E., Robert Korol, Anthony Szamboti, and Ted Walter. 2016. “15 Years Later: On the Physics of High-Rise Building Collapses.” *Europhysics* 47.4: 21–26.

Jonsson, Patrik. 2013. “JFK Assassination: Why Suspicions Still Linger about ‘Umbrella Man.’” *Christian Science Monitor* (November 22). http://www.csmonitor.com/USA/Politics/2013/1122/JFK-assassination-Why-suspicions-still-linger-about-Umbrella-Man

Joyner, James. 2006. “Move to Fire Professor for 9/11 Conspiracy Views.” *Outside the Beltway* (September 29).

Kean, Thomas, and Lee Hamilton. 2004. *The 9/11 Commission Report: Final Report of the National Commission on Terrorist Attacks upon the United States*. Washington, DC: National Commission on Terrorist Attacks upon the United States.

Kean, Thomas, and Lee Hamilton. 2006. *Without Precedent: The Inside Story of the 9/11 Commission*. New York: Vintage Books.

Kearns, Cristin E., Laura A. Schmidt, and Stanton A. Glantz. 2016. "Sugar Industry and Coronary Heart Disease Research: A Historical Analysis of Internal Industry Documents." *JAMA Internal Medicine* 176.11: 1680–1685.

Keeley, Brian L. 1999. "Of Conspiracy Theories." *Journal of Philosophy* 96.3 (March): 109–126. (Reprinted in Coady 2006a: 45–60.)

Keeley, Brian L. 2003. "Nobody Expects the Spanish Inquisition! More Thoughts on Conspiracy Theory." *Journal of Social Philosophy* 34.1 (Spring): 104–110. (Reprinted in Coady 2006a: 107–113.)

Keeley, Brian L. 2007. "God as the Ultimate Conspiracy Theory." *Episteme* 4.2: 135–149.

Keeley, Brian L. 2019. "The Credulity of Conspiracy Theorists: Conspiratorial, Scientific, and Religious Explanations Compared." In Uscinski 2019 (422–431).

Kirk, Michael. 2014. "United States of Secrets (Part One): The Program." *PBS Frontline Transcript*, Season 32, Episode 11. https://www.pbs.org/wgbh/frontline/film/united-states-of-secrets/transcript/

Klein, Naomi. 2007. *The Shock Doctrine: The Rise of Disaster Capitalism*. New York: Picador.

Kurtz, Howard. 1999. "Clinton Accuser's Story Aired." *Washington Post* (February 25). http://www.washingtonpost.com/wp-srv/politics/special/clinton/stories/broaddrick022599.htm

Lau, D. C. 1963. *Tao Te Ching*. New York: Penguin Books.

Layman, C. Stephen. 2007. *Letters to Doubting Thomas: A Case for the Existence of God*. New York: Oxford University Press.

Levy, Neil. 2007. "Radically Socialized Knowledge and Conspiracy Theories." *Episteme* 4.2: 181–192.

Lew, H. S., Richard W. Bukowski, and Nicholas J. Carino. 2005. "Design, Construction, and Maintenance of Structural and Life Safety Systems." *Federal Building and Fire Safety Investigation of the World Trade Center Disaster*. Washington, DC: NIST NCSTAR 1–1.

Lewandowsky, Stephen. 2019. "In Whose Hands the Future?" In Uscinski 2019 (149–177).

Lewandowsky, Stephan, John Cook, Klaus Oberauer, Elisabeth A. Lloyd, and Michael Marriott. 2015. "Recurrent Fury: Conspiratorial Discourse in the Blogosphere Triggered by Research on the Role of Conspiracist Ideation in Climate Denial." *Journal of Social and Political Psychology* 3: 142–78.

Lewandowsky, Stephan, John Cook, Klaus Oberauer, and Michael Marriott. 2013. "Recursive Fury: Conspiracist Ideation in the Blogosphere in Response to Research on Conspiracist Ideation." *Frontiers in Personality Science and Individual Differences* 4.73 (retracted).

Lewandowsky, Stephan, Elisabeth A. Lloyd, and Scott Brophy. 2018. "When THUNCing Trumps Thinking: What Distant Alternative Worlds Can Tell Us about the Real World." *Argumenta* 3.2 (issue 6): 217–231.

Lewis, Charles. 2014. *935 Lies: The Future of Truth and the Decline of America's Moral Integrity*. New York: Public Affairs.

Lewis, Neil A. 2005. "A Guide to the Memos on Torture." *New York Times*. http://www.nytimes.com/ref/international/24MEMO-GUIDE.html?_r=0

LoBianco, Tom. 2015. "Donald Trump on Terrorists: 'Take Out Their Families.'" *CNN* (December 3). https://www.cnn.com/2015/12/02/politics/donald-trump-terrorists-families/

Long, Tom. 2006. "A Muted Response from UNH: Professors' Right to Opinion Cited." *Boston Globe* (September 10).

MacQueen, Graeme. 2013. "Eyewitness Evidence of Explosions in the Twin Towers." In Gourley 2013 (173–196).

Mandik, Pete. 2007. "Shit Happens." *Episteme* 4.2: 205–218.

Mangan, Katherine. 2008. "Cass Sunstein to Leave U. of Chicago for Harvard." *Chronicle of Higher Education* (February 19). http://chronicle.com/article/Cass-Sunstein-to-Leave-U-of/40489/

Marantz, Andrew. 2017. "How a Liberal Scholar of Conspiracy Theories Became the Subject of a Right-Wing Conspiracy Theory." *New Yorker* (December 27).

Martin, Brian. 2019. "Technology and Evil." *Social Epistemology Review and Reply Collective* 8.2: 1–14.

Marx, Willem. 2006. "Misinformation Intern: My Summer as a Military Propagandist in Iraq." *Harper's Magazine* (September). https://harpers.org/archive/2006/09/misinformation-intern/

Matthews, Dylan. 2017. "The Rape Allegation against Bill Clinton, Explained." *Vox* (November 14). https://www.vox.com/2016/1/6/10722580/bill-clinton-juanita-broaddrick

McGovern, Ray, and Jason Leopold. 2011. "9/11 Not an 'Intelligence Failure.'" *The Real News Network* (September 11). https://therealnews.com/stories/rmcgoven0906-2

McHoskey, John W. 1995. "Case Closed? On the John F. Kennedy Assassination: Biased Assimilation of Evidence and Attitude Polarization." *Basic and Applied Social Psychology* 17.3: 395–409.

McKenzie-McHarg, Andrew. 2019. "Conspiracy Theory: The Nineteenth-Century Prehistory of a Twentieth-Century Concept." In Uscinski 2019 (62–81).

McLaren, Peter. 2003. "The Dialectics of Terrorism." In *Masters of War: Militarism and Blowback in the Era of American Empire*, ed. Carl Boggs, 149–190. New York: Routledge.

McMahon, Paula. 2016. "Feds Deceived Us about Billionaire Sex Offender's 'Sweetheart Deal,' Teen Victims Say." *Sun Sentinel* (February 11). http://www.sun-sentinel.com/local/palm-beach/fl-jeffrey-epstein-victims-20160210-story.html

McMurtry, John. 2015. "The Moral Decoding of 9–11. The Official Conspiracy Theory, the Free Press, and the 9–11 Turn." *Global Research* (October 1). https://www.globalresearch.ca/the-moral-decoding-of-9-11-the-official-conspiracy-theory-the-free-press-and-the-9-11-turn/5479046

McNeil, Donald B., Jr. 2010. "U.S. Apologizes for Syphilis Tests in Guatemala." *New York Times* (October 1). http://www.nytimes.com/2010/10/02/health/research/02infect.html?_r=0

Medsger, Betty. 2014. "Remembering an earlier time when a theft unmasked government surveillance." *Washington Post* (January 10). https://www.washingtonpost.com/lifestyle/style/remembering-an-earlier-time-when-a-theft-unmasked-government-surveillance/2014/01/10/30e9ee44-7963-11e3-8963-b4b654bcc9b2_story.html?utm_term=.90f9683841e0

Meisner, Jason, Jeff Coen, and Christy Gutowski. "Judge Calls Hastert 'Serial Child Molester,' Gives Him 15 Months in Prison." *Chicago Tribune* (April 28). http://www.chicagotribune.com/news/local/breaking/ct-dennis-hastert-sentencing-20160427-story.html

Moore, Alfred. 2019. "On the Democratic Problems of Conspiracy Politics." In Uscinski 2019 (111–121).

Moore, Jim. 1990. *Conspiracy of One*. Fort Worth, TX: Summit Publishing Group.

Ms. Smith [pseud.]. 2012. "Whistleblower Binney Says the NSA Has Dossiers on Nearly Every US Citizen." *Infowars* (July 15). (No longer available on the *Infowars* cite, the article can be found at: https://www.csoonline.com/article/2222780/hope-9--whistleblower-binney-says-the-nsa-has-dossiers-on-nearly-every-us-citizen.html)

Nagle, Gabrielle. 2017. "Dr. Whitehurst and the FBI Lab Scandal." *Whistleblower Protection Blog* (November 21). https://www.whistleblowersblog.org/2017/11/articles/intelligence-community-whistleblowers/dr-whitehurst-and-the-fbi-lab-scandal/

National Institute of Standards and Technology (NIST). 2008a. "Final Report on the Collapse of World Trade Center Building 7" (final version, NCSTAR 1A, November).

National Institute of Standards and Technology (NIST). 2008b. "Final Report on the Collapse of World Trade Center Building 7: Draft for Public Comment." April.

Neumann, Craig S., and Robert D. Hare. 2008. "Psychopathic Traits in a Large Community Sample: Links to Violence, Alcohol Use, and Intelligence." *Journal of Consulting and Clinical Psychology* 76.5: 893–899.

Newman, John. 2013. "Oswald, the CIA, and Mexico City." *PBS: Frontline* (November 19). http://www.pbs.org/wgbh/frontline/article/oswald-the-cia-and-mexico-city/

Noble, Holcomb B. 2013. *Cheney's War Crimes: The Reign of a De Facto President*. Bloomington, IN: AuthorHouse.

Oliver, J. Eric, and Thomas J. Wood. 2014. "Conspiracy Theories and the Paranoid Style(s) of Mass Opinion." *American Journal of Political Science* 58.4 (October): 952–966.

Olmsted, Kathryn S. 2009. *Real Enemies: Conspiracy Theories and American Democracy, World War I to 9/11*. New York: Oxford University Press.

Olmsted, Kathryn S. 2019. "Conspiracy Theories in U.S. History." In Uscinski 2019 (285–297).

Orr, Martin, and Ginna Husting. 2019. "Media Marginalization of Racial Minorities: 'Conspiracy Theorists' in U.S. Ghettos and on the 'Arab Street.'" In Uscinski 2019 (82–93).

Ostler, Catherine. 2015. "Jeffrey Epstein: The Sex Offender Who Mixes with Princes and Premiers." *Newsweek* (January 29). https://www.newsweek.com/2015/02/06/sex-offender-who-mixes-princes-and-premiers-302877.html

Palamara, Vincent. 2015. *JFK: From Parkland to Bethesda: The Ultimate Kennedy Assassination Compendium*. Walterville, OR: Trine Day.

Parry, Wynne. 2012. "Contradictions Don't Deter Conspiracy Theorists." *LiveScience* (January 27). http://www.livescience.com/18171-contradicting-conspiracy-theories-mistrust.html

Pepper, William F. 2003. *An Act of State: The Execution of Martin Luther King*. New York: Verso Books.

Pigden, Charles. 1995. "Popper Revisited, or What Is Wrong with Conspiracy Theories?" *Philosophy of the Social Sciences* 25.1: 3–34. (Reprinted in Coady 2006a: 17–43.)

Pigden, Charles. 2006. "Complots of Mischief." In Coady 2006a (139–166).

Pigden, Charles. 2007. "Conspiracy Theories and the Conventional Wisdom." *Episteme* 4.2: 219–232.

Pigden, Charles. 2017. "Are Conspiracy Theorists Epistemically Vicious?" In *A Companion to Applied Philosophy*, ed. Kasper Lippert-Rasmussen, Kimberley Brownlee, and David Coady, 120–132. Hoboken, NJ: Wiley-Blackwell.

Pigden, Charles. 2018. "Conspiracy Theories, Deplorables, and Defectibility: A Reply to Patrick Stokes." In Dentith 2018b (203–215).

Pigden, Charles. In press. "Conspiracy Theories and the Conventional Wisdom Revisited." In *Secrets and Conspiracies*, ed. Olli Loukola. Rodopi. https://philpapers.org/rec/PIGCTA-2

Pipes, Daniel. 1997. *Conspiracy: How the Paranoid Style Flourishes*. New York: Simon & Schuster.

Popper, Karl R. 1962. *The Open Society and Its Enemies*. Vol. 2. New York: Harper & Row.

Popper, Karl R. 1972. *Conjectures and Refutations*. 4th ed. London: Routledge Kegan Paul.

Popper, Karl R. 2006. "The Conspiracy Theory of Society." In *Conspiracy Theories: The Philosophical Debate*, ed. David Coady, 13–15. Burlington, VT: Ashgate.

Posner, Gerald. 1993. *Case Closed: Lee Harvey Oswald and the Assassination of JFK*. New York: Doubleday.

Prouty, Fletcher. 2011. *The Secret Team*. 2nd ed. New York: Skyhorse Publishing.

Räikkä, Juha. 2009a. "On Political Conspiracy Theories." *Journal of Political Philosophy* 17.2: 185–201.

Räikkä, Juha. 2009b. "The Ethics of Conspiracy Theorizing." *Journal of Value Inquiry* 43: 457–468.

Räikkä, Juha. 2018. "Conspiracies and Conspiracy Theories: An Introduction." *Argumenta* 3.2 (issue 6): 205–216.

Räikkä, Juha, and Lee Basham. 2019. "Conspiracy Theory Phobia." In Uscinski 2019 (178–186).

Ratner, Michael. 2008. *The Trial of Donald Rumsfeld: A Prosecution by Book*. New York: New Press.

Raymond, Terrence. 2012. *Evidence of Revision: The Assassination of America (Part 3—LBJ, Hoover and Others, What So Few Know Even Today)*. Conspiratus Ubiquitus. https://www.youtube.com/watch?v=VsXXE9qOA8s

Robertson, David G., and Asbjørn Dyrendal. 2019. "Conspiracy Theories and Religion: Superstition, Seekership, and Salvation." In Uscinski 2019 (411–421).

Rockefeller, David. 2002. *Memoirs*. New York: Random House.

Ronson, Jon. 2001. "Who Pulls the Strings? (Part 3)." *The Guardian* (March 10). https://www.theguardian.com/books/2001/mar/10/extract1

Roth, Richard. 2002. "A Look at the Annual Human Rights Watch International Film Festival." CNN.com/TRANSCRIPTS (aired June 15, 04:30 ET). http://www.cnn.com/TRANSCRIPTS/0206/15/dl.00.html

Ruppert, Michael C. 2004. *Crossing the Rubicon: The Decline of the American Empire at the End of the Age of Oil*. Gabriola Island, BC: New Society Publishers.

Russell, Bertrand. 1953. *The Impact of Science on Society*. New York: Simon and Schuster.

Russell, Bertrand. 1964. "16 Questions on the Assassination." (This can be found on various sites on the internet, including http://22november1963.org.uk/bertrand-russell-16-questions-on-the-assassination.)

Ryan, Kevin. 2013. *Another 19: Investigating Legitimate 9/11 Suspects*. Microbloom.

Ryan, Kevin, James R. Gourley, and Steven E. Jones. 2009. "Environmental Anomalies at the World Trade Center: Evidence for Energetic Materials." *The Environmentalist* 29.1: 56–63.

Sammon, Bill. 2002. *Fighting Back: The War on Terrorism—from Inside the Bush Whitehouse*. Washington, DC: Regnery Publishing.

Scaruffi, Piero. 2002. "Debunking 9/11 Conspiracy Theories." http://www.scaruffi.com/politics/911.html

Scott, Peter Dale. 1975. "Introduction." In *Accessories after the Fact: The Warren Commission, the Authorities, & the Report on the JFK Assassination*, by Sylvia Meagher, xiii–xvii. New York: Skyhorse Publishing.

Scott, Peter Dale. 2007. *The Road to 9/11: Wealth, Empire, and the Future of America*. Berkeley: University of California Press.

Scott, Peter Dale. 2013. "9/11 as a Deep Event: How CIA Personnel Helped Allow It to Happen." In Gourley (107–128).

Shane, Scott, and Mark Mazzetti. 2008. "Report Blames Rumsfeld for Detainee Abuse." *New York Times* (December 11).

Sheehan, Daniel. 2013a. *The People's Advocate: The Life and Legal History of America's Most Fearless Public Trust Lawyer*. Berkeley, CA: Counterpoint.

Sheehan, Daniel. 2013b. "JFK #17: Daniel Sheehan and Alternative Theories of the JFK Assassination Nov. 21, 2013." YouTube. https://www.youtube.com/watch?v=WnRd_bSTdoA

Shenon, Philip. 2008. *The Commission: The Uncensored History of the 9/11 Investigation*. New York: Twelve.

Shenon, Philip. 2013. *A Cruel and Shocking Act: The Secret History of the Kennedy Assassination*. New York: Henry Holt.

Shenon, Philip. 2015. "Yes, the CIA Director Was Part of the JFK Assassination Cover-Up: John McCone Was Long Suspected of Withholding Information from the Warren Commission. Now Even the CIA Says He Did." *Politico Magazine* (October 6). http://www.politico.com/magazine/story/2015/10/jfk-assassination-john-mccone-warren-commission-cia-213197

Simpich, Bill. 2014. *State Secret: Wiretapping in Mexico City, Double Agents, and the Framing of Lee Oswald*. Mary Ferrell Foundation. https://www.maryferrell.org/pages/State_Secret.html

Solomon, John. 2005. "Government Tested AIDS Drugs on Foster Kids: Children Not Provided with Basic Legal Protection, Review Finds." *NBC News.com* (May 4). https://www.nbcnews.com/health/health-news/government-tested-aids-drugs-foster-kids-flna1C9443062

Stokes, Patrick. 2016. "Between Generalism and Particularism about Conspiracy Theory: A Response to Basham and Dentith." *Social Epistemology Review and Reply Collective* 5.10: 34–39.

Stokes, Patrick. 2018a. "Conspiracy Theory and the Perils of Pure Particularism." In Dentith 2018b (29–37).

Stokes, Patrick. 2018b. "On Some Moral Costs of Conspiracy Theory." In Dentith 2018b (189–202).

Sunstein, Cass. 2014a. *Conspiracy Theories and Other Dangerous Ideas*. New York: Simon and Schuster.

Sunstein, Cass. 2014b. "Sunstein: The Conspiracy of Conspiracy Theories." *Newsday* (updated March 18).

Sunstein, Cass. 2015. "Conspiracy Theorists Have Suspicious, and Sometimes Paranoid Natures." *New York Times* (January 4). https://www.nytimes.com/roomfordebate/2015/01/04/are-conspiracy-theories-all-bad-17/conspiracy-theorists-have-suspicious-and-sometimes-paranoid-natures

Sunstein, Cass, and Adrian Vermeule. 2008. "Conspiracy Theories." *Social Science Research Network*. http://ssrn.com/abstract=1084585

Sunstein, Cass, and Adrian Vermeule. 2009. "Conspiracy Theories: Causes and Cures." *Journal of Political Philosophy* 17.2: 202–227.

Sutton, Robbie M., and Karen M. Douglas. 2014. "Examining the Monological Nature of Conspiracy Theories." In *Power, Politics, and Paranoia: Why People Are Suspicious of Their Leaders*, ed. Jan-Willem van Prooijen and Paul A. M. van Lange, 254–272. Cambridge: Cambridge University Press.

Swami, Viren, Tomas Chamorro-Premuzic, and Adrian Furnham. 2010. "Unanswered Questions: A Preliminary Investigation of Personality and Individual Difference Predictors of 9/11 Conspiracist Beliefs." *Applied Cognitive Psychology* 24: 749–761.

Swami, Viren, and Rebecca R. Coles. 2010. "The Truth Is Out There: Belief in Conspiracy Theories." *The Psychologist* 23.7 (July): 560–563.

Swami, Viren, Rebecca Coles, Stefan Stieger, Jakob Pietschnig, Adrian Furnham, Sherry Rehim, and Martin Voracek. 2011. "Conspiracist Ideation in Britain and Austria: Evidence of a Monological Belief System and Associations between Individual Psychological Differences and Real-World and Fictitious Conspiracy Theories." *British Journal of Psychology* 102: 443–463.

Swami, Viren, and Adrian Furnham. 2014. "Political Paranoia and Conspiracy Theories." In *Power, Politics, and Paranoia: Why People Are Suspicious of Their Leaders*, ed. Jan-Willem van Prooijen and Paul A. M. van Lange, 218–236. Cambridge: Cambridge University Press.

Swami, Viren, Martin Voracek, Stefan Stieger, Ulrich S. Tran, and Adriaan Furnham. 2014. "Analytic Thinking Reduces Belief in Conspiracy Theories." *Cognition* 133.3: 572–585.

Talbot, David. 2015. *The Devil's Chessboard: Allen Dulles, the CIA, and the Rise of America's Secret Government*. New York: HarperCollins.

Thomas, Stephen, and Sandra Quinn. 1991. "Public Health Then and Now: The Tuskegee Syphilis Study, 1932 to 1972: Implications for HIV Education and AIDS Risk Education Programs in the Black Community." *American Journal of Public Health* 81.11: 1498–1505. https://web.archive.org/web/20070701074701/http://minority-health.pitt.edu/archive/00000393/01/The_Tuskegee_Syphilis_Study_1932_to.pdf

Thompson, A. C., and Trevor Paglen. 2006. *Torture Taxi*. Hoboken, NJ: Melville House.

Turner, Nigel. 1991. *The Men Who Killed Kennedy* (Episode 3: "The Cover-up"). The History Channel.

US Department of Justice. 2015. "VII. King V. Jowers Conspiracy Allegations." https://www.justice.gov/crt/vii-king-v-jowers-conspiracy-allegations

US Senate. 1976a. "Foreign and Military Intelligence." In *Final Report of the Select Committee to Study Governmental Operations with Respect to Intelligence Activities*, Book I. https://archive.org/stream/finalreportofsel01unit/finalreportofsel01unit_djvu.txt

US Senate. 1976b. "Supplementary Detailed Staff Reports on the Intelligence

Activities and the Rights of Americans." In *Final Report of the Select Committee to Study Governmental Operations with Respect to Intelligence Activities*, Book III. https://archive.org/details/finalreportofsel03unit/page/n3

Uscinski, Joseph E. 2018. "The Study of Conspiracy Theories." *Argumenta* 3.2 (issue 6): 233–245.

Uscinski, Joseph E., ed. 2019. *Conspiracy Theories and the People Who Believe Them*. New York: Oxford University Press.

Uscinski, Joseph E., and Joseph M. Parent. 2014. *American Conspiracy Theories*. New York: Oxford University Press.

Valentine, Douglas. 1990. *The Phoenix Program*. Lincoln, NE: iUniverse.

van Prooijen, Jan-Willem, and Paul A. M. van Lange. 2014. "The Social Dimension of Belief in Conspiracy Theories." In *Power, Politics, and Paranoia: Why People Are Suspicious of Their Leaders*, ed. Jan-Willem van Prooijen and Paul A. M. van Lange, 237–253. Cambridge: Cambridge University Press.

Wagner-Egger, Pascal, Gérald Bronner, Sylvain Delouvée, Sebastian Dieguez, and Nicolas Gauvrit. 2019. "Why 'Healthy Conspiracy Theories' Are (Oxy) morons: Statistical, Epistemological, and Psychological Reasons in Favor of the (Ir)Rational View." *Social Epistemology Review and Reply Collective* 8.3: 50–67.

Walker, Jesse. 2014. "It's All a Conspiracy: The Limitations of Research into Paranoia and Conspiracy Theories." *Slate* (May 20). http://www.slate.com/articles/technology/future_tense/2014/05/conspiracy_theory_research_can_t_be_believed.single.html

Walker, Jesse. 2015. "What I Saw at the Conspiracy Theory Conference: When Tribes Collide." *Hit & Run Blog* (March 18). http://reason.com/blog/2015/03/18/what-i-saw-at-the-conspiracy-theory-conf

Welna, David. 2014. "Before Snowden: The Whistleblowers Who Tried to Lift the Veil." *NPR, Morning Edition* (July 22). https://www.npr.org/2014/07/22/333741495/before-snowden-the-whistleblowers-who-tried-to-lift-the-veil

West, Mick. 2018. *Escaping the Rabbit Hole: How to Debunk Conspiracy Theories Using Facts, Logic, and Respect*. New York: Skyhorse Publishing.

White House. 2009. "President Obama Announces Another Key OMB Post." White House Office of the Press Secretary website (April 20). https://obamawhitehouse.archives.gov/realitycheck/the-press-office/president-obama-announces-another-key-omb-post

White, John. 1967. "Rebuking President." *New Haven Register*. (John White's 1967 Letter to the Editor.) See http://www.connecticutmag.com/history/john-white-s-letter-to-the-new-haven-register/article_31a2c5d5-0796-5105-90ca-de6eea6bf45b.html

Wojcik, Daniel. 1997. *The End of the World as We Know It: Faith, Fatalism, and Apocalypse in America*. New York: New York University Press.

Wood, Michael J. 2016. "Some Dare Call It Conspiracy: Labeling Something a Conspiracy Theory Does Not Reduce Belief in It." *Political Psychology* 37.5: 695–705.

Wood, Michael J., and Karen Douglas. 2013. "'What about Building 7?' A Social Psychological Study of Online Discussion of 9/11 Conspiracy Theories." *Frontiers in Psychology* 4.409 (July): 1–9.

Wood, Michael J., and Karen M. Douglas. 2019. "Conspiracy Theory Psychology: Individual Differences, Worldviews, and States of Mind." In Uscinski 2019 (245–268).

Wood, Michael J., Karen M. Douglas, and Robbie M. Sutton. 2012. "Dead and Alive: Beliefs in Contradictory Conspiracy Theories." *Social Psychological and Personality Science* 3.6: 767–773.

Index

Printed and bound by CPI Group (UK) Ltd, Croydon, CR0 4YY
16/09/2026
14963489-0002